Tandem Repeat Polymorphisms

ADVANCES IN EXPERIMENTAL MEDICINE AND BIOLOGY

Tandem Repeat Polymorphisms

Genetic Plasticity, Neural Diversity and Disease

Edited by

Anthony J. Hannan, PhD

Florey Neuroscience Institutes, Melbourne Brain Centre, University of Melbourne, Parkville, Victoria, Australia

Springer Science+Business Media, LLC

Landes Bioscience

Springer Science+Business Media, LLC
Landes Bioscience

Printed in the USA.

Springer Science+Business Media, LLC, 233 Spring Street, New York, New York 10013, USA
http://www.springer.com

Please address all inquiries to the publishers:
Landes Bioscience, 1806 Rio Grande, Austin, Texas 78701, USA
Phone: 512/ 637 6050; FAX: 512/ 637 6079
http://www.landesbioscience.com

The chapters in this book are available in the Madame Curie Bioscience Database.
http://www.landesbioscience.com/curie

Tandem Repeat Polymorphisms: Genetic Plasticity, Neural Diversity and Disease, edited by Anthony J. Hannan. Landes Bioscience / Springer Science+Business Media, LLC dual imprint / Springer series: Advances in Experimental Medicine and Biology.

ISBN: 978-1-4614-5433-5

While the authors, editors and publisher believe that drug selection and dosage and the specifications and usage of equipment and devices, as set forth in this book, are in accord with current recommendations and practice at the time of publication, they make no warranty, expressed or implied, with respect to material described in this book. In view of the ongoing research, equipment development, changes in governmental regulations and the rapid accumulation of information relating to the biomedical sciences, the reader is urged to carefully review and evaluate the information provided herein.

Library of Congress Cataloging-in-Publication Data

Tandem repeat polymorphisms : genetic plasticity, neural diversity, and disease / edited by Anthony J. Hannan.
   p. ; cm. -- (Advances in experimental medicine and biology ; v. 769)
 Includes bibliographical references and index.
 ISBN 978-1-4614-5433-5 (alk. paper)
 I. Hannan, Anthony J., 1969- II. Series: Advances in experimental medicine and biology ; v. 769. 0065-2598
 [DNLM: 1. Polymorphism, Genetic. 2. Tandem Repeat Sequences. 3. Heredodegenerative Disorders, Nervous System--genetics. 4. Mutation. W1 AD559 v.769 2012 / QU 500]

  576.5'49--dc23
                    2012030285

DEDICATION

To all of those families who are suffering from tandem repeat expansion disorders

PREFACE

Tandem repeats of DNA sequences provide a unique and abundant source of genomic variability and recent evidence suggests they can modulate a range of biological processes in a wide variety of different species. These classes of repetitive DNA are variously referred to as simple sequence repeats, satellite DNA (microsatellites, minisatellites and satellites) or variable number tandem repeats. A key aspect of tandem repeats is that they represent highly polymorphic and uniquely mutable genomic components which can (depending on their sequence, length and location) affect the structure and function of DNA, RNA and protein.

This book addresses the role of tandem repeat polymorphisms (TRPs) in genetic plasticity, evolution, development, biological processes, neural diversity, brain function, dysfunction and disease. There are hundreds of thousands of unique tandem repeats in the human genome and their polymorphic distributions have the potential to greatly influence functional diversity and disease susceptibility. Recent discoveries in this expanding field are critically reviewed and discussed in a range of subsequent chapters, with a focus on the role of TRPs and their various gene products in evolution, development, diverse molecular and cellular processes, brain function and disease.

In the first chapter, I introduce these broad themes. This includes discussion of the specific proposal that TRPs could help solve the conundrum of 'missing heritability' produced by genome-wide association studies of various polygenic complex diseases which have only examined single nucleotide polymorphisms (SNPs). Subsequent chapters focus on key aspects of TRPs in health and disease. In the second chapter, David King shares his ideas regarding the role of simple sequence repeats in evolution, and provides a detailed discussion of repeat sequences as mutable sites providing genetic variability upon which natural selection can act. This theme is then extended by Noel Faux, who uses bioinformatic analyses of trinucleotide repeats encoding homopeptides to explore both the evolution and function of a wide variety of amino acid repeats located in diverse proteins across the phylogenetic spectrum. This bioinformatic exploration of the role of TRPs in normal biological functions is then extended by Sterling Sawaya and colleagues, who discuss evidence for the role of promoter microsatellites in modulating the expression of various human genes.

Expansions in tandem repeats ('dynamic mutations') are known to cause many disorders, which mainly affect the nervous system, including Huntington's disease (the most common polyglutamine disorder), spinocerebellar ataxias, Kennedy's disease (spinobulbar muscular atrophy), dentatorubral-pallidoluysian atrophy, Friedreich ataxia, polyalanine disorders, fragile X syndrome and related disorders. Robert Richards, who helped coin the term dynamic mutations, and his colleague Clare van Eyk, have provided an overview of this large and clinically significant area of research. Whilst the plasticity of these tandem repeats occurs at the DNA level, evidence for both 'gain of function' and 'loss of function' pathogenic effects of repeat expansions ('genetic stutters') at RNA and protein levels is discussed using specific examples of these monogenic disorders.

Danuta Loesch and Randi Hagerman review the exciting field that has evolved around the FMR1 gene, originally discovered due to its hosting of the large expansion of a 5'UTR trinucleotide (CGG) repeat which causes fragile X syndrome. Smaller 'premutation' repeat lengths have recently been shown by Hagerman and colleagues to cause fragile X tremor-ataxia syndrome (FXTAS), as well as contributing to other disorders. The focus then shifts to protein-coding trinucleotide repeats, with complementary chapters from Amy Robertson and Stephen Bottomley, as well as Saski Polling, Andrew Hill and Danny Hatters, exploring the biochemistry of expanded polyglutamine tracts and their roles in at least nine autosomal dominant neurodegenerative disorders. As Huntington's disease (HD), which was first described by George Huntington in 1872 and genetically mapped over a century later, is the most common of these so-called polyglutamine disorders it has been most intensively researched. Henry Waldvogel and colleagues review data on the neuropathology and related symptomatology of HD, linking molecular and cellular aspects to pathogenesis at the systems level. Another extraordinary polyglutamine disorder is Kennedy's disease (also known as spinobulbar muscular atrophy or SBMA). Jeffrey Zajac and Mark Tang discuss how polyglutamine polymorphism in the androgen receptor not only causes SBMA, but can contribute to other complex disorders via modulation of this sex hormone signaling system. A different neurodegenerative disease caused by a non-coding tandem repeat expansion is Friedreich ataxia. Corben and colleagues discuss how homozygosity of a GAA repeat expansion in an intron of the frataxin gene leads to downregulated expression and consequent neuropathology, motor and cognitive symptoms. In the final chapter, a unique group of disorders involving trinucleotide repeat expansions encoding polyalanine tracts, are reviewed by Cheryl Shoubridge and Jozef Gecz, providing insights into how expanded polyalanine in specific proteins leads to developmental abnormalities and neurocognitive dysfunction.

It is hoped that this book will help the reader to grasp the significance of TRPs in evolution, development, brain function and a variety of major clinical disorders. As we begin a new genetic revolution powered by next-generation sequencing, expanding knowledge of tandem repeats and their polymorphic variants will no doubt continue to enhance our understanding of genetic plasticity, neural diversity and disease.

Anthony J. Hannan, PhD
Florey Neuroscience Institute,
Melbourne Brain Centre
University of Melbourne
Parkville, Victoria, Australia

ABOUT THE EDITOR...

ANTHONY J. HANNAN is Head of the Neural Plasticity Laboratory, Florey Neuroscience Institutes and Associate Professor at the University of Melbourne, Parkville, Australia. Following undergraduate and PhD degrees at the University of Sydney, Anthony received postdoctoral neuroscience training at the University of Oxford, supported by a Nuffield Medical Fellowship. He currently holds an Australian Research Council (ARC) Future Fellowship (FT3) and an Honorary Senior Research Fellowship from the National Health and Medical Research Council (NHMRC). Main research interests include pathogenic mechanisms mediating Huntington's disease and related tandem repeat expansion disorders, as well as other cognitive and psychiatric illnesses. In his laboratory, experimental models of gene-environment interactions are used to explore experience-dependent plasticity in the healthy and diseased brain.

PARTICIPANTS

Andrew T. Bagshaw
Department of Pathology
University of Otago-Christchurch
Christchurch
New Zealand

Stephen P. Bottomley
Department of Biochemistry
   and Molecular Biology
Monash University
Clayton, Victoria
Australia

John L. Bradshaw
Experimental Neuropsychology
   Research Unit
School of Psychology and Psychiatry
Monash University
Clayton, Victoria
Australia

Emmanuel Buschiazzo
School of Natural Sciences
University of California – Merced
Merced, California
USA

Andrew J. Churchyard
Monash Neurology
Monash Medical Centre
Clayton, Victoria
Australia

Louise A. Corben
Bruce Lefroy Centre for Genetic
   Health Research
Murdoch Childrens Research Institute
The Royal Children's Hospital
Parkville, Victoria
Australia

Martin B. Delatycki
Department of Clinical Genetics
Austin Health
and
Department of Medicine
University of Melbourne at Austin Health
Heidelberg, Victoria
Australia

Marguerite V. Evans-Galea
Bruce Lefroy Centre for Genetic
   Health Research
Murdoch Childrens Research Institute
The Royal Children's Hospital
Parkville, Victoria
Australia

Richard L.M. Faull
Department of Anatomy with Radiology
Faculty of Medical and Health Sciences
and
Centre for Brain Research
University of Auckland
Auckland
New Zealand

Noel Faux
Mental Health Research Institute
The University of Melbourne
Parkville, Victoria
and
National Neuroscience Facility
Carlton, Victoria
Australia

Mark Ng Tang Fui
Department of Endocrinology
Austin Health
Heidelberg, Victoria
Australia

Jozef Gecz
Department of Genetics
    and Molecular Pathology
SA Pathology at the Women's
    and Children's Hospital
North Adelaide, South Australia
and
Department of Pediatrics
University of Adelaide
Adelaide, South Australia
Australia

Neil J. Gemmell
Department of Anatomy
    and Structural Biology
University of Otago
Dunedin
New Zealand

Nellie Georgiou-Karistianis
Experimental Neuropsychology
    Research Unit
School of Psychology and Psychiatry
Monash University
Clayton, Victoria
Australia

Randi Hagerman
Department of Pediatrics
MIND Institute
University of California at Davis
    Medical Center
Sacramento, California
USA

Anthony J. Hannan
Florey Neuroscience Institutes
Melbourne Brain Centre
University of Melbourne
Parkville, Victoria
Australia

Danny M. Hatters
Department of Biochemistry
    and Molecular Biology
University of Melbourne
Melbourne, Victoria
Australia

Andrew F. Hill
Department of Biochemistry
    and Molecular Biology
University of Melbourne
Melbourne, Victoria
Australia

Virginia Hogg
Lynette Tippett
Centre for Brain Research
and
Department of Psychology
University of Auckland
Auckland
New Zealand

David G. King
Department of Anatomy
and
Department of Zoology
Southern Illinois University Carbondale
Carbondale, Illinois
USA

Danuta Loesch
Department of Psychology
LaTrobe University
Melbourne, Victoria
Australia

Saskia Polling
Department of Biochemistry
    and Molecular Biology
University of Melbourne
Melbourne, Victoria
Australia

Robert I. Richards
Discipline of Genetics
School of Molecular
    and Biomedical Sciences
The University of Adelaide
Adelaide, South Australia
Australia

Amy L. Robertson
Department of Biochemistry
    and Molecular Biology
Monash University
Clayton, Victoria
Australia

Sterling M. Sawaya
Department of Anatomy
    and Structural Biology
University of Otago
Dunedin
New Zealand

Cheryl Shoubridge
Department of Genetics
    and Molecular Pathology
SA Pathology at the Women's
    and Children's Hospital
North Adelaide, South Australia
and
Department of Pediatrics
University of Adelaide
Adelaide, South Australia
Australia

Doris Thu
Brain Mind Institute
Ecole Polytechnique Federale de Lausanne
Lausanne
Switzerland

Lynette Tippett
Centre for Brain Research
and
Department of Psychology
University of Auckland
Auckland
New Zealand

Clare L. van Eyk
Discipline of Genetics
School of Molecular
    and Biomedical Sciences
The University of Adelaide
Adelaide, South Australia
Australia

Henry J. Waldvogel
Department of Anatomy with Radiology
Faculty of Medical and Health Sciences
and
Centre for Brain Research
University of Auckland
Auckland
New Zealand

Jeffrey D. Zajac
Department of Medicine
and
Department of Endocrinology
University of Melbourne at Austin Health
Heidelberg, Victoria
Australia

CONTENTS

7. MOLECULAR PATHWAYS TO POLYGLUTAMINE AGGREGATION..... 115

Amy L. Robertson and Stephen P. Bottomley

8. POLYGLUTAMINE AGGREGATION IN HUNTINGTON
AND RELATED DISEASES 125

Saskia Polling, Andrew F. Hill and Danny M. Hatters

9. SELECTIVE NEURODEGENERATION, NEUROPATHOLOGY
AND SYMPTOM PROFILES IN HUNTINGTON'S DISEASE 141

Henry J. Waldvogel, Doris Thu, Virginia Hogg, Lynette Tippett
and Richard L.M. Faull

ACKNOWLEDGMENTS

I would firstly like to thank the authors who have written chapters for this book, dedicating substantial amounts of precious time in their already hectic schedules. I would like to thank the dedicated staff at Landes Bioscience, particularly Celeste Carlton and Cynthia Conomos, for their assistance and patience during this long journey. I greatly appreciate ongoing discussions with past and present members of my laboratory, as well as other colleagues, at the Florey Neuroscience Institutes, University of Melbourne, as the intellectual stimulation they have provided has contributed to the development of my own thoughts in the introductory chapter. I thank Carolyn M. Hannan for her support with editing. My thanks also to the funding bodies that have allowed me to devote so much of my time to research, in particular the Australian Research Council and the National Health and Medical Research Council.

CHAPTER 1

TANDEM REPEAT POLYMORPHISMS
Mediators of Genetic Plasticity, Modulators of Biological Diversity and Dynamic Sources of Disease Susceptibility

Anthony J. Hannan

Florey Neuroscience Institutes, Melbourne Brain Centre, University of Melbourne, Parkville, Australia
Email: anthony.hannan@florey.edu.au

Abstract: Tandem repetitive DNA elements (tandem repeats), including microsatellites and simple sequence repeats, are extremely common throughout the genomes of a wide range of species. Tandem repeat expansions have been found to cause a range of monogenic diseases, such as Huntington's disease, various ataxias and other neurological diseases. The human genome contains hundreds of thousands of distinct tandem repeats, many of which appear to have evolved to regulate specific aspects of gene expression, RNA function and protein function. Tandem repeat polymorphisms (TRPs) provide a unique source of genetic variability that has an extended digital distribution, as opposed to the usual binary nature of single nucleotide polymorphisms. In this chapter I will review studies in which tandem repeats have been implicated in a multitude of molecular and cellular processes associated with the development, behavior and evolution of a variety of animal species, including mammals. Recent data suggesting that these repetitive sequences can increase the 'evolvability' of genomes provides further evidence that TRPs not only have functional consequences but also provide a rich source of genetic diversity that can facilitate evolutionary processes. I propose that a readily mutable subclass of tandem repeats may provide an important template for stochastic genetic variation, which could in turn generate diversity in epigenetics, development and organismal function, thus impacting upon evolution. Furthermore, the distinctive characteristics of TRPs also uniquely position them as contributors to complex polygenic disorders. Ultimately, there is much to be gained from systematic analysis of the 'repeatome', defined as the entire set of tandem repeats and other repetitive DNA in a genome, as well as their transcribed and translated expression products. Applying such approaches not only to the human genome but to other species will yield new insights into the genetic regulation of a wide range of biological processes in healthy and diseased states.

Tandem Repeat Polymorphisms: Genetic Plasticity, Neural Diversity and Disease,
edited by Anthony J. Hannan ©2012 Landes Bioscience and Springer Science+Business Media.

INTRODUCTION

DNA sequencing technologies have led to a recent revolution in our understanding of the human genome, as well as the genomes of other animals and many other species. Tandemly repeated DNA sequences, or tandem repeats, are increasingly being recognized as much more than 'genetic stutters', but rather key structural and functional elements of the human genome, as well as the genomes of other species.[1-4] The term tandem repeats encompasses satellite DNA (including minisatellites and microsatellites), simple sequence repeats (SSRs), as well as variable number tandem repeats (VNTRs). Tandem repeats are found commonly in exons, as well as introns and intergenic regions. Whilst the function of tandem repeats has only been explored in a relatively small number of genes, their abundance and locations suggest potentially widespread roles in the modulation of gene expression, RNA function, protein function and other molecular processes.[1,5-10] Furthermore, tandem repeats located in transcribed regions of the genome, which may constitute a large proportion of the human genome[11] have the capacity to alter the structure and function of both coding and noncoding RNA species. Those tandem repeats located in coding regions (and encoding amino acid repeats via trinucleotides, hexanucleotides, etc.) have additional potential roles in modulating the structure and function of the encoded proteins.

The importance of tandem repeats can be recognized in a number of different domains. The high degree of polymorphism in tandem repeats may confer a capacity to act as 'tuning knobs' for evolutionary processes.[1-6,12] This capacity derives from the fact that tandem repeats are far more mutable than single nucleotides, meaning that tandem repeat polymorphisms (TRPs) can have an extended digital distribution[2] as opposed to the binary possibilities presented by individual single nucleotide polymorphisms (SNPs). The mutability of tandem repeats has been proposed to add variability to brain development and function, thus providing a more dynamic template upon which natural selection may act.[1]

The capacity to compare the genomes of different individuals of a species has not only revealed high levels of conservation of many genes and intergenic regions, but also the extent of polymorphism.[13] While much attention has been focused on SNPs, recent studies have revealed other important polymorphic DNA sequences, including TRPs[2] and copy number variants (CNVs).[14]

EXPANSION OF TANDEM REPEATS IN RARE MONOGENIC DISORDERS WITH MENDELIAN INHERITANCE

The term 'dynamic mutations' has been used to describe the expansions of tandem repeats associated with a variety of diseases exhibiting Mendelian inheritance patterns.[15] TRPs, above a specific repeat length, have been shown to cause various autosomal dominant and recessive human disorders, including polyglutamine disorders (e.g., Huntington's disease and some spinocerebellar ataxias), Friedreich ataxia, fragile X syndrome and myotonic dystrophy.[16-18]

At least nine diseases with expansions in tracts of CAG repeats encoding polyglutamine tracts in different genes have been found to lead to neurodegeneration and consequent neurological (and in some cases also psychiatric) symptoms.[19-21] The recent discovery that spinocerebellar ataxia 8 (SCA8) and Huntington's disease-like

2 (HDL2) might be caused by expression of a CAG/glutamine tract expansion on the antisense strand suggests that there may now be at least 11 polyglutamine diseases,[22,23] although the potential roles of RNA toxicity and other nonpolyglutamine pathogenic mechanisms in these diseases have not been ruled out.

Huntington's disease (HD) is the most common of these polyglutamine diseases, and was first described by George Huntington in 1872. HD is caused by a CAG repeat expansion encoding an extended polyglutamine tract in the huntingtin protein and ultimately leading to a triad of cognitive, psychiatric and motor symptoms.[24] HD has been the most intensely studied polyglutamine disease, leading to major new insights into how the CAG/glutamine repeat expansion leads to pathogenesis in specific brain areas as well as some peripheral organs.[25-27] The other polyglutamine diseases include six spinocerebellar ataxias (SCA1,2,3,6,7,17), as well as dentatorubral pallidolysian atrophy (DRPLA) and spinal and bulbar muscular atrophy (SBMA).[18,28] The potential molecular mechanisms whereby abnormally expanded polyglutamine tracts induce cellular toxicity are reviewed elsewhere.[20,21]

A fascinating recent discovery of non-ATG-initiated translation associated with at least some tandem repeat expansions,[29] has raised unexpected possibilities regarding mechanisms of pathogenesis. This finding, along with knowledge of antisense transcripts suggests that a single tract of DNA could generate at least two transcripts, seven reading frames and 'potentially nine toxic entities!'.[30]

A separate class of human disorders involve polyalanine expansions in specific proteins and mainly present as abnormalities of development.[31,32] These polyalanine disorders do not appear to be neurodegenerative, thus setting them apart from those diseases caused by expanded polyglutamine, but rather seem to be caused by a disruption of the normal function of the polyalanine tracts within the respective proteins.[32]

Other disorders of tandem repeat expansion include some nonpolyglutamine spinocerebellar ataxias, Friedreich ataxia, fragile X syndrome, fragile X tremor-ataxia syndrome (FXTAS) and myotonic dystrophy.[33-37] These disorders involve tandem repeats located in noncoding genomic regions and are therefore associated with abnormal gene expression, RNA structure and/or function.[18,38-40]

Collectively, these tandem repeat expansion disorders constitute a major personal, medical and economic burden, and therefore expanding our understanding and developing effective therapeutic approaches represents a clinical priority. Each disorder may require its own tailored therapeutic strategy, however it could be imagined that a therapy targeting polyglutamine toxicity, for example, may have efficacy across that wider group of neurodegenerative diseases. Furthermore, the study of these unique disorders, involving tandem repeat lengths beyond the normal range, may continue to provide insights into why tandem repeats have evolved, both in coding and noncoding regions of the human genome, and thus illuminate molecular and cellular processes in health and disease.

TANDEM REPEATS AS DYNAMIC MODULATORS OF DEVELOPMENT, BRAIN FUNCTION AND BEHAVIOR

Evidence has been provided for tandem repeats encoding homopeptide repeats in transcription factors as key regulators of developmental processes and subsequent anatomical diversity amongst various breeds of domestic dogs.[41] This study and other evidence supports a role for homopeptide-repeat containing proteins in specific molecular and cellular processes, including transcriptional regulation.[1,7,42-44]

Tandem repeats, and their unique polymorphic contributions to genetic plasticity have been proposed to contribute to the modulation of brain development and function.[1] Putative roles for tandem repeats in various neurotransmitter and neuromodulatory systems, affecting brain development, behavioural modulation as well as affective and cognitive function have been recently reviewed.[2,16]

What is the evidence that tandem repeats are common in genes that are important for neural development and function? Firstly, the fact that the vast majority of known monogenic tandem repeat expansion disorders affect the nervous system provides indirect evidence, with the caveat being that many of those disease genes are also expressed in nonneural tissues. Nevertheless, the vulnerability of the nervous system to expansions of tandem repeats (both protein coding and noncoding) suggests that these tandem repeats may be particularly important in neural processes such as neurodevelopment and brain function. Secondly, bioinformatic approaches have indicated that many genes hosting tandem repeats are associated with neural functions.[1,45] Evidence for tandem repeats as modulators of behavioural processes has also been produced. For example, a TRP in the vasopressin receptor in prairie voles has been reported to modulate brain function and social behaviour.[46]

One implication of these and other studies is that TRPs could modulate normal human development, brain function, cognition and behavior.[1] A reasonable starting point for testing such a general hypothesis is to examine some of the many genes that have been implicated in monogenic human diseases.[16] The Huntington's disease (HD) gene, encoding a glutamine repeat in the huntingtin (Htt) protein as discussed above, provides one example. In the healthy range (nonHD families) Htt is known to have around 5-34 CAG/glutamine repeats. This TRP in Htt is known to be under selective pressure during evolution.[47] Furthermore, evidence from mice shows that Htt is important in development and brain function[48,49] and that removal of the CAG repeat has functional consequences.[50]

One specific hypothesis I therefore propose, as an extension of previous ideas,[1] is that healthy individuals who are polymorphic for their CAG/glutamine repeat in the Htt gene/ protein will show differences in brain development, structure and function. A corollary of this would be that individuals who are 'gene-positive' for a CAG-expansion causing HD may have abnormal brain development, due to the functional effects of the long polyglutamine in Htt, prior to any 'toxic gain of function' leading to neurodegenerative changes. Consequently, in order to fully understand HD we will need to comprehend the role of the polyglutamine tract in the spatiotemporally regulated functions of Htt. This could be extended to many more tandem repeats, and their polymorphic variants, located in and around neurally expressed genes.

A PROPOSED ROLE FOR TANDEM REPEAT POLYMORPHISMS IN COMMON POLYGENIC DISORDERS

Recent genome-wide association (GWA) studies do not fully account for the major genetic contributions to common polygenic disorders, and this has led to an active search for the 'missing heritability'.[51,52] The evidence for tandem repeat expansions as contributors to monogenic disorders of Mendelian inheritance has been outlined above. However, this may represent only a small fraction of tandem repeat contributions to human disease. It has been proposed that many common polygenic disorders may involve TRPs as major contributors to so-called 'missing heritability'.[2] TRPs could thus play a key role in modulating disease susceptibility for a range of common polygenic disorders, including various psychiatric and neurological disorders.[2] Recent evidence from studies of sporadic amyotrophic lateral sclerosis (ALS) supports the importance of TRPs as dynamic sources of genetic susceptibility in such complex polygenic diseases.[53-55]

The vast majority of GWA studies currently involve analysis of SNPs and therefore do not assay other major types of polymorphisms, such as TRPs. There are hundreds of thousands of distinct TRPs in the human genome, and this represents an extensive source of genetic variance possessing a dynamic and unique polymorphic range. SNPs are, with very few exceptions, binary in nature, whereas TRPs display extended digital distributions (multiallelic genotypes). An individual tandem repeat can thus exhibit a large array of polymorphic variants, which in turn extends the potential variety of genetic contributions to disease susceptibility.[2]

GWA studies will increasingly involve analyses of whole genome sequences, via the utilisation of next-generation sequencing technologies. It has been proposed that such new GWA studies involving whole-genome sequencing evaluate TRPs (using appropriate sequencing and bioinformatic protocols), as well as SNPs and other polymorphisms, for disease associations, so that the missing heritability may finally be found.[2] Rapid and accurate whole-genome sequencing has the potential to elucidate roles for TRPs in the dynamic modulation of biological evolution, development, function and dysfunction. It is therefore a priority to characterise the human 'repeatome', which I define here as the full set of tandem repeats and other repetitive DNA elements in the genome, as well as their transcribed and translated expression products. The repeatome would naturally encompass the full extent of TRPs throughout populations of a given species.

TANDEM REPEAT INSTABILITY AS A SOURCE OF CELLULAR HETEROGENEITY AND PHENOTYPIC DIVERSITY IN THE DEVELOPING AND MATURE ORGANISM

Instability of tandem repeats during meiosis, and consequent variability of repeat length between generations, is thought to be responsible for genetic anticipation in tandem repeat-expansion disorders such as HD. Meiotic instability of tandem repeat length could also contribute to common polygenic disorders, where variability in repeat lengths could modulate genetic susceptibility between generations.[2] Tandem repeats can also potentially change length during mitosis, and the most striking consequences of such repeat instability would be expected during development and may allow cellular selection to favour specific tandem repeat lengths in particular tissues and cell types.[1]

One additional potential source of tandem repeat length variability could involve postmitotic instability. Evidence for such tandem repeat instability in postmitotic neurons has recently been found.[56] Postmitotic instability of tandem repeats needs to be investigated in detail to establish its full extent, however it is already known that repeat instability (involving well described microsatellites) has been implicated in some sporadic cancers, which could result from postmitotic repeat length mutations.[57]

TANDEM REPEAT POLYMORPHISMS AND EVOLUTION

The abundance and broad distribution of tandem repeats in the human genome suggests that these sequences may arise randomly as 'genetic stutters' and that evolution actively selects for such repetitive DNA sequences.[3] Tandem repeats, and simple sequence repeats (SSRs) in particular, have been proposed to function as 'tuning knobs' during evolution.[3,12,16]

It has been observed that tandem repeat lengths are distributed in a digital manner, in contrast to the binary possibilities offered by SNPs; TRPs can thus act as 'digital genetic modulators' producing a continuously variable array of genotypes.[1] Furthermore, recent experimental evidence from a yeast model demonstrates that tandem repeats can enhance the 'evolvability' of specific promoter sequences.[58,59] Tandem repeats could thus be mediators of genetic plasticity, providing a diverse and dynamic genomic template, and associated phenotypic diversity, upon which natural selection may act.[1,12]

I propose a further possibility, that TRPs provide a dynamic source of stochastic genetic variation influencing organismal development, function and thus evolution. This would be analogous to the role of 'stochastic epigenetic variation' recently proposed by Feinberg and Irizarry.[60] More specifically, it was suggested that stochastic epigenetic variation might mediate phenotypic variability, without altering the mean phenotype, thus both enhancing evolutionary fitness whilst increasing susceptibility to disease of a population exposed to a changing environment.[60] These two hypotheses may in fact be interconnected, as evidence from repeat expansion disorders indicates that tandem repeats can modulate epigenetic processes, including chromatin remodelling.[34,39]

CONCLUSION

Tandem repeats are being increasingly recognised as a major potential source of genetic plasticity and associated cellular and organismal diversity, in humans and many other species.[1,3,61] Tandem repeat instability may mediate genomic plasticity and somatic variation, thus enhancing diversity at cellular, tissue and systems levels.[1] Tandem repeat length instability, occurring either meiotically, mitotically or postmitotically, thus has the capacity to mediate functional diversity at molecular, cellular, physiological and behavioural levels.

Technical and bioinformatic challenges need to be met to facilitate thorough analyses of tandem repeats and their polymorphic variants at a whole genome level.[62] Understanding genetic plasticity involving other types of repetitive DNA, such as specific retrotransposons,[63] will also be a priority. These approaches promise to shed new light on the genetic regulation of developmental, physiological and evolutionary processes. Furthermore, systematic characterisation of the 'human repeatome' may provide fundamental new insights into the genetic basis of disease.

ACKNOWLEDGMENTS

The NHMRC and an ARC Future Fellowship currently support the author's research. I thank past and present members of the Hannan Laboratory for useful discussions.

REFERENCES

1. Nithianantharajah J, Hannan AJ. Dynamic mutations as digital genetic modulators of brain development, function and dysfunction. Bioessays 2007; 29:525-535.
2. Hannan AJ. Tandem repeat polymorphisms: modulators of disease susceptibility and candidates for 'missing heritability'. Trends Genet 2010; 26:59-65.
3. King DG. Evolution of simple sequence repeats as mutable sites. In: Hannan AJ, ed. Tandem Repeat Polymorphisms: Genetic Plasticity, Neural Diversity and Disease. Austin/New York: Landes Bioscience/ Springer Science+Business Media, 2012:10-25.
4. Sawaya SM, Bagshaw ATB, Buschiazzo E et al. Promoter microsatellites as modulators of human gene expression. In: Hannan AJ, ed. Tandem Repeat Polymorphisms: Genetic Plasticity, Neural Diversity and Disease. Austin/New York: Landes Bioscience/Springer Science+Business Media, 2012:41-54.
5. Hamada H, Seidman M, Howard BH et al. Enhanced gene expression by the poly(dT-dG) poly(dC-dA) sequence. Mol Cell Biol 1984; 4:2622-2630.
6. Kashi Y, King D, Soller M. Simple sequence repeats as a source of quantitative genetic variation. Trends Genet 1997; 13:74-78.
7. Faux NG, Bottomley SP, Lesk AM et al. Functional insights from the distribution and role of homopeptide repeat-containing proteins. Genome Res 2005; 15:537-551.
8. Verstrepen KJ, Jansen A, Lewitter F et al. Intragenic tandem repeats generate functional variability. Nat Genet 2005; 37:986-990.
9. Roberts J, Scott AC, Howard MR et al. Differential regulation of the serotonin transporter gene by lithium is mediated by transcription factors, CCCTC binding protein and Y-Box binding protein 1, through the polymorphic intron 2 variable number tandem repeat. J Neurosci 2007; 27:2793-2801.
10. Salichs E, Ledda A, Mularoni L et al. Genome-wide analysis of histidine repeats reveals their role in the localization of human proteins to the nuclear speckles compartment. PloS Genet 2009; 5:e1000397.
11. Mattick JS, Taft RJ, Faulkner GJ. A global view of genomic information—moving beyond the gene and the master regulator. Trends Genet 2010; 26:21-28.
12. Kashi Y, King DG. Simple sequence repeats as advantageous mutators in evolution. Trends Genet 2006; 22:253-259.
13. Richard GF, Kerrest A, Dujon B. Comparative genomics and molecular dynamics of DNA repeats in eukaryotes. Microbiol Mol Biol Rev 2008; 72:686-727.
14. Conrad DF, Pinto D, Redon R et al. Origins and functional impact of copy number variation in the human genome. Nature 2010; 464:704-712.
15. Richards RI. Dynamic mutations: a decade of unstable expanded repeats in human genetic disease. Hum Mol Genet 2001; 10:2187-2194.
16. Fondon JW III, Hammock EA, Hannan AJ et al. Simple sequence repeats: genetic modulators of brain function and behavior. Trends Neurosci 2008; 31:328-334.
17. Usdin K. The biological effects of simple tandem repeats: lessons from the repeat expansion diseases. Genome Res 2008; 18:1011-1019.
18. Van Eyk CL, Richards RI. Dynamic mutations: where are they now? In: Hannan AJ, ed. Tandem Repeat Polymorphisms: Genetic Plasticity, Neural Diversity and Disease. Austin/New York: Landes Bioscience/ Springer Science+Business Media, 2012:55-77.
19. Bauer PO, Nukina N. The pathogenic mechanisms of polyglutamine diseases and current therapeutic strategies. J Neurochem 2009; 110:1737-1765.
20. Polling S, Hill AF, Hatters DM. Polyglutamine aggregation in Huntington and related diseases. In: Hannan AJ, ed. Tandem Repeat Polymorphisms: Genetic Plasticity, Neural Diversity and Disease. Austin/New York: Landes Bioscience/Springer Science+Business Media, 2012:125-140.
21. Robertson AL, Bottomley SP. Molecular pathways to polyglutamine aggregation. In: Hannan AJ, ed. Tandem Repeat Polymorphisms: Genetic Plasticity, Neural Diversity and Disease. Austin/New York: Landes Bioscience/Springer Science+Business Media, 2012:115-124.
22. Moseley ML, Zu T, Ikeda Y et al. Bidirectional expression of CUG and CAG expansion transcripts and intranuclear polyglutamine inclusions in spinocerebellar ataxia type 8. Nat Genet 2006; 38:758-769.
23. Wilburn B, Rudnicki DD, Zhao J et al. An antisense CAG repeat transcript at JPH3 locus mediates expanded polyglutamine protein toxicity in Huntington's disease-like 2 mice. Neuron 2011; 70:427-440.
24. Waldvogel H, Thu D, Hogg V et al. Selective neurodegeneration, neuropathology and symptom profiles in Huntington's disease. In: Hannan AJ, ed. Tandem Repeat Polymorphisms: Genetic Plasticity, Neural Diversity and Disease. Austin/New York: Landes Bioscience/Springer Science+Business Media, 2012:141-152.
25. Zuccato C, Valenza M, Cattaneo E. Molecular mechanisms and potential therapeutical targets in Huntington's disease. Physiol Rev 2010; 90:905-981.

26. Raymond LA, André VM, Cepeda C et al. Pathophysiology of Huntington's disease: time-dependent alterations in synaptic and receptor function. Neuroscience. 2011; 198:252-273.

27. Ross CA, Tabrizi SJ. Huntington's disease: from molecular pathogenesis to clinical treatment. Lancet Neurol 2011; 10:83-98.

28. Zajac JD, Fui MNT. Kennedy's disease: clinical significance of tandem repeats in the androgen receptor. In: Hannan AJ, ed. Tandem Repeat Polymorphisms: Genetic Plasticity, Neural Diversity and Disease. Austin/New York: Landes Bioscience/Springer Science+Business Media, 2012:153-168.

29. Zu T, Gibbens B, Doty NS et al. Non-ATG-initiated translation directed by microsatellite expansions. Proc Natl Acad Sci USA 2011; 108:260-265.

30. Pearson CE. Repeat associated non-ATG translation initiation: one DNA, two transcripts, seven reading frames, potentially nine toxic entities! PLoS Genet 2011; 7(3):e1002018.

31. Messaed C, Rouleau GA. Molecular mechanisms underlying polyalanine diseases. Neurobiol Dis 2009; 34:397-405.

32. Shoubridge C, Gecz J. Polyalanine tract disorders and neurocognitive phenotypes. In: Hannan AJ, ed. Tandem Repeat Polymorphisms: Genetic Plasticity, Neural Diversity and Disease. Austin/New York: Landes Bioscience/Springer Science+Business Media, 2012:185-204.

33. Brouwer JR, Willemsen R, Oostra BA. Microsatellite repeat instability and neurological disease. Bioessays 2009; 31:71-83.

34. Kumari D, Usdin K. Chromatin remodeling in the noncoding repeat expansion diseases. J Biol Chem 2009; 284:7413-7417.

35. Schmucker S, Puccio H. Understanding the molecular mechanisms of Friedreich's ataxia to develop therapeutic approaches. Hum Mol Genet 2010; 19:R103-R110.

36. Loesch D, Hagerman R. Unstable mutations in the FMR1 gene and the phenotypes. In: Hannan AJ, ed. Tandem Repeat Polymorphisms: Genetic Plasticity, Neural Diversity and Disease. Austin/New York: Landes Bioscience/Springer Science+Business Media, 2012:78-114.

37. Corben LA, Georgiou-Karistianis N, Bradshaw JL. Characterising the neuropathology and neurobehavioural phenotype in Friedreich ataxia: A systematic review. In: Hannan AJ, ed. Tandem Repeat Polymorphisms: Genetic Plasticity, Neural Diversity and Disease. Austin/New York: Landes Bioscience/Springer Science+Business Media, 2012:169-184.

38. Li LB, Bonini NM. Roles of trinucleotide-repeat RNA in neurological disease and degeneration. Trends Neurosci 2010; 33:292-298.

39. Nakamori M, Thornton C. Epigenetic changes and noncoding expanded repeats. Neurobiol Dis 2010; 39:21-27.

40. Todd PK, Paulson HL. RNA-mediated neurodegeneration in repeat expansion disorders. Ann Neurol 2010; 67:291-300.

41. Fondon JW III, Garner HR. Molecular origins of rapid and continuous morphological evolution. Proc Natl Acad Sci USA 2004; 101:18058-18063.

42. Gerber HP, Seipel K, Georgiev O et al. Transcriptional activation modulated by homopolymeric glutamine and proline stretches. Science 1994; 263:808-881.

43. Faux NG, Huttley GA, Mahmood K et al. RCPdb: An evolutionary classification and codon usage database for repeat-containing proteins. Genome Res. 2007; 17:1118-1127.

44. Matsushima N, Tanaka T, Kretsinger RH. Non-globular structures of tandem repeats in proteins. Protein Pept Lett 2009; 16:1297-1322.

45. Riley DE, Krieger JN. Embryonic nervous system genes predominate in searches for dinucleotide simple sequence repeats flanked by conserved sequences. Gene 2009; 429:74-79.

46. Hammock EAD, Young LJ. Microsatellite instability generates diversity in brain and sociobehavioral traits. Science 2005; 308:1630-1634.

47. Rubinsztein DC, Amos B, Cooper G. Microsatellite and trinucleotide-repeat evolution: evidence for mutational bias and different rates of evolution in different lineages. Philos Trans R Soc Lond B Biol Sci 1999; 354:1095-1099.

48. Zeitlin S, Liu JP, Chapman DL et al. Increased apoptosis and early embryonic lethality in mice nullizygous for the Huntington's disease gene homologue. Nat Genet 1995; 11:155-163.

49. Dragatsis I, Levine MS, Zeitlin S. Inactivation of Hdh in the brain and testis results in progressive neurodegeneration and sterility in mice. Nat Genet 2000; 26:300-306.

50. Zheng S, Clabough EB, Sarkar S et al. Deletion of the huntingtin polyglutamine stretch enhances neuronal autophagy and longevity in mice. PLoS Genet 2010; 6(2):e1000838.

51. Manolio TA, Collins FS, Cox NJ et al. Finding the missing heritability of complex diseases. Nature 2009; 461:747-753.

52. Singleton AB, Hardy J, Traynor BJ et al. Towards a complete resolution of the genetic architecture of disease. Trends Genet 2011; 26:438-442.

53. Elden AC, Kim HJ, Hart MP et al. Ataxin-2 intermediate-length polyglutamine expansions are associated with increased risk for ALS. Nature 2010; 466:1069-1075.

54. Dejesus-Hernandez M, Mackenzie IR, Boeve BF et al. Expanded GGGGCC hexanucleotide repeat in noncoding region of C9ORF72 causes chromosome 9p-Linked FTD and ALS. Neuron 2011; 72:245-256.
55. Renton AE, Majounie E, Waite A et al. A hexanucleotide repeat expansion in C9ORF72 is the cause of chromosome 9p21-linked ALS-FTD. Neuron 2011; 72:257-268.
56. Gonitel R, Moffitt H, Sathasivam K et al. DNA instability in postmitotic neurons. Proc Natl Acad Sci USA 2008; 105:3467-3472.
57. Haberman Y, Amariglio N, Rechavi G et al. Trinucleotide repeats are prevalent among cancer-related genes. Trends Genet 2008; 24:14-18.
58. Vinces MD, Legendre M, Caldara M et al. Unstable tandem repeats in promoters confer transcriptional evolvability. Science 2009; 324:1213-1216.
59. Gemayel R, Vinces MD, Legendre M et al. Variable tandem repeats accelerate evolution of coding and regulatory sequences. Annu Rev Genet 2010; 44:445-477.
60. Feinberg AP, Irizarry RA. Evolution in health and medicine Sackler colloquium: Stochastic epigenetic variation as a driving force of development, evolutionary adaptation and disease. Proc Natl Acad Sci USA 2010; 107 Suppl 1:1757-1764.
61. O'Dushlaine CT, Shields DC. Marked variation in predicted and observed variability of tandem repeat loci across the human genome. BMC Genomics 2008; 9:175.
62. Hannan AJ. TRPing up the genome: Tandem repeat polymorphisms as dynamic sources of genetic variability in health and disease. Discov Med 2010; 10:314-321.
63. Baillie JK, Barnett MW, Upton KR et al. Somatic retrotransposition alters the genetic landscape of the human brain. Nature 2011; 479:534-537.

CHAPTER 2

EVOLUTION OF SIMPLE SEQUENCE REPEATS AS MUTABLE SITES

David G. King

*Department of Anatomy and Department of Zoology, Southern Illinois University Carbondale,
Carbondale, Illinois, USA*
Email: dgking@siu.edu

Abstract: Because natural selection is commonly presumed to minimize mutation rates, the discovery of mutationally unstable simple sequence repeats (SSRs) in many functional genomic locations came as a surprise to many biologists. Whether such SSRs persist in spite of or because of their intrinsic mutability—whether they constitute a genetic burden or an evolutionary boon—remains uncertain. Two contrasting evolutionary explanations can be offered for SSR abundance. First, suppressing the inherent mutability of repetitive sequences might simply lie beyond the reach of natural selection. Alternatively, natural selection might indirectly favor SSRs at sites where particular repeat-number variants have provided positive contributions to fitness. Indirect selection could thereby shape SSRs into "tuning knobs" that facilitate evolutionary adaptation by implementing an implicit protocol of incremental adjustability. The latter possibility is consistent with deep evolutionary conservation of some SSRs, including several in genes with neurological and neurodevelopmental function.

INTRODUCTION

"No one expected that DNA sequences could be so unstable or behave as these do" (Jean-Louis Mandel, quoted in *Science*).[1] The initial discovery that triplet-repeat expansion was responsible for several neurological diseases surprised many geneticists. Perhaps even more surprising has been the subsequent discovery that repeat-number variation can also have nonpathological yet biologically significant effects[2-5]. Functional consequences attend repeat-number variation in a wide diversity of simple sequence repeats (SSRs; the term encompasses both microsatellite and minisatellite DNA, i.e., tandem repetitive

Tandem Repeat Polymorphisms: Genetic Plasticity, Neural Diversity and Disease,
edited by Anthony J. Hannan ©2012 Landes Bioscience and Springer Science+Business Media.

sequences with motifs ranging from mono-, di- and tri-nucleotides up to several tens of basepairs in length). These surprisingly unstable repetitive stretches are so profusely distributed throughout eukaryotic genomes that many genes, perhaps most, include one or more variable SSRs within regulatory and/or coding domains. Although most attempts to count SSRs have been restricted by motif length, number of repeats or functional domain, two recent reports of human[6] and *Daphnia pulex*[7] have catalogued hundreds of thousands of SSRs in each genome. The distribution of particular motif classes within a genome can vary substantially among different species.

The sheer abundance of SSRs raises an intriguing question. Why has evolution permitted such prolific sources of genetic instability to become so prevalent? Or, in slightly less teleological language, how do these highly mutable genetic patterns escape elimination or suppression by natural selection? Two contrasting answers can be suggested.

One intriguing possibility is that genetic patterns which confer special modes of mutability are serving an "evolutionary function." In this view, the unexpected prevalence, diversity and high mutation rates of SSRs support a hypothesis that appropriately constrained mutability can be evolutionarily beneficial. If so, then pathological expansion of SSRs is more than just a clinical curiosity. Just as other diseases throughout history have stimulated investigation of basic biological processes, repeat-expansion pathologies may be revealing a previously unsuspected role for a ubiquitous feature of normal genetic organization.

But a more conventional and apparently more parsimonious explanation is that natural selection has but limited ability to eliminate mutation. According to a widely accepted principle of evolutionary biology, mutations of any sort occur not because variation is necessary for adaptation but simply because total suppression of mutation is not feasible. Hence SSRs' surprising instability represents nothing more than an accidental consequence of the replication slippage that inevitably accompanies sequence repetition. Each particular example of repeat-number variation, as documented throughout this volume, may be interesting in itself for its effect on a particular gene. But there should be no reason to expect such mutable sites to provide any novel insight into evolutionary processes.

This chapter discusses both possibilities, beginning with a brief historical review of the conventional argument, sharpened repeatedly over the past century, that all mutations are essentially accidents. Some inadequacies of this argument will then be considered in light of the less familiar "evolutionary function" hypothesis. While evidence for each explanation remains inconclusive, this essay will advocate the proposal that SSRs are common precisely because their particular style of mutation facilitates evolutionary adaptation and has therefore been favored, albeit indirectly, by natural selection. The distinctive properties of SSRs, which initially appeared so surprising, accord neatly with this proposed evolutionary function. SSRs might even have a special role in behavioral evolution through their effects on genes involved in nervous system development and function.

A BRIEF HISTORY OF THE "MUTATION" CONCEPT

Several interrelated questions have concerned geneticists for much of the past century. Which aspects of genetic variation should be defined as "mutations"? What is the fundamental nature of mutational mechanisms? Are mutation rates optimized to ensure evolutionary adaptability? Or are all mutations essentially accidental errors in DNA replication? Satisfactory answers to these questions remain elusive, although as Darwin[8] himself noted, "Some authors believe it to be as much the function of the reproductive system to produce individual differences, or very slight deviations of structure, as to make the child like its parents."

The precise meaning of "mutation" has evolved as the word itself was assimilated into the language of genetics. For Hugo de Vries,[9] one of the pioneering rediscoverers of Mendel's laws at the start of the twentieth century, a mutation was a saltational jump leading to a new species. But by 1919 Calvin Bridges[10] was applying the term more broadly, with "no restrictions of degree, covering the most extreme as well as the slightest detectable inherited variation." Bridges,[10] who worked with Thomas Hunt Morgan at Columbia University's famous *Drosophila* laboratory, also shared with many modern geneticists an intuitive understanding that deleterious mutations must vastly outnumber beneficial ones:

> *Any organism as it now exists must be regarded as a very complex physico-chemical machine with delicate adjustments of part to part. Any haphazard change made in this mechanism would almost certainly result in a decrease of efficiency.... Only an extremely small proportion of mutations may be expected to improve a part or the interrelation of parts in such a way that the fitness of the whole organism for its available environments is increased.*

Bridges simply presumed that mutations are "haphazard," with the extreme unlikelihood of beneficial mutations being a self-evident corollary. But by 1937, Alfred H. Sturtevant (another member of Morgan's *Drosophila* group at Columbia) had confirmed "accidental" as a defining attribute of mutation. Sturtevant[11] reasoned that selection should favor the lowering of mutation rates to reduce the loss of reproductive potential due to deleterious mutation. He then considered a possible tendency in the opposite direction based on the necessity of mutations for evolutionary adaptation:

> *It seems at first glance that there should be a counter-selection, due to the occurrence of favorable mutations. It is true that favorable mutations furnish the only basis for improvement of the race and must be credited with being the only raw material for evolution. It would evidently be fatal for a species, in the long run, if its mutation rate fell to zero, for adjustment to changing conditions would then not long remain possible.*

But Sturtevant[11] rejected this possibility:

> *While this effect may occur, it is difficult to imagine its operation. It is clear that the vast majority of mutations are unfavorable.... [F]or every favorable mutation, the preservation of which will tend to increase the number of genes in the population that raises the mutation rate, there are hundreds of unfavorable mutations that will tend to lower it. Further, the unfavorable muta-tions are mostly highly unfavorable and will be more effective in influencing the rate than will the relatively slight improvements that can be attributed to the rare favorable mutations.*

Sturtevant[11] then asked, rhetorically, "why does the mutation rate not become reduced to zero?" To this critical question, he gave a famous reply: "No answer seems possible at present, other than the surmise that the nature of genes does not permit such a reduction. In short, **mutations are accidents and accidents will happen**" (emphasis added).

Three decades later, in his classic 1966 text *Adaptation and Natural Selection*, evolutionary theorist George C. Williams[12] responded to what was still a frequent assertion, "that natural selection will not produce too low a mutation rate because that would reduce the evolutionary plasticity of the species," with a conclusion even stronger than Sturtevant's:

> *[N]atural selection of mutation rates has only one possible direction, that of reducing the frequency of mutation to zero. That mutations should continue to occur...requires no special explanation. It is merely a reflection of the unquestionable principle that natural selection can often produce mechanisms of extreme precision, but never of perfection.... Evolution has probably reduced mutation rates to far below species optima, as the result of unrelenting selection for zero mutation rate in every population. Mutation is, of course, a necessary precondition to continued evolutionary change. So evolution takes place, not so much because of natural selection, but to a large degree in spite of it.*

This same basic argument continues to be re-iterated into our present century. For example, Sniegowski et al[13] write:

> *[I]t can be appealing to suppose that the genomic mutation rate is adjusted to a level that best promotes adaptation. Most mutations with phenotypic effects are harmful, however and thus there is relentless selection within populations for lower genomic mutation rates. Selection on beneficial mutations can counter this effect by favoring alleles that raise the mutation rate, but the effect of beneficial mutations on the genomic mutation rate is extremely sensitive to recombination and is unlikely to be important in sexual populations.*

As Sniegowski et al[13] explain, it is the cost of accurate DNA replication, not a need for evolutionary plasticity, that determines mutation rates:

> *The physiological cost of reducing mutation below the low level observed in most populations may be the most important factor in setting the genomic mutation rate in sexual and asexual systems, regardless of the benefits of mutation in producing new adaptive variation. Maintenance of mutation rates higher than the minimum set by this 'cost of fidelity' is likely only under special circumstances.*

A recent authoritative review of mutation rate evolution (see ref. 14) again echoed Sturtevant's[11] argument and re-affirmed that "the cost of fidelity is the generally accepted explanation for nonzero mutation rates in multicellular eukaryotes."

This prevailing view of mutation, as exemplified by the quotations above, has changed little over the past century in spite of a tremendous increase in our understanding of DNA metabolism with its associated diversity of mutational mechanisms. Mutations continue to be regarded as accidental errors such that the vast majority must be deleterious, albeit with some acknowledged exceptions (below).

A BRIEF CRITIQUE OF MUTATIONS AS ACCIDENTS

One reason why subsequent authors still rehearse the essentials of Sturtevant's argument lies in the tenacity of a contrary narrative in which higher-than-minimal mutation rates really are maintained because of their past contribution to adaptive evolution. Although Sturtevant's and Williams' arguments have dominated genetics for several decades, this contrary view is resurging: "Increasing numbers of biologists are invoking 'evolvability' to explain the general significance of genomic and developmental phenomena affecting genetic variation" (see ref. 15).

Early interpretations, naively attributing evolvability (also variously called "evolutionary plasticity," "evolutionary potential," etc.) to selection for the future good of the species, can be dismissed on grounds that "natural selection has no foresight" (e.g., ref. 16). But just as implications of "design" are hard to avoid when discussing the function of complex adaptive structures, so also are implications of "foresight" hard to avoid when speaking of genomic patterns that generate novel variation.[17,18] The critical issue here lies not with foresight but with the production of hereditary variation as a proper biological function, i.e., as an advantageous trait that selection has favored over preceding generations. The philosophical foundations for evolutionary theory do not require that mutations be "accidental" or "haphazard," only that they be "random" with respect to current adaptive needs. Otherwise mutation itself rather than natural selection would direct the process of adaptation.[19]

The conviction that mutations are haphazard persists largely through repeated assertion in textbooks and prominent publications—e.g., "It is common sense that most mutations that alter fitness at all will lower it;"[20] "the vast majority of mutations with observable effects are deleterious."[14] Yet although many studies have measured the accumulation of deleterious mutations, there remains even now remarkably little experimental evidence regarding the proportion of mutations that increase or decrease fitness to some degree[21] and none that effectively distinguishes among different classes of mutation. In the absence of such evidence, the classic "mutations are accidents" argument becomes essentially circular: Because mutations are accidental, if they affect fitness at all they must mostly be deleterious. Because fitness-affecting mutations are mostly deleterious, selection cannot favor mutability. Because selection cannot favor mutability, mutations must occur only as accidents.

But if mutation is defined simply and broadly—i.e., any change in inherited genetic information, with "no restrictions to degree"[10]—then it clearly embraces the consequences of several highly organized processes that are hardly accidental. The most familiar example is meiotic recombination, whereby novel gene sequences can be created by precise reciprocal exchange between alleles that differ at more than one site. Sexual reproduction normally assures that every gamete has a unique haploid genotype, randomly generated from a vast number of viable possibilities. Yet even though any particular genotype is an unpredictable chance event, the label "accidental error" is inappropriate (except for inviable aneuploids). Although the selection pressures responsible for maintaining sex and recombination in most plant and animal populations remain controversial, most theories nevertheless recognize variation in one form or another as the principal overriding advantage.[22] Reconciling these well known facts with the "mutations are accidents" argument has necessitated, as routine practice, that the products of recombination be explicitly excluded from the definition of "mutation" (e.g., ref. 23: "**Mutation**: An error in replication of a nucleotide sequence or any other alteration of the genome that is not manifested as reciprocal recombination").

Also often set apart from the "mutation" category is "programmed gene re-arrangement," a source for highly structured variation used by parasitic trypanosomes to alter expression of surface antigens as they reproduce within a host, thereby facilitating evasion of the host's immune response.[24] Additional strategies for the active generation of internally organized variation are known in prokaryotic organisms. For example, mutation-prone "contingency genes"[25] are recognized among microbial geneticists as having a legitimate evolutionary role, predictably generating mutations of particular types that help assure survival of some descendents even if current conditions change. Mutations produced by contingency genes are still "random" (i.e., they occur whether needed or not and only in chance individuals), but they are no more accidental errors than are particular results from shuffling cards or rolling dice in an orderly game of chance. Such mutational mechanisms are presumably shaped by recurring shifts in selection pressure over many preceding generations. Explaining genetic patterns that have such evolutionary functions "requires a change in our attitude towards the sources of genetic variation, which until recently have largely been thought to rely on errors and accidents happening to DNA."[26]

Routinely excluding such manifestly non-accidental sources of genetic variation from consideration as mutation has hindered recognition that the concept of heritable genetic change embraces several highly constrained mechanisms in addition to those "errors" that are patently accidental. The resulting semantic confusion is exacerbated by common reference to "the genomic mutation rate" as if this were a single parameter characterizing a well-defined unitary process. Even apart from the special exceptions above, "mutation" remains a composite concept that encompasses a number of disparate mechanisms (e.g., nucleotide substitution, replication slippage, transposable element activity, etc.). This diversity needs to be disaggregated. "Mutation rate" should never be described by a single statistic.[27] Instead, each separate source of DNA sequence modification should be analyzed on its own terms.

Once the significance of several distinct sources of hereditary variation is acknowledged, the simplistic conclusion of relentless selection for lower mutation rates becomes far less compelling. Furthermore, classical analyses of mutation rate evolution[11-13] have generally assumed that particular genes determine an average genome-wide mutation rate by influencing the overall fidelity of nucleotide base-pairing. Any "mutator allele" that increases this mutation rate must reduce fitness by causing haphazard errors throughout the genome. Meanwhile, "the effect of beneficial mutations on the genomic mutation rate is extremely sensitive to recombination,"[13] and in sexually reproducing populations any fortuitous beneficial mutant would have only a small probability of close linkage to the mutator. Standing in sharp contrast to such analyses are the parameters that could allow indirect selection to favor increased mutability—i.e., a relatively low likelihood for deleterious fitness effects together with reliable linkage between beneficial mutant alleles and a cause for increased mutability. Remarkably, these are exactly the parameters that characterize the mutability of SSRs.

SSRs AS SOURCES OF "TUNING KNOB" VARIATION

Although low background rates for nucleotide substitution appear consistent with Williams'[12] "unrelenting selection for zero mutation rate," rates for repeat-number mutations at SSR sites can be several orders of magnitude higher. The resulting variation in repeat number is so pervasive it can be used for DNA fingerprinting. Because phenotypic effects are seldom evident, such variation has been "generally assumed to evolve neutrally."[28] But an unqualified assumption that repeat number variants have no significant effect on evolutionary fitness can no longer be justified. Evidence has been accumulating for almost three decades that variation in repeat number can and does exert small-scale, quantitative effects on many aspects of gene function.[2-5] Even if the percentage of SSRs that do influence phenotype is quite small, SSRs are so numerous that repeat-number variants must still make a substantial contribution to overall phenotypic variation.

Functional effects of repeat-number variation are not limited to rare cases of pathological expansion, nor even to SSRs that directly encode amino acid repeats. So-called "noncoding" SSRs with a variety of different motifs are also found in introns, in UTRs and in upstream and downstream regulatory regions of many genes. (The adjective "noncoding" is potentially misleading, as it typically refers to any sequence that does not directly encode peptide sequences with canonical triplet codons. As ironically noted in a recent article in a prominent journal,[29] "many functions are encoded...in the noncoding portion of the genome"). Early on, evidence that such mutation-prone SSR sites could provide an abundant supply of small-scale quantitative genetic variation led to speculation that these sites function as "evolutionary tuning knobs."[30,31] SSRs would thus embody an "implicit protocol"[32] for incremental adjustability.

In fact, classical evolutionary theory has long held that "mutations of small effect" can improve fitness with a probability approaching fifty percent,[33] especially in natural conditions where selection pressures vary over space and time. And as early as the 1960s Levins[34,35] had demonstrated how changing or heterogenous environments can lead to increased mutation rates. But prior to discovery of SSRs, hardly anyone had imagined how unstable DNA sequences such as SSRs could evade Bridges'[10] intuitive expectation for a very low proportion of beneficial mutations. SSRs demonstrate how readily a simple mechanism can preferentially yield mutations whose characteristically small effect on phenotype could carry a nonnegligible probability of selective advantage as well as a low probably for harm. Although any newly-arising allele of small effect, even if beneficial, can be readily lost through genetic drift before weak selection can increase its prevalence in a population,[36] nevertheless high rates of repeat-number mutation guarantee a continued resupply of such alleles.

Thus the presumption that deleterious mutations must vastly outnumber beneficial ones has become quite doubtful for repeat-number variants at SSR sites. Even at those sites associated with repeat expansion diseases, pathological expansion arises only from rare "premutation" alleles at one extreme of a normal, nonpathogenic range. Most variation falls within this relatively safe range, but if selection should favor a shift in repeat number then the high mutation rate of SSRs assures that appropriate new variation will be quickly forthcoming.

Empirical evidence for nonpathological phenotypic effects of naturally occurring SSR variation, especially for noncoding SSRs, remains quite limited relative to the multitude of SSRs found throughout most eukaryotic genomes. Nevertheless, several cases already include circumstantial evidence that SSR variants have supported adaptive differentiation among natural populations (ref. 37; more recently refs. 38-40). Vinces et al[41] have provided the strongest experimental evidence to date that SSRs can indeed serve an evolutionary function in eukaryotes, reporting not only functional effects of repeat-number variation within promotor regions but also establishing that this variation could be the basis for evolutionary adaptation in laboratory populations of yeast (*Saccharomyces cerevisiae*).

Not only should the characteristic mutability of SSRs carry a fair probability for adaptive advantage, this mutability is also inextricably associated with sequence repetition at each individual SSR site. At any SSR, each repeat-number allele retains the inherent site-specific mutability by which it arose. Thus selection favoring any advantageous repeat-number variant also favors the site's potential for incremental adjustability. Since recombination cannot separate cause (i.e., the site-specific adjustability protocol) from consequence (particular alleles), this intrinsic linkage should strongly dispose SSRs toward indirect site-by-site selection for adaptively appropriate mutation rates.[37,42-45]

Selective shaping of local mutation rates requires some hereditary variation in those rates. In the case of SSRs, the rate for repeat-number mutation can be lower or higher depending on the presence or absence of interruptions or imperfect repeats.[46] Once mutation rate variation exists at a particular site, indirect selection acts through direct selection upon individual repeat-number mutant alleles. Indirect selection against a higher rate is relatively inefficient, since any copy of an allele with a higher mutation rate can be eliminated only after that copy gives rise to a deleterious repeat-number mutant. In contrast, indirect selection favoring a high-mutation-rate allele can be much more effective. Once a beneficial repeat-number mutation appears at a single copy of such an allele, direct selection that increases the frequency of the beneficial mutant necessarily increases the frequency of the high mutation rate as well, since this mutation rate is retained by each copy of the beneficial mutant. Thus, as long as any beneficial variants appear within a population before every copy of a high-mutation-rate site is eliminated by direct selection against a long series of deleterious mutants, the higher rate (i.e., the "tuning knob" protocol of incremental adjustability) will prevail at that site.

ORIGIN AND MAINTENANCE OF SSRs

As long as all SSRs were believed to lie in nonfunctional intergenic domains, neither their mutability nor their abundance posed any special theoretical difficulty. Since their prevalence in functional domains has become more widely appreciated, however, simply dismissing them as meaningless genetic junk is no longer adequate. A satisfactory explanation should answer two separate questions. First, by what mechanisms do SSRs originate de novo? Second, once any particular SSR has arisen, how is its presence maintained over time?

Explanations for the origins of SSRs remain, at best, incomplete.[46] Minisatellite SSRs require some mechanism for initial duplication of a lengthy motif. In contrast, microsatellite SSRs with their shorter motifs can arise either by chance nucleotide substitution in previously nonrepetitive DNA or by short insertions that duplicate adjacent sequence.[47] Microsatellite SSRs of several different motifs can also be created in abundance through the action of transposable elements (TEs). Moreover, microsatellite SSRs can in turn promote the activity of TEs (for examples, see refs. 43,46). This association suggests the intriguing possibility of a synergistic co-evolution between SSRs and TEs, especially since TEs have also been proposed as major contributors to evolvability.[48,49]

Regardless of how SSRs originate, a complete explanation for their abundance in functional domains should consider not only the speculative evolutionary "tuning knob" function (above) but also two alternative hypotheses of "adaptive function" and "mutation pressure."

Adaptive Function

If sequence repetition is necessary for some essential adaptive function, then natural selection might retain SSRs in spite of the intrinsic mutability that attends sequence repetition. For SSRs in each of several motif classes and functional domains, this hypothesis requires that sequence repetition offers sufficient immediate functional advantage to offset the presumed liability of frequent mutations. Yet it is far from obvious why sequences with greater stability could not function equally well at most SSR loci. For example, a repeating amino acid stretch can be encoded by a DNA sequence in which codon repetition is interrupted by alternative codon usage, thereby reducing the propensity toward replication slippage. Just such stabilizing interruptions are indeed found in some sequences that encode amino acid repeats. Presumably an immediate adaptive role for any other SSR class could also be served by functionally equivalent but nonrepetitive sequences. Thus, although this "adaptive function" hypothesis may apply in special cases, it seems doubtful that all roles occupied by SSRs require essential sequence repetition. The only function that is plainly shared by all SSRs is that of mutability itself.

Mutation Pressure

SSRs might also be self-perpetuating through their own intrinsic mutability, if this mutation pressure were sufficient to resist spontaneous degradation of sequence repetition by nucleotide substitition.[46] By this hypothesis, once a repetitive sequence exceeds a threshhold number of repeats at which replication slippage becomes frequent, then repeat expansion can reverse any reduction in repeat number. (The threshhold length for replication slippage remains inadequately characterized for most motif classes.)[46,47,50] At the same time, sequential bouts of expansion and contraction can purge mutations that would otherwise interrupt motif repetition. Thus replication slippage alone might explain the persistence of "junk" SSRs in nonfunctional intergenic regions. Nevertheless, its adequacy for explaining SSRs in functional domains remains open to question. For this "mutation pressure" hypothesis by itself to explain the persistence of functional SSRs, one must presume for a wide range of distinct SSRs with differing motifs and functional roles that selection pressure against mutability at each SSR site is too weak to overcome the mutation pressure. This hypothesis also requires a tacit assumption that cost-effective molecular mechanisms for suppressing replication slippage have proven altogether inaccessible to the evolutionary process.

Evolutionary Function

Unlike either the "adaptive function" or the "mutation pressure" hypothesis on its own, the "evolutionary function" hypothesis imposes no requirement for overcoming the putative cost of deleterious accidental mutations. Indeed, the "evolutionary function" hypothesis proposes that SSRs persist in functional domains because of their advantageous mutations rather than in spite of deleterious ones. This hypothesis is supported primarily by a close correspondence between the peculiar properties of SSRs and the special conditions needed to sustain indirect selection of mutability (see SSRs as sources of "tuning knob" variation, above). By this hypothesis, SSRs are selected for their "tuning knob" role as efficient suppliers of potentially adaptive variation, providing an abundant and practically inexhaustible supply of reversible, quantitative variation that can facilitate evolutionary adaptation.[37,42-45]

Nevertheless, even though the "evolutionary function" hypothesis directly contradicts the conventional view that "mutations are accidents," it remains compatible with both "adaptive function" and "mutation pressure" hypotheses. If sequence repetition should be directly advantageous for any particular SSR, then any indirect benefit from incremental adjustability would simply reinforce direct selection. And if mutation pressure can promote the abundance of nonfunctional SSRs, it should also assist the evolutionary function of SSRs by maintaining these sequences through periods when they are not yielding beneficial variants. But without an evolutionary function for SSRs, both of these more conventional hypotheses lack persuasive power for SSRs in functional domains. Thus acknowledging an "evolutionary function" for SSRs may create a more robust explanation for the prevalence of SSRs across their full range of motif classes and genomic locations.

Even though many questions regarding SSR origins and maintenance remain to be addressed by future research, a scenario such as the following may be readily imagined based on the above considerations. Once an SSR appears at a particular site (by whatever mechanism), repeat-number variation will begin to accumulate. If the SSR resides in a truly nonfunctional region of the genome, the ensuing variation should have no impact on fitness. The SSR may then shrink or grow at the whim of replication slippage and genetic drift, perhaps thereby maintaining itself over an indefinite number of generations[46] while incidentally preserving its low-risk potential for some future contribution to adaptation. If an SSR initially emerges at a site where its sequence fits into a pre-existing functional role, or if a novel role becomes established in its region of influence, then repeat-number variation will inevitably have some effect on that role. If that variation happens to be consistently deleterious, selection will favor mutations that suppress replication slippage by shortening or interrupting the repeat, eventually eliminating sequence repetition at the site. But if incremental adjustability is at least occasionally advantageous, then indirect selection will preserve the beneficial variants and with them the site-specific mutational mechanism by which they arose. The mere presence of a variable SSR at any particular functional location would then imply that current or recent adaptation had exploited repeat-number variation at that site.

EVOLUTIONARY CONSERVATION OF SSRs

Buschiazzo and Gemmel[6] have recently analyzed microsatellite SSRs in alignments of the human genome with genomes of 16 other vertebrate species. They report that the extent of SSR conservation between human and other species declines exponentially with increasing phylogenetic distance, paralleling the declining proportion of alignable genome sequence. This is consistent with prior observation that SSRs at particular locations are often not shared among related species.[51] But Buschiazzo and Gemmel[6] also report a surprisingly high level of conservation over deep evolutionary time. Almost 200,000 microsatellite SSRs are shared between human and at least one nonprimate species, with over 10,000 conserved between human and opossum (*Monodelphus domestica*). Chicken (*Gallus gallus*), frog (*Xenopus tropicalis*), zebrafish (*Danio rerio*) and pufferfish (*Tetraodon nigroviridis*) each share with human over 1000 microsatellites. These latter numbers represent SSRs enduring for several hundred million years.

Unfortunately, little can be safely inferred from sequence conservation alone, without additional information or assumptions. Long term sequence conservation is commonly taken as evidence of an important sequence function, at least for protein-coding sequences. However the reliability of such inference for SSRs, especially for noncoding SSRs, should not be presumed without further analysis. Exponential decline in the number of SSRs conserved over time since evolutionary divergence could be consistent with either of two quite different interpretations.

One possibility is that SSRs persist even without any enduring function, simply because replication slippage can plausibly shield an SSR (to an unknown extent) from routine degradation by nucleotide substitution and genetic drift (see Mutation pressure, under Origin and maintenance of SSRs, above). They would arise spontaneously and then eventually disappear, with a "life cycle" whose duration or "half life" is determined by the distribution over time of competing types of mutations.[46] In this case, exponential decline in SSR conservation, including the appearance of exceptional conservation for a few SSRs, could be nothing more than a simple statistical expectation of random decay across a very large array of SSRs. In other words, the phylogenetic lability of SSRs among related species might simply reinforce the conventional assumption[28] that these sequences "evolve neutrally" under the influence of their intrinsic mutability.

On the other hand, this same pattern of exponential decline in SSR conservation might obtain because specific selection pressures vary extensively across a phylogeny. If SSRs are preserved by indirect selection (see Evolutionary function, under Origin and maintenance of SSRs, above), then the observed decline in proportion of conserved SSRs could result from a decreasing fraction of adjustable sites that are shared over time by diverging species. After all, increasing phylogenetic distance between species is commonly accompanied by increasing divergence of adaptive traits, which must be accomplished through patterns of sequence divergence that remain largely unexplored. In this case, exceptional sequence conservation would indeed indicate an important function. But lack of conservation would not necessarily indicate any absence of function. An SSR could serve a temporarily important function during a particular episode of adaptation, only to be superseded by other SSRs as adaptive divergence continued.

Buschiazzo and Gemmel[6] also reported that the extent of conservation declines more rapidly for noncoding SSRs than for those located in exons. If SSRs do serve an evolutionary function, then this observation suggests that noncoding SSRs experience weaker functional constraint and hence may be serving more labile "tuning knob" roles. This in turn appears consistent with current understanding that much adaptive evolution, at least at the level of morphology and behavior, occurs through changes in regulatory sequences where many noncoding SSRs are found. Deep conservation of any particular non-exonic SSR would then suggest an especially persistent locus for regulatory adjustment. Unfortunately, in such domains we currently have little basis for predicting either the type or the degree of sequence divergence that accompanies adaptive divergence.

A peculiar style of SSR conservation, shared by a small set of 22 human genes, was recently discovered by Riley and Krieger.[52,53] The transcript for each of these genes includes in its untranslated region a long uninterrupted dinucleotide SSR whose upstream flanking sequence is highly conserved between human and opossum (*Monodelphis domestica*). Alignments of these genes with homologues in 17 nonhuman vertebrate species revealed that the human dinucleotide SSRs were frequently replaced by other SSRs with alternative motifs. Thus these sites reveal an evolutionary history during which each site's character as an SSR has been retained even while its specific sequence has been extensively remodelled. Something more than simple mutation pressure has evidently constrained evolution at these sites, to retain a basic SSR pattern in spite of mutational churning sufficient to transform the sites. A constraint based on immediate adaptive function should be expected to minimize the extent of sequence remodelling, while mutation pressure sufficient to remodel the site should not, by itself, readily recreate a different SSR. Thus an evolutionary function that constrains the site as an SSR while exploiting the mutational flexibility of simple sequence repetition appears especially plausible for these sites.

Apart from these few intriguing examples, patterns of SSR conservation in both coding and noncoding domains remain poorly characterized. Future research that associates conserved SSRs with particular functional domains and gene ontologies may help discriminate among alternative explanations for their abundance.

SSRs WITH NEUROLOGICAL SIGNIFICANCE

For the purpose of this volume, the relationship between SSR variation and nervous system function has special relevance. An intuitive expectation that the evolution of adaptive behavior must require exquisite adjustment of innumerable parameters of neuronal anatomy and physiology suggests, to this writer at least, the possibility that incremental adjustability supplied by SSRs may play a special role in nervous system evolution. Several observations appear consistent with such a possibility.

First of all is the remarkable predominance of neurological disorders in the list of human repeat expansion pathologies (e.g., ref. 54, and other chapters in this volume). Whether such a functional bias is meaningful or just a statistical fluke is not yet clear. Nevertheless, at least until evidence comes to light that other systems are equally prone to pathological triplet repeat expansions, one might entertain an ad hoc speculation that relatively recent and rapid human evolution not only has utilized triplet repeats in many genes with neurological function but also has pushed several of these to the limit of their functional capacity, dangerously near the edge of the "premutation" repeat-number range where further expansion can become progressively pathological.

Prompted by early reports of an association between triplet repeats and neurological disorders, many laboratories began seeking additional examples of trinucleotide repeats in protein-coding domains. As early as 1994, Gerber et al[3] reported homopeptide stretches (which can be encoded by either perfect or imperfect triplet DNA repeats) in several transcription factors. A few years later, a search by Karlin and Burge[55] for proteins containing multiple homopeptide stretches found a preponderance of developmental proteins, including many involved in nervous system development. Huntley et al[56] confirmed that SSRs are overrepresented in developmental proteins but also found that apart from some polyhistidine sequences SSRs are not especially enriched in genes expressed in brain and nervous system. Łabaj et al[57] recently reported that polyleucine is overrepresented in signal peptides, transient regions soon cleaved and degraded from growing protein chains. Phenotypic effects of repeat variation at most such sites have yet to be demonstrated. Nevertheless, a few recent studies have emphasized the possibility that such SSRs play evolutionary roles involving a variety of regulatory mechanisms. For example, Huntley and Clark,[58] analyzing amino acid repeats in 12 species of the fly genus *Drosophila*, found such sequences to be especially common in genes encoding developmental, signaling and regulatory factors. They also report that "the presence of repeats is associated with an increase in evolutionary rate upon the entire sequence in which they are embedded." In a different fly species (*Teleopsis dalmanni*, "stalk-eyed" flies with bizarre head shape), Birge et al[59] found that genes encoding glutamine repeats were overrepresented among genes expressed in developing head, including nervous tissues, with several of these genes showing repeat-number variation that was correlated with variation in head shape.

Coding SSRs with minisatellite motifs (i.e., motif sequences longer than six basepairs) have also been implicated in nervous system evolution. Tompa[60] reports that evolution by SSR expansion has shaped a number of intrinsically unstructured proteins, including at least four with known neurological function: Neural zinc finger factor-1 (with a repeating motif of 44 basepairs), neuromodulin bt (an 11 basepair repeat), neurofilament-H (a hexanucleotide repeat, at the upper end of the microsatellite range) and prion protein (an octanucleotide repeat). Tompa[60] concludes, "these repeat regions carry important functions and, thus, their inherent genetic instability and the structurally/functionally permissive nature of unstructured proteins provide a unique combination for rapid and advantageous evolutionary changes." Tyedmers et al[61] have independently hypothesized the prion PSI[+] "as a capacitor to promote evolvability," because of its ability to reveal cryptic genetic variation (at least in yeast) and thus promote survival in fluctuating environments.

As the sample above indicates, most studies associating SSRs with gene functions have concentrated on those, especially triplet repeats, that occur in exons. Although many genes also contain SSRs in noncoding domains, ontologies for such genes remain poorly characterized. One exception is the set of human genes found by Riley and Krieger[52,53] to contain transcribed but untranslated dinucleotide SSRs flanked by deeply conserved sequences (above). Of these 22 genes, 19 have known functions. Remarkably, all but one of these 19 genes have critical roles in the embryonic nervous system. Thus this newly described genomic pattern, possessing both a highly conserved feature (SSRs in transcribed but untranslated regions) and a highly variable feature (motif patterns in the SSR sites), appears essential for several neurodevelopmental functions that evidently entail repeated evolutionary remodelling of the included SSR.

This miscellany of observations (also see ref. 54 as well as other chapters in this volume) implicates a wide variety of mutationally variable SSRs in neurological as well as other functions. But apart from the evident role of repeat expansion in several human neurological disorders, definitive conclusions regarding a special or widespread role for SSRs in behavioral evolution remain tantalizingly out of reach. Nevertheless such observations are surely sufficient to warrant some attention to the possibility of an evolutionary "tuning knob" role for any SSR that is found anywhere near a gene with any neurological function.

CONCLUSION

George C. Williams, who argued so strongly in 1966, that "natural selection of mutation rates has only one possible direction, that of reducing the frequency of mutation to zero,"[12] also admitted in the same volume that "our current picture of evolutionary adaptation is, at best, oversimplified and naive."[62] SSRs, by exemplifying how high mutation rates may prevail when the probability of deleterious variation is sufficiently low and beneficial mutants are directly linked to the mutational mechanism, may thus guide our understanding of mutation beyond Sturtevant's[11] dismissive dictum that "accidents will happen."

At the very least, it has become evident that mutationally unstable SSRs can have important biological functions. The characteristic properties, abundant distribution and phylogenetic conservation of SSRs are consistent with multiple roles for these surprisingly mutable sequences, including the production of potentially advantageous variation. A complete explanation for SSRs will surely include an evolutionary role, with their mutability shaped by indirect selection to provide an implicit "tuning knob" protocol of incremental adjustability. If so, then we should be attentive to the possibility that repeat-number variation is influencing the function of practically any gene, including most especially those that guide the development and function of nervous tissue.

REFERENCES

1. Morell V. The puzzle of the triplet repeats. Science 1993; 260:1422-1423.
2. Hamada H, Seidman M, Howard BH et al. Enhanced gene expression by the poly(dT-dG) ·poly(dC-dA) sequence. Mol Cellular Biol 1984; 4:2622-2630.
3. Gerber HP, Seipel K, Georgiev O et al. Transcriptional activation modulated by homopolymeric glutamine and proline stretches. Science 1994; 263:808-811.
4. Kashi Y, King DG, Soller M. Simple sequence repeats as a source of quantitative genetic variation. Trends Genet 1997; 13:74-78.
5. Li YC, Korol AB, Fahima T et al. Microsatellites within genes: Structure, function and evolution. Mol BiolEvol 2004; 21:991-1007.
6. Buschiazzo E, Gemmel NJ. Conservation of Human Microsatellites across 450 Million Years of Evolution. Genome Biol Evol 2010; 2:153-165.
7. Sung W, Tucker A, Bergeron DR et al. Simple sequence repeat variation in the Daphnia pulex genome. BMC Genomics 2010; 11:691. doi:10.1186/1471-2164-11-691.
8. Darwin CR. On the Origin of Species by Means of Natural Selection. London: John Murray, 1859:131 (Facsimile edition by Harvard University Press, Cambridge, Massachusetts, 1964).
9. de Vries H. The origin of species by mutation. Science 1902; 15:721-729.
10. Bridges CB. Specific modifiers of eosin eye color in Drosophila melanogaster. J Exp Zool 1919; 28:337-384.
11. Sturtevant AH. Essays on evolution. I. On the effects of selection on mutation rate. Q Rev Biol 1937; 12:464-467.
12. Williams GC. Adaptation and Natural Selection. Princeton: Princeton University Press, 1966:139-141.
13. Sniegowski PD, Gerrish PJ, Johnson T et al. The evolution of mutation rates: separating causes from consequences. BioEssays 2000; 22:1057-1066.
14. Baer CF, Miyamoto MM, Denver DR. Mutation rate variation in multicellular eukaryotes: causes and consequences. Nature Rev Genet 2007; 8:619-631.
15. Sniegowski PD, Murphy HA. Evolvability. Current Biology 2006; 16:R831-R834.
16. Ayala FJ. One hundred fifty years without Darwin are enough! Genome Res 2009; 19:693-699.
17. Darlington CD. A diagram of evolution. Nature 1978; 276:447-452.
18. Lennox JG. Teleology. In: Keller EF, Lloyd EA, eds. Keywords in Evolutionary Biology. Cambridge, Massachusetts: Harvard University Press, 1992:324-333.
19. Gould SJ. Darwinism and the expansion of evolutionary theory. Science 1982; 216:380-387.
20. Maynard Smith J. Evolutionary Genetics. Oxford: Oxford University Press, 1989:55.

21. Shaw RG, Shaw FH, Geyer C. What fraction of mutations reduces fitness? Evolution 2003; 57:686-689.
22. Hadany L, Comeron JM. Why are sex and recombination so common? Ann NY Acad Sci 2008; 1133:26-43.
23. Futuyma DJ. Evolutionary Biology, 3rd ed. Sunderland, Massachusetts: Sinauer Associates, 1998: 769.
24. Barry JD. Implicit information in eukaryotic pathogens as the basis of antigenic variation. In: Caporale LH, ed. The Implicit Genome. Oxford: Oxford University Press, 2006:91-106.
25. Bayliss CD, Moxon ER. Repeats and variation in pathogen selection. In: Caporale LH, ed. The Implicit Genome. Oxford: Oxford University Press, 2006:54-76.
26. Arber W. Gene products with evolutionary functions. Proteomics 2005; 5:2280-2284.
27. King DG, Kashi Y. Mutation rate variation in eukaryotes: evolutionary implications of site-specific mechanisms. Nature Rev Genet 2007; 8. doi:10.1038/nrg2158-c1.
28. Ellegren H. Microsatellites: Simple sequences with complex evolution. Nat Rev Genet 2004; 5:435-445.
29. Parker SCJ, Hansen L, Abaan HO et al. Local DNA topography correlates with functional noncoding regions of the human genome. Science 2009; 324:389-392.
30. Trifonov EN. The multiple codes of nucleotide sequences. Bull Math Biol 1989; 51:417-432.
31. King DG, Soller M, Kashi Y. Evolutionary tuning knobs. Endeavour 1997; 21:36-40.
32. Doyle J, Csete M, Caporale L. An engineering perspective: The implicit protocols. In: Caporale LH, ed. The Implicit Genome. Oxford: Oxford University Press, 2006:294-298.
33. Fisher RA. The Genetical Theory of Natural Selection. Oxford: Oxford University Press, 1930.
34. Levins R. Theory of fitness in a heterogeneous environment. VI. The adaptive significance of mutation. Genetics 1967; 56:163-178.
35. Levins R. Evolution in Changing Environments: Some Theoretical Explorations. Princeton: Princeton University Press, 1968.
36. Orr HA. The population genetics of adaptation: The distribution of factors fixed during adaptive evolution. Evolution 1998; 52:935-949.
37. Kashi Y, King DG. Simple sequence repeats as advantageous mutators in evolution. Trends Genet 2006; 22:253-259.
38. Sawyer LA, Sandrelli F, Pasetto C et al. The period gene Thr-Gly polymorphism in Australian and African Drosophila melanogaster populations: Implications for selection. Genetics 2006; 174:465-480.
39. Lindqvist C, Laakkonen L, Albert VA. Polyglutamine variation in a flowering time protein correlates with island age in a Hawaiian plant radiation. BMC Evol Biol 2007; 7:105. doi:10.1186/1471-2148-7-105.
40. Johnsen A, Fidler AE, Kuhn S et al. Avian Clock gene polymorphism: evidence for a latitudinal cline in allele frequencies. Mol Ecol 2007; 16:4867-4880.
41. Vinces MD, Legendre M, Caldara M et al. Unstable tandem repeats in promoters confer transcriptional evolvability. Science 2009; 324:1213-1216.
42. King DG, Soller M. Variation and fidelity: The evolution of simple sequence repeats as functional elements in adjustable genes. In: Wasser SP, ed. Evolutionary Theory and Processes: Modern Perspectives. Dordrecht: Kluwer Academic Publishers, 1999:65-82.
43. King DG, Trifonov EN, Kashi Y. Tuning knobs in the genome: Evolution of simple sequence repeats by indirect selection. In: Caporale LH, ed. The Implicit Genome. Oxford: Oxford University Press, 2006:77-90.
44. Kashi Y, King DG. Has simple sequence repeat mutability been selected to facilitate evolution? Isr J Ecol Evol 2006; 52:331-342.
45. King DG, Kashi Y. Indirect selection for mutability. Heredity 2007; 99:123-124.
46. Buschiazzo E, Gemmel NJ. The rise, fall and renaissance of microsatellites in eukaryotic genomes. BioEssays 2006; 28:1040-1050.
47. Zhu Y, Strassmann JE, Queller DC. Insertions, substitutions, and the origin of microsatellites. Genet Res Camb 2000; 76:227-236.
48. Jurka J, Kapitonov VV, Kohany O et al. Repetitive sequences in complex genomes: Structure and evolution. Annu Rev Genomics Hum Genet 2007; 8:241-259.
49. Oliver KR, Green WK. Transposable elements: powerful facilitators of evolution. BioEssays 2009; 31:703-714.
50. Kelkar YD, Strubczewski N, Hile SE et al. What is a microsatellite: A computational and experimental definition based upon repeat mutational behavior at A/T and GT/AC repeats. Genome Biol Evol 2010; 2:620-635.
51. Barbará T, Palma-Silva C, Paggi GM et al. Cross-species transfer of nuclear microsatellite markers: potential and limitations. Molec Ecol 2007; 16:3759-3767.
52. Riley DE, Krieger JN. Embryonic nervous system genes predominate in searches for dinucleotide simple sequence repeats flanked by conserved sequences. Gene 2009; 429:74-79.
53. Riley DE, Krieger JN. UTR dinucleotide simple sequence repeat evolution exhibits recurring patterns including regulatory sequence motif replacements. Gene 2009; 429:80-86.
54. Fondon JW, Hammock EAD, Hannan A et al. Simple sequence repeats: Genetic modulators of brain function and behavior. Trends Neurosci 2008; 31:328-334.

55. Karlin S, Burge C. Trinucleotide repeats and long homopeptides in genes and proteins associated with nervous system disease and development. Proc Natl Acad Sci USA 1996; 93:1560-1565.
56. Huntley MA, Mahmood S, Golding BG. Simple sequence in brain and nervous system specific proteins. Genome 2005; 48:291-301.
57. Łabaj PP, Leparc GG, Bardet AF et al. Single amino acid repeats in signal peptides. FEBS Journal 2010; 277:3147-3157.
58. Huntley MA, Clark AG. Evolutionary analysis of amino acid repeats across the genomes of 12 Drosophila species. Mol Biol Evol 2007; 24:2598-2609.
59. Birge LM, Pitts ML, Baker RH et al. Length polymorphism and head shape association among genes with polyglutamine repeats in the stalk-eyed fly, Teleopsis dalmanni. BMC Evol Biol 2010; 10:227. doi:10.1186/1471-2148-10-227.
60. Tompa P. Intrinsically unstructured proteins evolve by repeat expansion. BioEssays 2003; 25:847-855.
61. Tyedmers J, Madariaga ML, Lindquist S. Prion switching in response to environmental stress. PLoS Biol 2008; 6(11): e294. doi:10.1371/journal.pbio.0060294.
62. Williams GC. Adaptation and Natural Selection. Princeton: Princeton University Press, 1966:270.

CHAPTER 3

SINGLE AMINO ACID AND TRINUCLEOTIDE REPEATS

Function and Evolution

Noel Faux

Mental Health Research Institute, The University of Melbourne, Parkville, Victoria, Australia; and National Neuroscience Facility, Carlton, Victoria, Australia
Email: nfaux@unimelb.edu.au

Abstract: The most well known effect of single amino acid repeat expansion, beyond a certain threshold, is the development of a specific disease, depending on the protein in which the expansion has occurred. For example, the expansion of the glutamine repeat in huntingtin leads to the debilitating neurodegenerative disease, Huntington's disease. Similarly, there are a range of other disorders caused by trinucleotide repeat expansions encoding polyglutamine or polyalanine tracts. The age of onset of the polyglutamine-induced neurodegenerative diseases is usually negatively correlated with the length of expanded CAG/glutamine repeat. However, recent studies have given evidence that single amino acid repeats may also play critical roles in normal protein function and that changes in the length of single amino acid repeats is likely to play a beneficial role in evolution. This chapter will look at the prevalence, function and possible role single amino acid repeats have in evolution and other biological processes.

INTRODUCTION

Repetitive deoxyribonucliec acid (DNA) makes up ~50% of the human genome[1] and a substantial proportion of other eukaryote genomes. Repetitive elements can be divided into two groups: interspersed repeats or transposable elements, and tandem repeats, which include microsatellites, minisatellites, satellites and simple sequence repeats, such as trinucleotide repeats (TNRs).[2] Approximately 3% of the human genome contains simple sequence repeats.[1] Trinucleotide repeats are of particular interest as they are able

---

Tandem Repeat Polymorphisms: Genetic Plasticity, Neural Diversity and Disease,
edited by Anthony J. Hannan ©2012 Landes Bioscience and Springer Science+Business Media.

to undergo rapid expansion and contractions between generations. As will be discussed in this chapter, coding TNRs have both negative and positive phenotypic effects. The expansion of TNRs beyond certain lengths can lead to a number of severely debilitating diseases. Conversely, there is growing evidence that changes in the length of TNRs may have positive effects via evolutionary processes.

FREQUENCY OF HOMOPEPTIDES AND THE FORCES INFLUENCING REPEAT LENGTH

Trinucleotide repeats present within coding regions of the genome encode single amino acid repeats (homopeptides). As the selective pressures applied upon genes compared to the rest of the genome are different, the frequencies and types of TNR within coding regions are likely to be different compared to the rest of the genome and thus will influence the frequency of homopeptides. Moreover, these pressures are likely to lead to differing prevalence of homopeptides across species. Recent studies have given some insight into the influences and selective pressures which influence the frequency and types of homopeptides in an organism.

Prevalence of Homopeptides in the Proteomes of All Organisms

The first global surveys of homopeptides[3,4] highlighted an apparent lack of homopeptides in prokaryotes and an abundance in eukaryotes. This observation was latter confirmed in a larger number of organisms[5,6] (Table 1). The predominant homopeptides in eukaryotes are primarily those of the uncharged polar amino acids glutamine (Q), asparagine (N), serine (S), threonine (T), and histidine (H); the small residues G and A; the nonpolar residue P; or the acidic residues aspartic acid (D) and glutamic acid (E). It is important to note that between species the order of prevalence differs.[3,4] The prevalence of these homopeptides was latter confirmed by analysis of larger datasets.[5,7-11] Further, not only are Q, N, E, D common as homopeptides, their frequency as long homopeptides is notable, especially compared to the other common homopeptides, such as A, P and G, which while common, have a larger preponderance as short homopeptides[5] (Fig. 1). One of the largest differences across taxonomic groups is the paucity of asparagine homopeptides in vertebrates compared to the other eukaryotic species.[3,5,8,9,12] This difference was suggested to be as a result of strong selection against asparagine homopeptides in mammals.[8] Moreover, it was postuleted that the other differences between the species might be based on the underlying genome content. One such difference is the nucleotide make up of the codons between the species, i.e., *C. elegans*, *S. cerevisiae* and *A. thaliana* favour adenine and thymine (A+T) rich codons whereas *H. sapiens* and *D. melanogaster* favour cytosine and guanine (C+G) rich codons.

Influence of Nucleotide Content on Homopeptide Prevalence

The class III POU genes *Brain-1*, *Brain-2* and *Scip* in mammals all contain alanine (A), proline (P), glycine (G), histidine (H) and glutamine (Q) homopeptides. Studies of these genes have shown a positive correlation between the ratio of homopeptide length to the amino acid content of the protein (i.e., hAAR = hAA/nAA. Where hAA is the number of amino acids in the homopeptide—min length of 3 amino acids—and nAA is the total

Table 1. The percentage of homopeptides and RCPs in GENPEPT (as of 2005, GENPEPT is the nonredundant protein database from NCBI [ftp.ncbi.nih.gov]) and then broken down by three major taxa groups, eukaryotes, prokaryotes and viruses and other environmental sequences, adapted from reference 5 ©2005, with copyright permission of Cold Spring Harbor Laboratory Press.

	GENPEPT		Eukaryote		Prokaryote		Other (Viruses/ Environmental Sequences)	
	Repeats	Proteins	Repeats	Proteins	Repeats	Proteins	Repeats	Proteins
Alanine	11.24%	13.51%	11.22%	13.56%	15.62%	18.01%	9.75%	11.08%
Valine	0.27%	0.31%	0.19%	0.25%	0.56%	0.65%	1.08%	0.75%
Leucine	3.00%	4.29%	2.97%	4.37%	4.36%	5.04%	2.86%	3.17%
Isoleucine	0.10%	0.15%	0.07%	0.10%	0.19%	0.22%	0.47%	0.60%
Proline	8.86%	10.52%	8.54%	10.22%	13.50%	13.26%	10.85%	12.40%
Methionine	0.05%	0.06%	0.04%	0.06%	0.00%	0.00%	0.19%	0.12%
Phenylalanine	0.36%	0.50%	0.36%	0.53%	0.06%	0.07%	0.47%	0.39%
Tryptophan	0.01%	0.01%	0.01%	0.01%	0.00%	0.00%	0.00%	0.00%
Glycine	10.96%	13.44%	10.27%	12.77%	19.29%	20.24%	15.67%	17.10%
Serine	11.70%	14.62%	11.14%	14.53%	23.52%	18.59%	13.61%	13.87%
Threonine	5.49%	6.46%	5.12%	6.08%	3.92%	4.25%	10.36%	11.14%
Cystine	0.12%	0.14%	0.08%	0.12%	0.00%	0.00%	0.61%	0.42%
Asparagine	13.06%	9.99%	14.30%	11.02%	1.93%	2.09%	3.12%	3.14%
Glutamine	15.27%	15.26%	16.48%	16.75%	3.24%	3.67%	6.09%	5.51%
Tyrosine	0.10%	0.14%	0.08%	0.12%	0.25%	0.29%	0.30%	0.27%
Aspartic acid	3.36%	4.57%	3.19%	4.45%	2.12%	2.45%	5.79%	6.65%
Glutamic acid	8.76%	11.52%	8.90%	11.99%	4.17%	4.39%	8.86%	9.85%
Lysine	3.81%	5.16%	3.94%	5.44%	1.56%	1.80%	3.19%	3.80%
Arginine	1.38%	1.91%	0.95%	1.36%	3.73%	4.11%	5.37%	6.41%
Histidine	2.09%	2.84%	2.15%	2.98%	1.99%	2.31%	1.38%	1.74%
Total	54,566	37,355	48,691	32,628	1,607	1,388	4,268	3,339

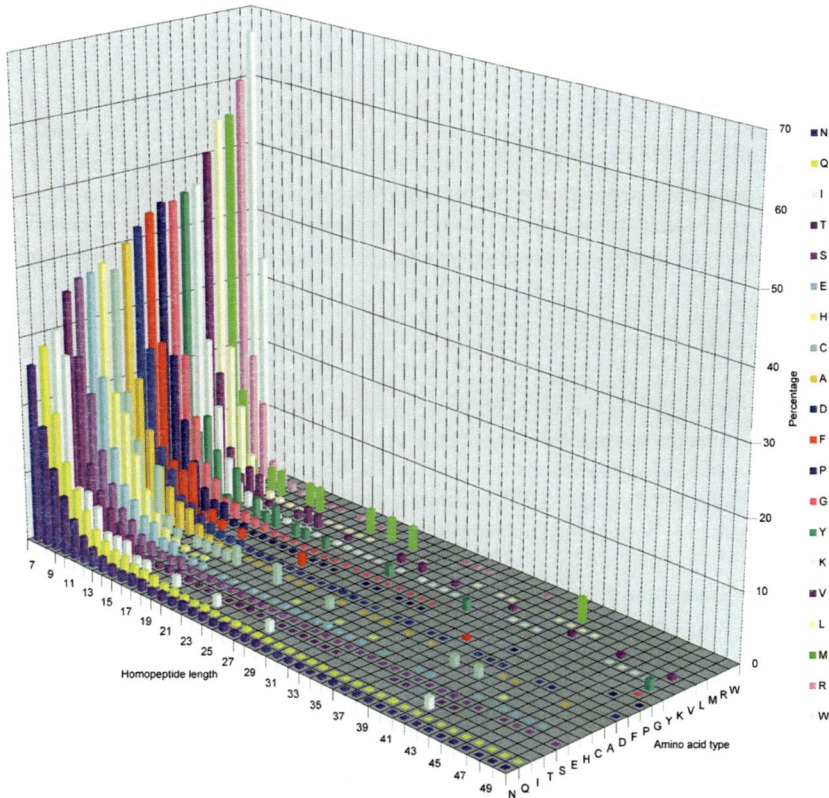

Figure 1. Length of homopeptide repeats. Three-dimensional plot showing repeat length (x-axis) versus amino acid type (y-axis; also highlighted in key) versus percentage of each repeat class of a particular length limited to those repeats <51 amino acids in length. A blank square indicates that no repeat of that length and type exists in GENPEPT. There are repeats >50 amino acids in length; however, these are infrequent, with lengths up to 410 amino acids and a sporadic distribution. Reproduced from reference 5 ©2005, with copyright permission of Cold Spring Harbor Laboratory Press.

number of amino acids in the repeat containing protein [RCP]) and the G+C content at the third codon position.[13,14] This observation was not seen for the nonvertebrate homologues (which do not contain the aforementioned homopeptides) nor for *Brain-4* in mammals and invertebrates (*Brain-4* is a homologue of *Brain-1*, *Brain-2* and *Scip*). Furthermore, hAAR was shown to be positively correlated to the G+C content at the third codon position in 34 transcription factors which contain A, G, and P homopeptides.[14] Moreover, the class III POU genes, which contain homopeptides, in *Homo sapiens*, *Mus musculus*, *Rattus norvegicus*, *Xenopus laevis*, *Brachydanio rerio*, *Drosophila melanogaster*, *Bombyx mri*, *Dugesia japonica* and *Caenorhabitis elegans* have elevated G+C levels (not attributed to the presence of the homopeptides). While those which do not contain homopepides do not have elevated G+C levels.[14] This G+C rich observation was extended to the alanine, glycine and proline RCPs in *H. Sapiens* and *Danio rerio*. These studies showed a positive correlation between the alanine, glycine and proline content of the RCPs and the G+C content at the third codon position of the encoding genes, compared to a random reference set of proteins from the respective species.[15,16]

The results of the above four studies,[13-16] lead to the hypothesis that the apparent prevalence of certain homopeptides is due to the selectively neutral (at the protein level) pressure towards G+C richness (G+C pressure)[13-15] within a genome region such as in different isochores. This is especially true in vertebrates, as their genomes have a mosaic pattern of G+C content across their chromosomes (reviewed in ref. 17). The G+C rich background in conjunction with the enrichment of homo- and hetero-dicodons in the homopeptide rich regions suggests that a specific codon organisation may be required for the generation of homocodons, and that the di-codons may act as primers for slip strand structures.[18]

However, it seems that these observations[13-16] may only be applicable to alanine, glycine and proline RCPs. In particular, the analysis of poly-Q RCPs across orthologous genes in *H. sapiens*, *M. musculus* and *R. norvegicus* revealed that while the encoding genes of these RCPs are G+C rich, this elevation is directly related to the presence of the homopeptides themselves rather than the rest of the amino acids.[9] Moreover, analysis of *H. sapiens* poly-Q RCPs did not reveal an elevation of G+C content at the third codon position,[19,20] thus argue against the hypothesis of G+C pressure driving the prevalence of specific homopeptide types on two levels. Firstly, once the amino acid composition of the RCP is factored in, the third codon position within the RCP genes is not found to be G+C enriched.[20] Secondly, the hypothesis of G+C pressure assumed selection at the protein level is neutral, i.e., it is assumed that homopeptides are neutral structures. A number of studies have shown that this is not the case.[19,21]

In particular, *Saccharomyce cerevisiae* contains only a subset of homopeptides (i.e., not all amino acids form homopeptides in this organism). Moreover, there is a bias towards certain encoding TNRs (correlating with the observed homopeptide bias)[10] indicating that selection at the protein level plays a direct role in the prevalence of homopeptides. Furthermore, RCPs in eukaryotes have been shown to have distinct selective pressure patterns across their RCPs.[19,22] Overall, the flanking regions of the homopeptides are under weaker selective constraints in comparison to the nonhomopeptide proportion of the protein.[19,22] Moreover, the flanking regions of conserved homopeptides have been shown to be under stronger selective pressure than those which are not conserved.[19,22] Further, glutamine homopeptides whose lengths are polymorphic between the different *Mus* species are under weaker selective pressure than those that are monomorphic in length, i.e., the homopeptide length is conserved.[21] These larger studies clearly indicate that, at least for glutamine homopeptides in eukaryotes, homopeptides do not lay within neutral regions and are thus not neutral themselves, and that selection plays an important role in the prevalence of homopeptide type and homopeptide length.

The hypothesis of G+C pressure driving the prevalence of homopeptide types implies that the codons of predominant homopeptides would be G+C rich, and that the prevalent homocodons would also be G+C rich. However, the results of four studies challenge this hypothesis.[10,22-24] These studies suggest that the mechanism of homopeptide development differed between species and homopeptide types. For example, in *S. cerevisiae* the aspartic acid (GAC and GAT), asparagine (ACC and AAT), glutamic acid (GAA) and glutamine (CAG) homopeptides contain a prevalence of long homocodons, most of which are not G+C rich.[10] The glutamine homopeptides in *H. sapiens*, *M. musculus* and *S. cerevisiae*, particularly the mammalian species, contain significantly longer CAG homocodons (in comparison with the expected homocodon length based the codon frequencies in the homopeptide).[10,23,24] In contrast, *D. melanogaster*'s glutamine homopeptides contain a high frequency of short CAG and CAA homocodons, and *Arabidopsis thaliana* glutamine homopeptides contain a high frequency of long CAG and CAA homocodons.[24] Looking at a large range of species, it has been shown that any apparent nucleotide bias within a homopeptide, can be explained by the species' transcriptome.[22]

Taken together these data suggest that homopeptide generation is different between species, i.e., those homopeptides which contain long homocodons may have been generated by slip strand mechanisms, whereas those which do not contain long homocodons may have been generated by an accumulation of point mutations.[10,22]

Forces Driving Homopeptide Prevalence

It has been proposed that G+C pressure drives the amino acid content and the prevalence of homopeptides in mammals.[13-15] This idea is based on the observation that alanine, proline and glycine homopeptides are the most prevalent homopeptide type in mammals, and also the positive correlation between alanine, proline and glycine content and the G+C content of the third codon position of these RCPs.

However, Hancock and Simon[20] argue that selective pressure at the protein level is the primary force driving the homopeptide prevalence and that the species' genetic backgrounds explain the apparent G+C elevation and GC3 correlation in RCPs.

In support of this latter hypothesis, the analysis of the patterns of selective pressure acting upon the RCPs in *H. sapiens*, *M. musculus* and *R. norvegicus*,[22] have shown that the 33 amino acids flanking the homopeptides are under less selective pressure than the rest of the protein (excluding the homopeptide). Also, the flanking regions of conserved homopeptides are under stronger selective pressure than those surrounding nonconserved homopeptides. This is in agreement with the earlier analysis of glutamine homopeptides in *H. sapiens* and *M. musculus*.[19] In addition, analysis of the G+C content of RCPs, after taking into account the organism's transcriptome, revealed that RCPs are not G+C rich.[22] As further evidence of selection being applied at the protein level, the global survey of homopeptides[5] revealed that polar and acidic amino acids are frequently present as homopeptides, while nonpolar and basic homopeptides occur at much lower frequencies. While there are some exceptions (alanine (nonpolar) homopeptides occur as frequently as glutamine (polar) homopeptides; and tyrosine (polar) homopeptides occur infrequently, it is clear that, in general, hydrophobic homopeptides are frequently short, while hydrophilic homopeptides are longer, as are (to a lesser extent) acidic and basic homopeptides.[5] These data suggest that hydrophobic amino acids are strongly selected against as long homopeptides and, in some cases, strongly selected against as whole homopeptides. This strong selection towards polar homopeptides is most likely due to hydrophobic collapse, as suggested by Oma[25] and Dorsman,[26] leading to protein aggregation and fibrillogenesis and hence cellular toxicity and cell death.

Therefore the above data, suggest that selection at the protein level influences the prevalence and length of homopeptides within a protein, and that the species' genome influences the codons used to encode the homopeptides.

HUMAN DISEASES CAUSED BY AN EXPANSION OF A HOMOPEPTIDE

As there are apparent strong selective pressures being applied upon homopeptides, changes to the presence of and/or length of the homopeptide within a RCP may have detrimental effects on the function of the protein and possibly on to the organism. To date, twenty heritable diseases have been identified that are caused by the expansion of a single homopeptide within the associated protein (see Table 2). Like the noncoding TNR

Table 2. Diseases caused by an expansion of a single amino acid repeat within the protein. The disease and homopeptide lengths were obtained from reference 31. The expression details were obtained from the Human Protein Reference Database.[32]

OMIM	Protein	Disease	Amino Acid Repeat	Repeat Length Normal	Repeat Length Disease	Aggregation	Location of Expression
143100	Huntingtin[27]	Huntington's disease	Glutamine	6-35	36-121	Y	Ubiquitous
64400	Ataxin-1	Spinocerebellar Ataxia-1	Glutamine	6-38	39-82	Y	Brain, neurons
183090	Ataxin-2	Spinocerebellar Ataxia-2	Glutamine	15-34	36-52	Y	Ubiquitous
109150	Ataxin-3	Ataxin-3 (Machado-Joseph disease)	Glutamine	12-41	55-84	Y	Ubiquitous
183086	Voltage-dependent calcium channel α-1A subunit	Spinocerebellar Ataxia-6	Glutamine	5-20	21-25	Y	Brain
164500	Ataxin-7	Spinocerebellar Ataxia-7	Glutamine	7-35	38-240	Y	CNS
607136	TATA box-binding protein	Spinocerebellar Ataxia-17	Glutamine	7-35	38-240	Y	Ubiquitous
125370	Atrophin1	Dentatorubral-pallidoluysian atrophy (Haw river syndome)	Glutamine	7-23	49-88	Y	Brain, heart, lung, kidney, placenta, muscle, testis, nervous system, cornea
313200	Androgen receptor[28]	Spinal and bulbar muscular atrophy	Glutamine	22-35	40-54	Y	Prostate, adrenal cortex, epididymis, endometrium, corpus luteum, thyroid follicular cell, skin fibroblast, liver

continued on next page

Table 2. Continued

OMIM	Protein	Disease	Amino Acid Repeat	Repeat Length		Aggregation	Location of Expression
				Normal	Disease		
606438	Junctophilin 3 or open reading frame on the anit-sense.	Huntington disease-like 2	Alanine* or leucine* or glutamine†	15A	49-60	Unclear	Brain
164300	Poly(A)-binding protein-2	Oculopharyngeal muscular dystrophy	Alanine	6-7	8-13	Y	Superior cervical ganglion, sympathetic (chain) ganglionic trunk, peripheral nervous system, centeral nervous system
209880	Paired-like homeobox 2B	Congenital central hypoventilation syndrome	Alanine	20	25-29	N/E	Nervous system
186000	Homeobox D13	Syndactyly, Type II	Alanine	15	22-29	Y	Foetus, heart, brain, liver, skeletal muscle
309510	Aristaless-related homeobox	Partington X-linked mental retardation syndrome	Alanine	10-16	17-23	Y	Bone marrow stroma cell, osteopblasts, artery, calvaria
119600	runt-related transcription factor 2	Cleidocranial dysplasia (minor form)	Alanine	11-17	27	Y	Brain
609637#	Zinc finger protein of cerebellum, 2	Holoprosencephaly 5	Alanine	15	25	N/E	Uterus, nervous system, cervix

continued on next page

Table 2. Continued

OMIM	Protein	Disease	Amino Acid Repeat	Repeat Length		Aggregation	Location of Expression
				Normal	Disease		
140000	Homeobox A13	Hand-foot-uterus syndrome	Alanine	18	24-27	Y	Ovary and eye
110100	Forkhead transcription factor FOXL2	Blepharophimosis, ptosis, and epicanthus inversus Type II	Alanine	14	19-24	Y	
300123	SRY-BOX 3	Mental retardation, X-linked, with isolated growth hormone deficiency	Alanine	15	26	Y	Fetus
	SRY-BOX 3	Anterior pituitary hypoplasia	Alanine	15	22 (29)	Y	Fetus
137215	G1- TO S-phase transition 1	Cancer susceptibility	Glycine	8-11	12 (30)	N/E	Ubiquitous

*Due to alternative splice sites the reading frames for this repeat can either be part of an exon (leading to different reading frames) or in the 3' UTR.
#Putative mode of expansion: somatic recombination.[33]
†The evidence to-date suggests that the molecular cause of the HDL2 phenotype is multi modal. The transcript from the sense strand for the splice variant which puts the repeat in the 3' UTR, leads to RNA aggregation or RNA foci:[34] Recently it has been shown in mouse models and some evidence in humans, that transcription can occur on the antisense strand leading to a peptide product which contains a poly-glutamine, and that this peptide maybe present in the intranuclear inclusions in HDL2 patients.[35]
Abbreviation: N/E, no evidence.

diseases, the majority of these diseases are neurological in nature. These diseases may cause deficits in learning and memory, diminution of executive function, reduced verbal span, alteration in complex visuoperceptual and visuoconstructive abilities, disorientation, psychiatric symptoms and loss of motor function.

The expansion of a glutamine homopeptide in a number of proteins leads to a number of diseases. For example, such an expansion event in Huntingtin leads to Huntington's disease. Normal individuals have between 6-35 glutamines in the homopeptide, whereas diseased individuals have >36 glutamines in the homopeptide. This expansion results in progressive and selective neuronal cell death, leading to convulsive movements, dementia and psychiatric symptoms with an age of onset usually during mid-life (OMIM:143100,[31]). The expansion in Ataxin-1 and 2 leads to spinocerebellar ataxia 1 and 2 respectively. Normal individuals have between 6-38 and 15-34 glutamines in the homopeptides in Ataxin-1 and 2 respectively, whereas individuals with the disease have between 39-82 and 36-52 glutamines in the homopeptide in Ataxin-1 and 2, respectively. Both diseases, spinocerebellar ataxia 1 and 2, are characterised by progressive degeneration of the cerebellum, brainstem, and spinal cord resulting in paralysis of the motor nerves of the eyes, peripheral neuropathy and dementia (OMIM:164400). There are at least six other polyglutamine diseases, most of which are spinocerebellar ataxia (Table 2).

The expansion of an alanine homopeptide in a number of proteins also leads to several diseases.[36] For example, such an expansion event in Poly(A)-Binding Protein-2 leads to oculopharyngeal muscular dystrophy. Normal individuals have six or seven alanines in the homopeptide, whereas, individuals with the disease have between 8-13 alanines in the homopeptide. This disease is characterised by painful and difficult swallowing and progressive paralysis of the eyelids (OMIM:164400 and 183090). The expansion in Runx2 leads to cleidocranial dysplasia. Normal individuals have between 11-17 alanines in the homopeptide in Runx2, whereas, individuals with the disease have at least 27 alanines in the homopeptide. This expansion leads to shortness of the hands and feet and minor craniofacial features of leidocranial dysplasia (OMIM:119600).

There are two characteristics common to all these diseases. Firstly, the majority of homopeptide expansions lead to the formation of fibrils and/or to protein aggregates (Table 2) which are the predominant component of insoluble inclusion bodies. Secondly, like for the noncoding TNR diseases, the length of the repeat is negatively correlated with the age of onset.[37] However, the age of disease onset is not solely based on the length of the homopeptide. A recent study showed that the variability in the age of onset for Huntington's disease is predominantly determined by the environment (60%) with only 40% being attributed to the genetic component, not including Huntingtin itself.[38] This followed on from the first evidence for environmental modulation of HD, which was produced in a transgenic mouse model.[39] The above examples and the list of diseases in Table 2 highlight that the length of expansion required for disease progression depends on the amino acid type. For example, for glutamine homopeptides the length of expansion is >20 amino acids while for alanine homopeptides a length of >8 amino acids results in disease.

THE STRUCTURAL ROLES OF HOMOPEPTIDES

The exact functional or structural role(s) homopeptides have within the protein is unclear. One hypothesis is that they are simply protein spacer elements between functional domains.[3] Another, suggests that specific homopeptides play roles in mediating or modulating protein—protein interactions[40-42] or regulating transcription.[43,44] Examples included:

- Proline-rich regions are known to be involved in protein—protein interactions and bind Src homology 3 (SH3) domains[45]
- The poly-proline tract in Huntingtin which binds the SH3-containing growth factor receptor bound protein 2 (Grb2)-like protein[46,47]
- The glycine homopeptide in the 75 kDa translocon (Toc-75) protein at the outer-envelope-membrane of chloroplasts has been shown to be important in targeting of the protein to the outer envelope of the chloroplast[48]
- A number of poly-arginine rich proteins have been shown to be involved in the binding of RNA[49,50]
- Both glutamine and aspartic acid homopeptides have been shown to inhibit ribosome-activated adenosine triphosphotase (ATPase) activity of the yeast translation elongation factor 3 (EF-3)[51]
- The arginine-rich regions of human immunodeficiency virus 1 (HIV-1) proteins regulator of virion (Rev) and trans-activator of transcription (Tat) bind RNA species and when these regions were modified to arginine homopeptides they still bound RNA[49,50]
- Free poly-glutamine peptides have been shown to bind the transcriptional co-activator TATA box binding protein (TBP)-associated factor, RNA polymerase II (TAF(II)130) which suppresses cyclic adenosine monophosphate (cAMP) response element-binding (CREB)-dependent transcriptional activation[52]

These studies point to the possibility that homopeptides all have a similar general role, in that they are involved in the binding of either protein or nucleic acids, and may regulate this binding.

ROLE OF HOMOPEPTIDES IN EVOLUTION

Homopeptides are not neutral structures and they play a number of roles in important cellular processes. Changes in their length are therefore likely to influence the function of the RCP.

The analysis of amino acid repeat loci in 96 dog breeds revealed a high level of purity, i.e., infrequently interrupted repeats.[53] Changes in the length of the repeat in two particular proteins were associated with gross morphological differences. There was a direct correlation between the absence of the PQ repeat in Alx-4 and the presence of an extra digit in the rear paw of the Great Pyrenees dog breed. A correlation was also shown between the length ratio of the alanine and glutamine homopeptides in Runx-2 and the level of clinorhynchy and mid-face length across 90 dog breeds (as well as within the breed over time).[53]

As TNR expansion and contraction is up to 100,000 times faster than point mutations,[54,55] this form of mutation and its effect may represent a molecular mechanism for rapid morphological changes during the evolution of a species, which natural selection can act upon.[1] In support of this hypothesis, the evolution of the alanine homopeptide has been shown to confer transcriptional repression in the ultrabithorax protein in the insects phyla, and is believed to have contributed to the increased specialisation of the abdominal segments in insects.[43]

The majority of RCPs belong to three main functional groups: transcription, translation and signalling. Numerous studies have shown that homopeptides in transcription factors act as either repressors or activators of transcription.[43,44] Thus, any changes in the length of these homopeptides is likely to impact on the expression of the transcribed gene(s), which may in turn lead to changes in specific gene and protein networks. In addition, it has been suggested that homopeptides may also influence protein-protein interactions, especially in large protein—protein complexes,[5] and changes in homopeptide length are also likely to change the kinetics of these interactions and possibly the actual binding partners themselves. Changes in homopeptide length in RCPs involved in signal cascades are likely to have a direct effect on the protein-protein interaction network.[20] Together, these changes may not only lead to morphological changes,[53] but also changes in the general biochemistry of the cell, such as the rate of signaling and ion transport across the cell membrane.

To prevent continual change in homopeptide length via uncontrolled contractions and expansions of homocodons, it is suggested that mechanisms may exist that reduce/ prevent such events from occurring. The stability of TNRs (i.e., their propensity to expand or contract) is influenced by a number of *cis*-factors, such as DNA methylation[56,57] which may play a role in the regulation of TNR expansion and contraction during evolution of eukaryotes.

DNA methylation occurs in a large number of eukaryotes such as vertebrates, insects, *S. cerevisiae*, plants, fungi, algae and certain ciliated protozoa.[58-68] It has been shown that the methylation state of DNA changes in response to environmental stress[69,70] and genomic stress in the form of chromosomal remodeling in species hybrids.[71] Thus, it is plausible that during times of stress, changes in the DNA methylation may facilitate TNR expansion and contraction[72] leading to changes in homopeptide length. These changes might then generate new gene expression patterns and protein-protein interactions, which, in turn, may lead to changes in morphology and biochemistry which selection can act upon. If these stressors/selective pressures were maintained over time, the repeat length would stabilise through synonymous point mutations, most likely being enriched for transversion events, thus disrupting the TNR and stabilising the homopeptide.

CONCLUSION

During the life cycle of DNA, TNRs have several opportunities to expand or contract in length. The hyper-expansion of TNRs within the noncoding regions of the human genome results in a number of severely debilitating diseases. A change in the length of a TNR within a coding region of a genome alters the length of the encoded amino acid tract (homopeptide). This change in length may have either a beneficial role or a harmful effect on the organism. The expansion of homopeptides in certain proteins leads to a number of severely debilitating diseases. In contrast, the contraction and expansion of homopeptides can lead to morphological changes of an organism upon which selection can then act. Therefore, in evolutionary terms, changes in the length of homopeptides can have a beneficial effect in allowing a species to rapidly adapt to new environments.

The evidence is now becoming stronger in favour of the hypothesis that selection at the protein level is the main driving mechanism, which dictates the prevalence and length of homopeptides within an organism. Further, that point mutations within a homocodon tract, particularly that of transversions, may act as mechanism for homopeptide length stabilisation.

REFERENCES

1. Lander ES, Linton LM, Birren B et al. Initial sequencing and analysis of the human genome. Nature 2001; 409(6822):860-921.
2. Cooper DN. Nature encyclopedia of the human genome. Nature Publ Group; 2003.
3. Karlin S, Burge C. Trinucleotide repeats and long homopeptides in genes and proteins associated with nervous system disease and development. Proc Natl Acad Sci U S A 1996; 93(4):1560-1565.
4. Green H, Wang N. Codon reiteration and the evolution of proteins. Proc Natl Acad Sci U S A 1994; 91(10):4298-4302.
5. Faux NG, Bottomley SP, Lesk AM et al. Functional insights from the distribution and role of homopeptide repeat-containing proteins. Genome Res 2005; 15(4):537-551.
6. Jorda J, Kajava A V. Protein homorepeats sequences, structures, evolution, and functions. Adv Protein Chem Struct Biol 2010; 79:59-88.
7. Huntley MA, Golding GB. Neurological proteins are not enriched for repetitive sequences. Genetics 2004; 166(3):1141-1154.
8. Karlin S, Brocchieri L, Bergman A et al. Amino acid runs in eukaryotic proteomes and disease associations. Proc Natl Acad Sci U S A 2002; 99(1):333-338.
9. Albà MM, Guigó R. Comparative analysis of amino acid repeats in rodents and humans. Genome Res 2004; 14(4):549-554.
10. Mar Alba M, Santibanez-Koref MF, Hancock JM. Amino acid reiterations in yeast are overrepresented in particular classes of proteins and show evidence of a slippage-like mutational process. J Mol Evol 1999; 49(6):789-797.
11. Holtzman JL. Amyloid-β vaccination for Alzheimer's dementia. The Lancet 2008; 372(9647):1381-1381.
12. Kreil D, Kreil G. Asparagine repeats are rare in mammalian proteins. Trends Biochem Sci 2000; 25(6):270-271.
13. Sumiyama K, Washio-Watanabe K, Saitou N et al. Class III POU genes: generation of homopolymeric amino acid repeats under GC pressure in mammals. J Mol Evol 1996; 43(3):170-178.
14. Nakachi Y, Hayakawa T, Oota H et al. Nucleotide compositional constraints on genomes generate alanine-, glycine-, and proline-rich structures in transcription factors. Mol Biol Evol 1997; 14(10):1042-1049.
15. Cocquet J, de Baere E, Caburet S et al. Compositional biases and polyalanine runs in humans. Genetics 2003; 165(3):1613-1617.
16. Veitia R. Amino acids runs and genomic compositional biases in vertebrates. Genomics 2004; 83(3):502-507.
17. Bernardi G. The compositional evolution of vertebrate genomes. Gene 2000; 259(1-2):31-43.
18. Caburet S, Vaiman D, Veitia R. A genomic basis for the evolution of vertebrate transcription factors containing amino acid runs. Genetics 2004; 167(4):1813-1820.
19. Hancock J, Worthey E, Santibanez-Koref M. A role for selection in regulating the evolutionary emergence of disease-causing and other coding CAG repeats in humans and mice. Mol Biol Evol 2001; 18(6):1014-1023.
20. Hancock JM, Simon M. Simple sequence repeats in proteins and their significance for network evolution. Gene 2005; 345(1):113-118.
21. Ogasawara M, Imanishi T, Moriwaki K et al. Length variation of CAG/CAA triplet repeats in 50 genes among 16 inbred mouse strains. Gene 2005; 349:107-119.
22. Faux N, Huttley G, Mahmood K et al. RCPdb: An evolutionary classification and codon usage database for repeat-containing proteins. Genome Res 2007; 17(7):1118-1127.
23. Alba MM, Santibanez-Koref MF, Hancock JM. Conservation of polyglutamine tract size between mice and humans depends on codon interruption. Mol Biol Evol 1999; 16(11):1641-1644.
24. Alba MM, Santibanez-Koref MF, Hancock JM. The comparative genomics of polyglutamine repeats: extreme differences in the codon organization of repeat-encoding regions between mammals and Drosophila. J Mol Evol 2001; 52(3):249-259.
25. Oma Y, Kino Y, Sasagawa N et al. Intracellular localization of homopolymeric amino acid-containing proteins expressed in mammalian cells. J Biol Chem 2004; 279(20):21217-21222.
26. Dorsman J, Pepers B, Langenberg D et al. Strong aggregation and increased toxicity of polyleucine over polyglutamine stretches in mammalian cells. Hum Mol Genet 2002; 11(13):1487-1496.
27. Waldvogel HJ, Thu D, Hogg V et al. Selective neurodegeneration, neuropathology and symptom profiles in Huntington's disease. In: Hannan AJ, ed. Tandem Repeat Polymorphisms: Genetic Plasticity, Neural Diversity and Disease. Austin/New York: Landes Bioscience/Springer Science+Business Media, 2012:141-152.
28. Zajac JD, Fui MNT. Kennedy's disease: clinical significance of tandem repeats in the androgen receptor. In: Hannan AJ, ed. Tandem Repeat Polymorphisms: Genetic Plasticity, Neural Diversity and Disease. Austin/New York: Landes Bioscience/Springer Science+Business Media, 2012:153-168.
29. Woods KS, Cundall M, Turton J et al. Over- and underdosage of SOX3 is associated with infundibular hypoplasia and hypopituitarism. Am J Hum Genet 2005; 76(5):833-849.

30. Brito M, Malta-Vacas J, Carmona B et al. Polyglycine expansions in eRF3/GSPT1 are associated with gastric cancer susceptibility. Carcinogenesis 2005; 26(12):2046-2049.
31. McKusick-Nathans Institute for Genetic Medicine J. Online Mendelian Inheritance in Man, OMIM (TM).
32. Peri S, Navarro J, Kristiansen T et al. Human protein reference database as a discovery resource for proteomics. Nucleic Acids Res 2004; 32(Database issue):D497-D501.
33. Brown LY, Odent S, David V et al. Holoprosencephaly due to mutations in ZIC2: alanine tract expansion mutations may be caused by parental somatic recombination. Hum Mol Genet 2001; 10(8):791-796.
34. Rudnicki DD, Holmes SE, Lin MW et al. Huntington's disease-like 2 is associated with CUG repeat-containing RNA foci. Ann Neurol 2007; 61(3):272-282.
35. Wilburn B, Rudnicki DD, Zhao J et al. An antisense CAG repeat transcript at JPH3 locus mediates expanded polyglutamine protein toxicity in Huntington's disease-like 2 mice. Neuron 2011; 70(3):427-440.
36. Shoubridge C, Gecz J. In: Hannan AJ, ed. Tandem Repeat Polymorphisms: Genetic Plasticity, Neural Diversity and Disease. Austin/New York: Landes Bioscience/Springer Science+Business Media, 2012:185-204.
37. Zoghbi HY, Orr HT. Glutamine repeats and neurodegeneration. Annu Rev Neurosci 2000; 23:217-247.
38. Wexler NS, Lorimer J, Porter J et al. Venezuelan kindreds reveal that genetic and environmental factors modulate Huntington's disease age of onset. Proc Natl Acad Sci U S A 2004; 101(10):3498-3503.
39. van Dellen A, Blakemore C, Deacon R, York D, Hannan AJ. Delaying the onset of Huntington's in mice. Nature 2000; 404(6779):721-722.
40. Mitchell PJ, Tjian R. Transcriptional regulation in mammalian cells by sequence-specific DNA binding proteins. Science 1989; 245(4916):371-378.
41. Perutz MF, Johnson T, Suzuki M, Finch JT. Glutamine repeats as polar zippers: their possible role in inherited neurodegenerative diseases. Proc Natl Acad Sci U S A 1994; 91(12):5355-5358.
42. Kazemi-Esfarjani P, Trifiro MA, Pinsky L. Evidence for a repressive function of the long polyglutamine tract in the human androgen receptor: possible pathogenetic relevance for the (CAG)n-expanded neuronopathies. Hum Mol Genet 1995; 4(4):523-527.
43. Galant R, Carroll SB. Evolution of a transcriptional repression domain in an insect Hox protein. Nature 2002; 415(6874):910-913.
44. Gerber H-P, Seipel K, Georgiev O et al. Transcriptional activation modulated by homopolymeric glutamine and proline stretches. Science 1994; 263(5148):808-811.
45. Ren R, Mayer B, Cicchetti P et al. Identification of a ten-amino acid proline-rich SH3 binding site. Science 1993; 259(5098):1157-1161.
46. Liu YF, Deth RC, Devys D. SH3 domain-dependent association of huntingtin with epidermal growth factor receptor signaling complexes. J Biol Chem 1997; 272(13):8121-8124.
47. Sittler A, Wälter S, Wedemeyer N et al. SH3GL3 associates with the Huntingtin exon 1 protein and promotes the formation of polygln-containing protein aggregates. Mol Cell 1998; 2(4):427-436.
48. Inoue K, Keegstra K. A polyglycine stretch is necessary for proper targeting of the protein translocation channel precursor to the outer envelope membrane of chloroplasts. Plant J 2003; 34(5):661-669.
49. Calnan B, Tidor B, Biancalana S et al. Arginine-mediated RNA recognition: the arginine fork. Science 1991; 252(5010):1167-1171.
50. Nam YS, Petrovic A, Jeong KS et al. Exchange of the basic domain of human immunodeficiency virus type 1 Rev for a polyarginine stretch expands the RNA binding specificity, and a minimal arginine cluster is required for optimal RRE RNA binding affinity, nuclear accumulation, and trans-activation. J Virol 2001; 75(6):2957-2971.
51. Uritani M, Nakano K, Aoki Y et al. Polyamino acids that inhibit the interaction of yeast translational elongation factor-3 (EF-3) with ribosomes. J Biochem (Tokyo) 1994; 115(5):820-824.
52. Shimohata T, Nakajima T, Yamada M et al. Expanded polyglutamine stretches interact with TAFII130, interfering with CREB-dependent transcription. Nat Genet 2000; 26(1):29-36.
53. Fondon J, Garner H. Molecular origins of rapid and continuous morphological evolution. Proc Natl Acad Sci U S A 2004; 101(52):18058-18063.
54. Ellegren H. Microsatellite mutations in the germline: implications for evolutionary inference. Trends Genet 2000; 16(12):551-558.
55. Ellegren H. Microsatellites: simple sequences with complex evolution. Nat Rev Genet 2004; 5(6):435-445.
56. Cleary J, Pearson C. The contribution of cis-elements to disease-associated repeat instability: clinical and experimental evidence. Cytogenet Genome Res 2003; 100(1-4):25-55.
57. Robertson KD. DNA methylation and human disease. Nat Rev Genet 2005; 6(8):597-610.
58. Gutierrez JC, Callejas S, Borniquel S et al. DNA methylation in ciliates: implications in differentiation processes. Int Microbiol 2000; 3(3):139-146.
59. Gowher H, Leismann O, Jeltsch A. DNA of Drosophila melanogaster contains 5-methylcytosine. EMBO J 2000; 19(24):6918-6923.
60. Lyko F, Ramsahoye BH, Jaenisch R. DNA methylation in Drosophila melanogaster. Nature 2000; 408(6812):538-540.

61. Marhold J, Rothe N, Pauli A et al. Conservation of DNA methylation in dipteran insects. Insect Mol Biol 2004; 13(2):117-123.
62. Hattman S, Kenny C, Berger L et al. Comparative study of DNA methylation in three unicellular eucaryotes. J Bacteriol 1978; 135(3):1156-1157.
63. Vanyushin BF. Enzymatic DNA methylation is an epigenetic control for genetic functions of the cell. Biochemistry (Mosc) 2005; 70(5):488-499.
64. Rogers SD, Rogers ME, Saunders G et al. Isolation of mutants sensitive to 2-aminopurine and alkylating agents and evidence for the role of DNA methylation in Penicillium chrysogenum. Curr Genet 1986; 10(7):557-560.
65. Zhang X, Yazaki J, Sundaresan A et al. Genome-wide high-resolution mapping and functional analysis of DNA methylation in arabidopsis. Cell 2006; 126(6):1189-1201.
66. Wang Y, Jorda M, Jones PL et al. Functional CpG methylation system in a social insect. Science 2006; 314(5799):645-647.
67. Varriale A, Bernardi G. DNA methylation and body temperature in fishes. Gene 2006; 385:111-121.
68. Varriale A, Bernardi G. DNA methylation in reptiles. Gene 2006;385:122-127.
69. Steward N, Ito M, Yamaguchi Y et al. Periodic DNA methylation in maize nucleosomes and demethylation by environmental stress. J Biol Chem 2002; 277(40):37741-37746.
70. Wada Y, Miyamoto K, Kusano T et al. Association between up-regulation of stress-responsive genes and hypomethylation of genomic DNA in tobacco plants. Mol Genet Genomics 2004; 271(6):658-666.
71. O'Neill RJ, O'Neill MJ, Graves JA. Undermethylation associated with retroelement activation and chromosome remodelling in an interspecific mammalian hybrid. Nature 1998; 393(6680):68-72.
72. Ruden DM, Garfinkel MD, Xiao L et al. Epigenetic regulation of trinucleotide repeat expansions and contractions and the "biased embryos" hypothesis for rapid morphological evolution. Current Genomics 2005; 6:145-155.

CHAPTER 4

PROMOTER MICROSATELLITES AS MODULATORS OF HUMAN GENE EXPRESSION

Sterling M. Sawaya,*,[1] Andrew T. Bagshaw,[2]
Emmanuel Buschiazzo[3] and Neil J. Gemmell[1]

[1]Department of Anatomy and Structural Biology, University of Otago, Dunedin, New Zealand; [2]Department of Pathology, University of Otago-Christchurch, Christchurch, New Zealand; [3]School of Natural Sciences, University of California-Merced, Merced, California, USA
*Corresponding Author: Sterling M. Sawaya—Email: sterlingsawaya@gmail.com

Abstract: Microsatellites in and around genes have been shown to modulate levels of gene expression in multiple organisms, ranging from bacteria to humans. Here we will discuss promoter microsatellites known to modulate gene expression, with a few key examples related to the human brain. Many of the microsatellites we discuss are highly conserved in mammals, indicating that selection may favor their retention as "tuning knobs" of gene expression. We will also discuss the mechanisms by which microsatellites in promoters can alter gene expression as they expand and contract, with particular attention to secondary structures like Z-DNA and H-DNA. We suggest that promoter microsatellites, especially those that are highly conserved, may be an important source of human phenotypic variation.

INTRODUCTION

Microsatellites, i.e., tandem repeats with a motif length less than 7 base pairs, are a common occurrence in both eukaryotic[1] and prokaryotic genomes.[2] When found near a gene's transcription start site, i.e., in the "promoter," microsatellites can increase or decrease levels of gene expression by expanding and contracting in length (herein referred to as modulation of gene expression). In this chapter we review evidence that human promoter microsatellites can modulate gene expression. Microsatellites in these regions can regulate gene expression by forming unusual DNA secondary structures.[3,4]

Tandem Repeat Polymorphisms: Genetic Plasticity, Neural Diversity and Disease,
edited by Anthony J. Hannan ©2012 Landes Bioscience and Springer Science+Business Media.

Promoter Microsatellites across the Tree of Life

In bacteria, microsatellites are often studied in "contingency loci," genes that switch between functional and nonfunctional states as microsatellites within coding regions expand and contract.[2] Bacteria tend to have small genomes, a large proportion of which codes for proteins, thus limiting the size of their regulatory regions. Nevertheless, even in these tiny organisms, microsatellites in promoter regions can modulate gene expression.[5-8]

In yeast, promoter microsatellites can also be sources of variation in gene expression. Around 25% of *Saccharomyces cerevisiae* genes contain a tandem repeat in their promoter. Expansion and contraction of these microsatellites alters chromatin structure, resulting in changes in levels of gene expression.[9] Microsatellites found in yeast promoters can modify nucleosome binding.[9,10] For example, a microsatellite with the motif CCGNN motif has been shown to affect both gene expression and recombination by altering histone positioning.[11]

Promoter microsatellites have been investigated in multiple different rodent species. One of the most notable studies was done in the voles *Microtus ochrogaster* and *M. montannus*, where the length of a microsatellite upstream of the vasopressin 1a receptor gene was shown to affect mating behavior and parental care.[12] In humans, microsatellites upstream of the vasopressin 1a receptor gene are also associated with behavior, although the positions and compositions of these microsatellites differ from the microsatellites in the rodent gene.[13-17] In another example, the rat nucleolin gene, the length of a microsatellite with the motif AC/GT found just upstream of the transcription start site is associated with levels of gene expression.[18] Other AC/GT microsatellites in rat promoters have been investigated as general inhibitors of expression, but were not tested for a relationship between length and expression.[19-21] These AC/GT microsatellites are very common in the human genome and have been postulated to be important modulators of gene expression in mammals.[22,23]

Microsatellites have also been shown to modulate expression in non-model organisms.[24,25] In tilapia, an AC microsatellite upstream of the prolactin gene affects salt tolerance.[24] In chicken, the malic enzyme gene contains a CT/AG microsatellite that affects expression in a length dependent manner.[25] The evidence that microsatellites modulate gene expression in such a wide range of organisms suggests that this phenomenon is common to all forms of life. Importantly, good support that microsatellites modulate gene expression comes from research on the human genome.

HUMAN PROMOTER MICROSATELLITES

Support for the importance of promoter microsatellites can be found in molecular genetics literature.[14,27-49] These studies have investigated the ability of promoter microsatellites to modulate expression using reporter constructs,[27-47] as well as quantitative reverse-transcription PCR.[14,48,49] These studies also found associations between phenotypes and the lengths of the microsatellites examined. Other studies have found associations between promoter microsatellites and human phenotypes but have not directly tested for the ability of these microsatellites to modulate expression (e.g., refs. 13, 15-17). Because of the extensive number of these studies, we will not attempt to cite them all here.

Additional support for the importance of promoter microsatellites comes from our recent work in bioinformatics, in which we examined the conservation of mammalian

Table 1. Conserved microsatellites shown in Figures 1 and 2. The table contains the name of the gene, standardized motif of the microsatellites, position of the microsatellite relative to the transcription start site, change in gene expression for expanded microsatellites and reference number.

Gene	Motif	Relative Position	Δ Expression for Long Microsatellites	Ref.
nos1	(AC)n	−92	Increase	29,30
pax6	(AC)n-(AG)n	−955	Increase	31,41
fli1	(AG)n	49	Decrease	90,91
hmga2	(AG)n	279	Increase	93

microsatellites using phylogenetics.[26] Human promoters have recently been noted to contain a high density of microsatellites,[9] and our work has demonstrated that microsatellites in these regions are often highly conserved.[26] In fact, the more proximal a microsatellite is to the transcription start site, the more likely it is to be conserved.[26] This suggests that microsatellites are present in human promoters because they are favored by selection.

Here we discuss specific human promoter microsatellites known to modulate gene expression as they change in length (Table 1). The high level of conservation of most of these microsatellites invokes the intriguing hypothesis that they may be conserved as a source of genetic variation.

nos1 Promoter Microsatellite

nos1 is an excellent example of a neurological gene that contains a highly conserved promoter microsatellite (Fig. 1). *nos1* encodes the neural nitric-oxide synthase, which generates nitric oxide (NO), an important signaling molecule. NO is a vasorelaxant with a short half-life, which, in the brain, acts as a rapid paracrine signal produced in response to increased calcium levels (reviewed in ref. 50).

Two independent studies have shown that an AC/GT microsatellite 92 base pairs upstream of a *nos1* transcription start site modulates gene expression, with longer repeats inducing higher expression levels in reporter constructs.[29,30] Carrying a short *nos1* promoter microsatellite is strongly associated with impulsivity in humans, presumably because of reduced NO signaling in the brain.[51,52] These associations are influenced by environmental factors[51] as well as interactions with levels of platelet monoamine oxidase-A.[53] Additionally, knock-out mice lacking *nos1* display aggressive, impulsive behaviors in certain environments.[54] Taken together, this evidence implicates NO signaling as an important determinant of impulsivity, and that the highly conserved microsatellite in the *nos1* promoter can modulate the strength of this pathway, at least in humans.

The conservation of this microsatellite provides further support that it serves an important function (Fig. 1). Intriguingly, in the opossum genome the AC/GT motif is replaced by the trinucleotide motif GTA and an adjacent poly-A run has been replaced by a poly-G run in opossum. Both the poly-A and poly-G runs are expected to be hyper-mutable,[55] but have never been investigated for their potential effects on gene expression. These microsatellites may also play a role modulating *nos1* gene expression in mammals.

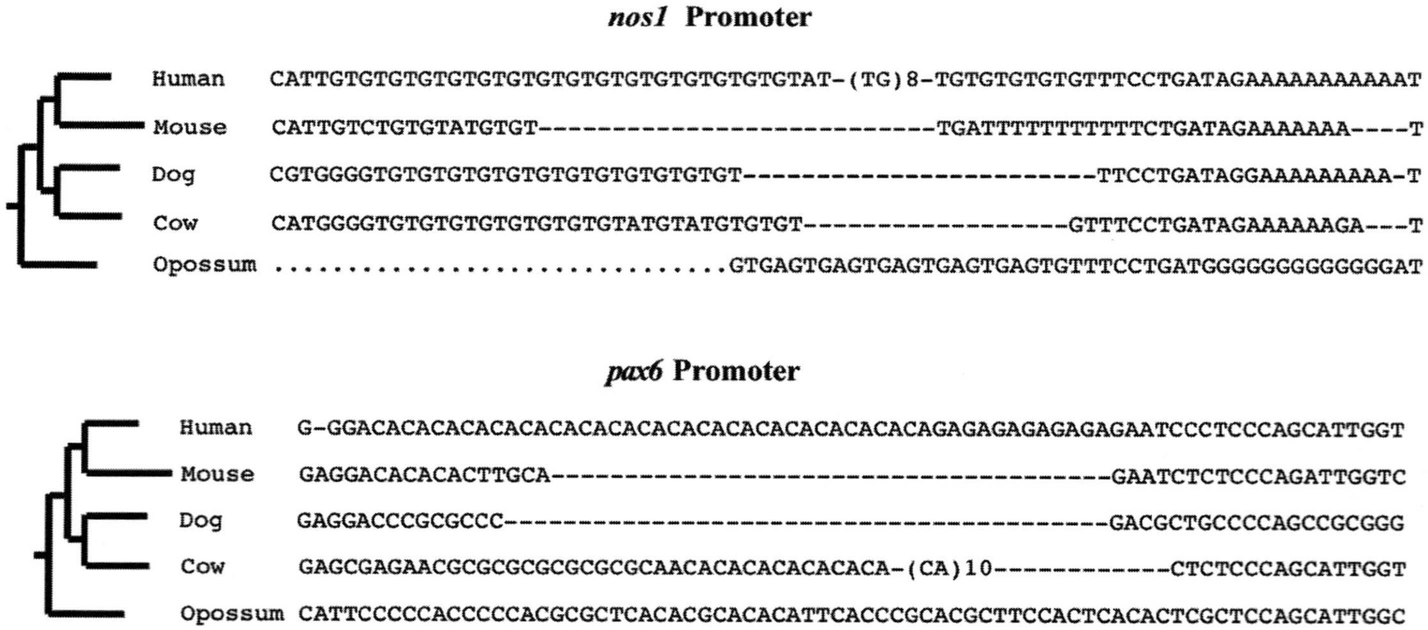

nos1 Promoter

```
Human    CATTGTGTGTGTGTGTGTGTGTGTGTGTGTGTGTAT-(TG)8-TGTGTGTGTGTGTTCCTGATAGAAAAAAAAT
Mouse    CATTGTCTCTGTGTATGTGT-------------------------------------TGATTTTTTTTTTCTGATAGAAAAAAA----T
Dog      CGTGGGTGTGTGTGTGTGTGTGTGTGTGTGT--------------------------TTCCTGATAGGAAAAAAAA--T
Cow      CATGGGGTGTGTGTGTGTGTGTGTGTATGTATGTGT----------------------GTTTCCTGATAGAAAAAGA----T
Opossum  ..............GTGAGTGAGTGAGTGAGTGAGTGTGTTCCTGATGGGGGGGGGGGGAT
```

pax6 Promoter

```
Human    G--GGACACACACACACACACACACACACACACAGAGAGAGAGAGAATCCTCCCAGCATTGGT
Mouse    GAGGACACACACTTGCA-------------------------------------------GAATCTCTCCCAGATTGGTC
Dog      GAGGACCCGCGCCC------------------------------------GACGCTGCCCCAGCCGCGGG
Cow      GAGCGAGAACGCGCGCGCGCGCAACACACACACACACA-(CA)10------------------CTCTCCCAGCATTGGT
Opossum  CATTCCCCACCCCACCCGGCTCACACGCCACACCATTCACCCGCACGCTTCCACTCACACTCGCTCCAGCATTGGC
```

Figure 1. Alignment of the NOS1 and PAX6 promoter microsatellites in 5 mammals. Note the motif replacements in opossum for NOS1.

pax6 Promoter Microsatellites

pax6 is a homeobox gene that is essential for eye and nose development and mice lacking *pax6* fail to develop lenses in their eyes and lack nasal cavities.[56] *pax6* is also expressed at high levels in proliferative zones in the fetal brain and is involved in neural development.[57] *pax6* is important for dorsoventral patterning in the forebrain[58] by limiting cell migration across the cortical-striatal boundary.[59]

Transcription of *pax6* is initiated from two distinct promoters, one of which is used primarily in the cerebral cortex.[60] Approximately 1 kb upstream of this promoter is a compound $(AC)_n(AG)_n$ microsatellite.[41] While this microsatellite is not highly conserved in mammals, it can be found in species distantly related to humans, such as cow (Fig. 1).

This *pax6* promoter microsatellite modulates expression, with longer repeats displaying higher levels of gene expression in reporter gene assays.[31,41] Altering either the $(AC)_n$ or the $(AG)_n$ motif individually alters expression levels[31] and carriers of high expressing microsatellite alleles are predisposed to myopia.[31] *pax6* is a strong candidate for myopia because of its role in eye development.[56] Inducing high expression of *pax6* causes eye abnormalities in mice[61] and SNPs in *pax6* have been associated with myopia in humans.[62,63] Intriguingly, this microsatellite may affect myopia in other mammals.

MECHANISMS OF EXPRESSION MODULATION

A detailed understanding of the mechanisms by which promoter microsatellites modulate gene expression remains elusive. Theoretically, there are many ways in which a change in the length of a microsatellite can result in a change in gene expression. We will discuss all of the various possibilities but will focus on the ability for microsatellites to form DNA secondary structures. DNA is commonly found in the standard, Watson and Crick double helix, B-DNA. Non-B-DNA structures are known to regulate gene expression and can form in microsatellites composed of various motifs.[3,4]

The AC/TG Motif and Z-DNA Formation

Z-DNA is a left-handed double helix structure that can form in sequences composed of alternating purine-pyrimidine bases. There are multiple motifs that, when repeated in a microsatellite, are predicted to form Z-DNA. Of these motifs, only one is found commonly in the human genome, AC/GT.[3] AC/GT microsatellites are the 2nd most common microsatellites in the human genome,[64] and are well known to form Z-DNA in vivo (reviewed in ref. 3). Other motifs with Z-DNA forming potential, such as the motif CG/GC, are very uncommon in the genomes of mammals.[65]

Although Z-DNA is a higher energy conformation than canonical Watson-Crick B-DNA, it can be induced by negative torsional strain ("untwisting") in supercoiled plasmids[66] and strong evidence indicates that it can form on chromosomal DNA in vivo.[67,68] There are thousands of regions in the human genome that are predicted to form Z-DNA, and these sequences appear to be enriched around human promoters.[69,70]

Z-DNA can form in promoters because promoters commonly encounter negative torsional strain induced in the wake of a moving polymerase[4] or by chromatin remodeling proteins-like BRG1.[71,72] Formation of Z-DNA expels bound nucleosomes,[71-74] opening chromatin and allowing the binding of regulatory elements,[72-74] such as the TATA-binding protein.[73] Z-DNA can also block the movement of RNA polymerase when it occurs downstream of the transcription start site.[75]

Although promoters contain a high density of sequences with Z-DNA forming potential, conclusive evidence that all of these sequences are functional as Z-DNA forming elements has been somewhat elusive. Some of the earliest in vivo evidence for Z-DNA formation in promoters comes from non-microsatellite Z-DNA forming sequences.[67,68] These remarkable studies probed active nuclei with Z-DNA binding antibodies and found Z-DNA formation is induced by transcription.[67,68,76]

Other studies have provided less direct evidence. For example, in rat, removal of AC/GT promoter microsatellites increases levels of gene expression.[18-21,77,78] However, none of these studies tested for a relationship between length and expression. Therefore, although removing these microsatellites increases expression, expansion of the microsatellite may not necessarily result in decreased expression. In fact, the relationship between AC/GT microsatellite length and expression is not necessarily so simple.

Of the 24 promoter microsatellites examined for their ability to modulate gene expression, 17 contain the motif AC/GT.[14,27,29-31,33,34,37-44,47,49] All of these studies have shown that repeat expansion alters transcription levels, but longer repeats do not necessarily increase transcription. In fact, some studies showed that although these sequences modulate expression, there is no simple relationship between length and expression levels.[33,38,44,47] Therefore, the relationship between microsatellite length and gene expression depends on the gene and the sequence context.

While Z-DNA formation remains the most widely accepted function of these AC/GT microsatellites, some studies have provided evidence that protein binding may also be involved.[38,40,47] These two explanations are not mutually exclusive, and both may be occurring.[3] However, there are only a few Z-DNA binding proteins found in humans, none of which have been shown to act as transcription factors.[3] Additionally, Z-DNA formation is transitive and very brief.[79] So although Z-DNA formation can alter local protein binding,[72-74] Z-DNA is unlikely to be a common binding site for transcription factors in human promoters. However, this does not rule out the GT/CA microsatellites as potential transcription factor binding sites, and proteins may bind these sequences as they transition between B-DNA and Z-DNA.

In summary, the alteration of Z-DNA forming potential as AC/GT microsatellites expand and contract is a widely accepted mechanism by which gene expression can be modulated.[22,23] Because Z-DNA blocks RNA-polymerase[75] increased Z-DNA formation downstream of the transcription start site is expected to inhibit levels of expression. When found upstream, Z-DNA formation can alter the accessibility of transcription factor binding sites,[71,72,74] which could theoretically increase or decrease transcription, depending on the context. Because of their frequency in the human genome, especially in promoter regions,[64] these AC/GT microsatellites may be responsible for a significant proportion of human expression variation.[22]

The CT/AG Motif and H-DNA Formation

DNA sequences composed of poly-purine/poly-pyrimidine mirror repeats are capable of forming H-DNA, a DNA triplex structure that, like Z-DNA, can form under negative torsional strain.[80] When H-DNA forms in promoters, it can act as a functional element and affect gene expression.[81-84]

Microsatellites with the motif CT/AG are capable of forming H-DNA.[4] Additionally, when in its B-DNA form, CT/AG microsatellites are a binding site for the GAGA factor, a chromatin remodeling transcription factor originally discovered in Drosophila.[85] Replacing Drosophila CT/AG microsatellites with other sequences that have H-DNA forming potential does not induce similar levels of expression, suggesting that GAGA binding is the primary function of these sequences.[86] Although a vertebrate homolog for the GAGA factor has recently been discovered,[87] the extent to which it binds CT/AG microsatellites is yet undetermined.

In vertebrates, CT/AG microsatellites can function as H-DNA forming elements.[25,88-90] As these microsatellites expand they increase DNAse sensitivity, likely by increasing H-DNA forming potential.[25,88,90] By altering chromatin structure, these microsatellites can alter levels of expression when found in promoters.[25,88,90] For example, a CT/AG microsatellite 200 base pairs upstream of the smooth muscle myosin light chain kinase gene induces high levels of gene expression when it expands.[90] A similar change in expression is seen when a CT/AG microsatellite upstream of the chicken malic enzyme gene expands.[25]

A well studied example of a regulatory CT/AG microsatellite is in the 5' UTR of the *fli1* gene. This microsatellite is believed to form H-DNA,[88] and is conserved between humans and mice[91,92] as well as other mammals (Fig. 2). Expressed in white blood cells, *fli1* regulates immune activity, and short microsatellites induce the highest levels of expression.[91,92] High levels of *fli1* expression cause renal disease in mice.[93] There is no clear benefit to the modulation of *fli1* expression, and the evolutionary function of this microsatellite is unclear. Nevertheless, this microsatellite is well conserved in mammals (Fig. 2).

Another example of a conserved microsatellite with H-DNA forming potential is found in the 5' UTR of the gene *hmga2*.[89] Changes in the length of this microsatellite alter gene expression.[94] In mice, knocking out *hmga2* results in a pygmy phenotype,[95] and the length of this microsatellite is associated with height in humans.[96] Additionally, SNPs in this gene are very strongly associated with height and *hmga2* is one of the top candidate genes for height according to a recent meta-analysis of genome-wide association studies.[97] This gene appears to be a major determinant of human growth,[97] but more work needs to be done to determine the extent to which its 5' UTR microsatellite modifies height and weight.

The *hmga2* microsatellite has different motifs in different mammals (Fig 2). Intriguingly, all of these motifs are poly-pyrimidine and have the potential to form H-DNA. For example, the cow *hmga2* gene contains a trinucleotide CTT/GAA motif and the opossum contains the tetranucleotide CTTT/GAAA motif. The presence of these motifs supports the hypothesis that H-DNA serves a functional purpose in this region.

In summary, H-DNA forms under negative torsional strain, and modifies the surrounding chromatin structure. Longer CT/AG microsatellites form H-DNA more readily,[25,88,90] and expansion of these microsatellites may also affect the rate at which H-DNA can form. Increased H-DNA formation in the 5' UTR may decrease expression if it interferes with transcription, as is seen in the *fli1* promoter.[91,92] Conversely, expression may increase if H-DNA formation induces an open chromatin structure that is favorable for transcription.

The binding of GAGA factor proteins to these microsatellites further complicates the relationship between CT/AG length and gene expression. More work in necessary to determine the exact mechanisms by which promoter CT/AG microsatellites modulate gene expression in the human genome. Furthermore, researching other H-DNA forming microsatellites, such as CTT/GAA, which do not bind GAGA factor, will help elucidate the function of H-DNA in promoters.

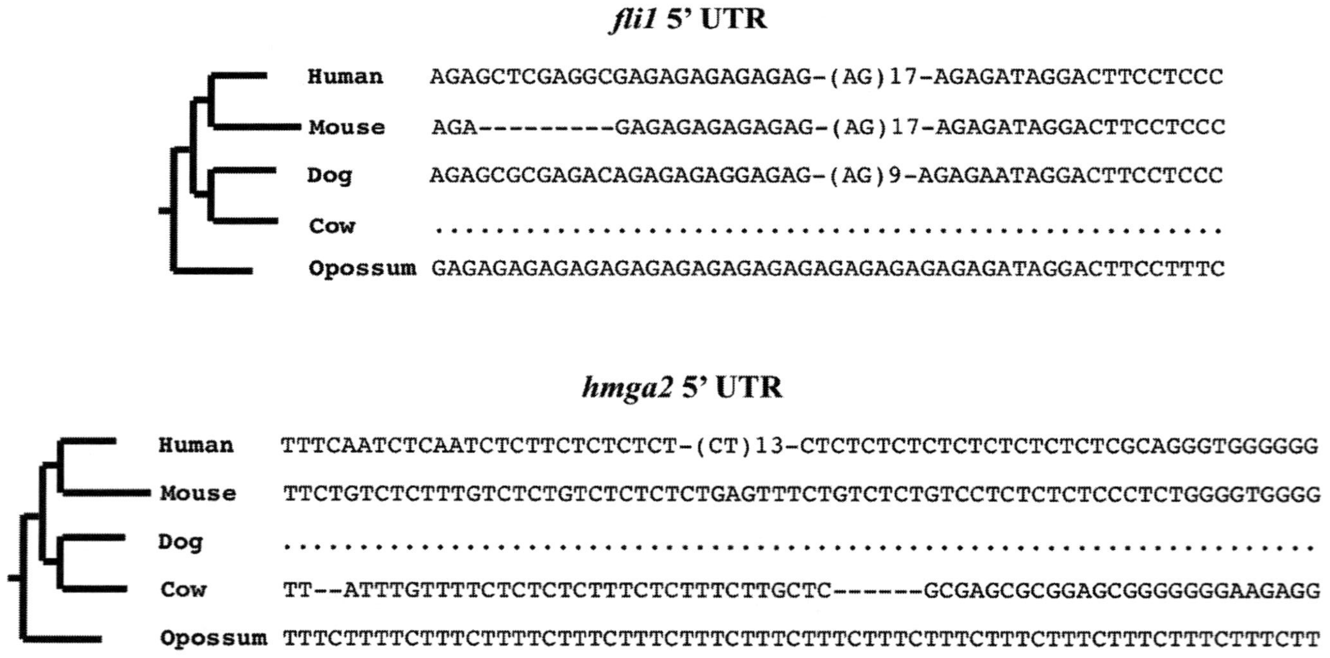
fli1 5' UTR
Human AGAGCTCGAGGCGAGAGAGAGAGAG-(AG)17-AGAGATAGGACTTCCTCCC
Mouse AGA---------GAGAGAGAGAGAG-(AG)17-AGAGATAGGACTTCCTCCC
Dog AGAGCGCGAGACAGAGAGAGGAGAG-(AG)9-AGAGAATAGGACTTCCTCCC
Cow ....
Opossum GAGAGAGAGAGAGAGAGAGAGAGAGAGAGAGAGAGAGAGATAGGACTTCCTTTC
hmga2 5' UTR
Human TTTCAATCTCAATCTCTTCTCTCTCT-(CT)13-CTCTCTCTCTCTCTCTCTCGCAGGGTGGGGGG
Mouse TTCTGTCTCTTTGTCTCTGTCTCTCTCTGAGTTTCTGTCTCTGTCCTCTCTCTCCCTCTGGGGTGGGG
Dog ....
Cow TT--ATTTGTTTTCTCTCTCTTTCTCTTTCTTGCTC------GCGAGCGCGGAGCGGGGGGGAAGAGG
Opossum TTTCTTTTCTTTCTTTTCTTTCTTTCTTTCTTTCTTTCTTTCTTTCTTTCTTTCTTTCTTTCTTTCTT

Figure 2. Alignments for two potential H-DNA forming microsatellites in the 5′ UTR of FLI1 and HMGA2. Note that although the motif differs between mammals for the HMGA2, all of these motifs have H-DNA forming potential.

The A/T Motif and Nucleosome Positioning

Microsatellites composed of the motif A/T (herein poly-A repeats) are the most common microsatellite in the human genome[64] and are determinants of nucleosome positioning.[98-101] Poly-A repeats form rigid DNA structures[98] that, when wrapped around a histone octamer, disrupt nucleosome binding.[99] This disruption increases the probability that poly-A repeats will be nucleosome free.[100] While the majority of work on poly-A repeats has been done in yeast, these general results apply to all eukaryotes including humans.[101]

When found in yeast promoters, poly-A repeats increase gene expression bi-directionally.[102] This effect is more pronounced for longer repeats.[10] Their effect on gene expression is independent of transcription factor binding,[103] making poly-A repeats a general mechanism by which gene expression can be modulated. However, we are unaware of any research on the effect that these microsatellites have on human gene expression. Given the prevalence of this motif in the human genome, including promoters,[64] these microsatellites are potentially an important source of unexplored variation in human gene expression.

Other Motifs and Mechanisms of Expression Modulation

There are many microsatellite motifs found in the human genome, and so far the discussion has focused on common motifs that have been well studied. There are many motifs that have not been studied in depth, but that can be found in promoters. For example, microsatellites with the motif "AT" are relatively common in the human genome, but we can only find one example where this motif is shown to modulate human gene expression.[104] In this example, expression levels cycled between high and low levels with a period of approximately 10 base pairs, suggesting that this microsatellite modulates expression by altering the orientation of, and space between, surrounding binding sites.[104] This highlights an important general mechanism by which all microsatellites can modulate gene expression, as minor changes in spacing between different binding sites can affect transcription rates.[105,106]

Microsatellites with longer motifs can also function as promoter elements.[107,108] For example, a pentanucleotide repeat upstream of the PIG3 gene interacts directly with p53, an important tumor suppressor.[107] Also, the length of a tetra-nucleotide repeat upstream of the human cholecystokinin gene is associated with variation in gene expression and panic disorder.[108] Although microsatellites with longer motifs have lower rates of expansion and contraction,[55] they may nevertheless serve as sources of phenotypic variation in the human genome.

Other motifs are discussed in detail in other chapters in this book, and will only be mentioned here briefly. Of particular interest is the motif CCG/CGG, which is involved in fragile-X syndrome.[109] This motif can form secondary structures that involve Hoogsteen base pairing, such as G-quadruplex (G-4) DNA.[110] These structures can act as activators or repressors of gene expression, depending on their context,[111] and some transcription factors target G-4 DNA structures as binding sites.[112] This motif has also been shown to modulate levels of expression when found upstream of a gene's transcription start site.[45]

Another motif capable of forming G-4 DNA is GGA/TCC.[113] In the *c-myb* gene, repression of transcription in a reporter construct is alleviated when the number of GGA/TCC repeats falls below the threshold required for quadruplex formation, indicating that G-4 DNA formation is important for expression.[114] This motif is an important regulator of gene expression in at least one human gene[115] and is capable of binding the transcription factor Sp1.[116] However, while GGA/TCC microsatellites appear to serve as functional promoter elements, more work needs to be done to determine whether naturally occurring polymorphisms in these microsatellites are responsible for variation in gene expression.

Interaction between Motifs

Because torsional strain is responsible for secondary structure formation in microsatellites, the interaction between adjacent microsatellites should not be ignored. Absorption of negative torsional strain by one microsatellite can prevent secondary structure formation in the surrounding region. For example, a long microsatellite with the motif $(CCTG/CAGG)_n$ can form slipped strand structures that interfere with transcription and are associated with myotonic dystrophy. These harmful structures are prevented with the expansion of an adjacent AC/GT microsatellite because Z-DNA formation can absorb the torsional strain that causes the strands to slip.[117] Theoretically, absorption of torsional strain by Z-DNA may also affect the formation of other secondary structures, such as H-DNA and G-4 DNA.

When microsatellites are directly adjacent to each other, like the compound $(AC)_n$-$(AG)_n$ microsatellite found in the *pax6* promoter, the effect of the expansion of one motif may affect the structure and function of the adjacent motif.[31,41] Investigating compound microsatellites, or microsatellites that are near other microsatellites will improve our understanding of structural interactions. However, the potential for motif interaction complicates interpretation of experimental results, making non-compound microsatellites a better choice for association studies.

CONCLUSION

Microsatellites have been shown, in multiple species, to modulate levels of gene expression as they expand and contract. This evolutionary tuning of phenotypes may serve a beneficial role.[118,119] Some microsatellites have been conserved in vertebrates over long evolutionary time periods,[120] and many promoter microsatellites are conserved in mammals,[26] perhaps for their ability to modulate gene expression. Here we have provided multiple examples of well-studied, conserved promoter microsatellites that appear to be modulating phenotypes in humans. Investigating other conserved promoter microsatellites will help determine whether microsatellites can be conserved as sources of variation in gene expression.

REFERENCES

1. Toth G, Gaspari Z, Jurka J. Microsatellites in different eukaryotic genomes: survey and analysis. Genome Res 2000; 10:967-981.
2. Moxon R, Bayliss C, Hood D. Bacterial contingency loci: the role of simple sequence DNA repeats in bacterial adaptation. Annu Rev Genet 2006; 40:307-333.
3. Wang G, Vasquez KM. Z-DNA, an active element in the genome. Front Biosci 2007; 12:4424-4438.
4. Kouzine F, Sanford S, Elisha-Feil Z et al. The functional response of upstream DNA to dynamic supercoiling in vivo. Nat Struct Mol Biol 2008; 15:146-154.
5. Martin P, Makepeace K, Hill SA et al. Microsatellite instability regulates transcription factor binding and gene expression. Proc Natl Acad Sci U S A 2005; 102:3800-3804.

6. Sarkari J, Pandit N, Moxon ER et al. Variable expression of the Opc outer membrane protein in Neisseria meningitidis is caused by size variation of a promoter containing poly-cytidine. Mol Microbiol 1994; 13:207-217.
7. Sawaya R, Arhin FF, Moreau F et al. Mutational analysis of the promoter region of the porA gene of Neisseria meningitidis. Gene 1999; 233:49-57.
8. Carson SD, Stone B, Beucher M et al. Phase variation of the gonococcal siderophore receptor FetA. Mol Microbiol 2000; 36:585-593.
9. Vinces MD, Legendre M, Caldara M et al. Unstable tandem repeats in promoters confer transcriptional evolvability. Science 2009; 324:1213-1216.
10. Iyer V, Struhl K. Poly(dA:dT), a ubiquitous promoter element that stimulates transcription via its intrinsic DNA structure. EMBO J 1995; 14:2570-2579.
11. Kirkpatrick DT, Wang YH, Dominska M et al. Control of meiotic recombination and gene expression in yeast by a simple repetitive DNA sequence that excludes nucleosomes. Mol Cell Biol 1999; 19:7661-7671.
12. Hammock EA, Young LJ. Microsatellite instability generates diversity in brain and sociobehavioral traits. Science 2005; 308:1630-1634.
13. Walum H, Westberg L, Henningsson S et al. Genetic variation in the vasopressin receptor 1a gene (AVPR1A) associates with pair-bonding behavior in humans. Proc Natl Acad Sci U S A 2008; 105:14153-14156.
14. Knafo A, Israel S, Darvasi A et al. Individual differences in allocation of funds in the dictator game associated with length of the arginine vasopressin 1a receptor RS3 promoter region and correlation between RS3 length and hippocampal mRNA. Genes Brain Behav 2008; 7:266-275.
15. Meyer-Lindenberg A, Kolachana B, Gold B et al. Genetic variants in AVPR1A linked to autism predict amygdala activation and personality traits in healthy humans. Mol Psychiatry 2009; 14:968-975.
16. Bachner-Melman R, Dina C, Zohar AH et al. AVPR1a and SLC6A4 gene polymorphisms are associated with creative dance performance. PLoS Genet 2005; 1:e42.
17. Ukkola LT, Onkamo P, Raijas P et al. Musical aptitude is associated with AVPR1A-haplotypes. PLoS ONE 2009; 4:e5534.
18. Rothenburg S, Koch-Nolte F, Rich A et al. A polymorphic dinucleotide repeat in the rat nucleolin gene forms Z-DNA and inhibits promoter activity. Proc Natl Acad Sci U S A 2001; 98:8985-8990.
19. Hayes TE, Dixon JE. Z-DNA in the rat somatostatin gene. J Biol Chem 1985; 260:8145-8156.
20. Thomas MJ, Freeland TM, Strobl JS. Z-DNA formation in the rat growth hormone gene promoter region. Mol Cell Biol 1990; 10:5378-5387.
21. Naylor LH, Clark EM. d(TG)n.d(CA)n sequences upstream of the rat prolactin gene form Z-DNA and inhibit gene transcription. Nucleic Acids Res 1990; 18:1595-1601.
22. Rockman M, Wray G, Wray G. Abundant raw material for cis-regulatory evolution in humans. Molecular Biology and Evolution 2002; 19:1991-2004.
23. Rothenburg S, Koch-Nolte F, Haag F. DNA methylation and Z-DNA formation as mediators of quantitative differences in the expression of alleles. Immunol Rev 2001; 184:286-298.
24. Streelman JT, Kocher TD. Microsatellite variation associated with prolactin expression and growth of salt challenged tilapia. Physiol Genomics 2002; 9:1-4.
25. Xu G, Goodridge AG. A CT repeat in the promoter of the chicken malic enzyme gene is essential for function at an alternative transcription start site. Arch Biochem Biophys 1998; 358:83-91.
26. Sawaya S, Lennon D, Buschiazzo E et al. Measuring microsatellite conservation in mammals with a phylogenetic birth death model. Submitted to Gen Biol and Evol 2011.
27. Chen YH, Lin SJ, Lin MW et al. Microsatellite polymorphism in promoter of heme oxygenase-1 gene is associated with susceptibility to coronary artery disease in type 2 diabetic patients. Hum Genet 2002; 111:18.
28. Contente A, Dittmer A, Koch MC et al. A polymorphic microsatellite that mediates induction of PIG3 by p53. Nat Genet 2002; 30:315-320.
29. Rife T, Rasoul B, Pullen N et al. The effect of a promoter polymorphism on the transcription of nitric oxide synthase 1 and its relevance to Parkinson's disease. J Neurosci Res 2009; 87:2319-2325.
30. Reif A, Jacob CP, Rujescu D et al. Influence of functional variant of neuronal nitric oxide synthase on impulsive behaviors in humans. Arch Gen Psychiatry 2009; 66:41-50.
31. Ng TK, Lam CY, Lam DS et al. AC and AG dinucleotide repeats in the pax6 P1 promoter are associated with high myopia. Mol Vis 2009; 15:2239-2248.
32. Albanese V, Biguet NF, Kiefer H et al. Quantitative effects on gene silencing by allelic variation at a tetranucleotide microsatellite. Hum Mol Genet 2001; 10:1785-1792.
33. Gao PS, Heller NM, Walker W et al. Variation in dinucleotide (GT) repeat sequence in the first exon of the STAT6 gene is associated with atopic asthma and differentially regulates the promoter activity in vitro. J Med Genet 2004; 41:535-539.
34. Yim JJ, Lee HW, Lee HS et al. The association between microsatellite polymorphisms in intron II of the human Toll-like receptor 2 gene and tuberculosis among Koreans. Genes Immun 2006; 7:150-155.
35. Agarwal AK, Giacchetti G, Lavery G et al. CA-Repeat polymorphism in intron 1 of HSD11B2 : effects on gene expression and salt sensitivity. Hypertension 2000; 36:187-194.

36. Akai J, Kimura A, Hata RI. Transcriptional regulation of the human type I collagen alpha2 (COL1A2) gene by the combination of two dinucleotide repeats. Gene 1999; 239:65-73.
37. Itokawa M, Yamada K, Yoshitsugu K et al. A microsatellite repeat in the promoter of the N-methyl-D-aspartate receptor 2A subunit (GRIN2A) gene suppresses transcriptional activity and correlates with chronic outcome in schizophrenia. Pharmacogenetics 2003; 13:271-278.
38. Searle S, Blackwell JM. Evidence for a functional repeat polymorphism in the promoter of the human NRAMP1 gene that correlates with autoimmune versus infectious disease susceptibility. J Med Genet 1999; 36:295-299.
39. Yamada N, Yamaya M, Okinaga S et al. Microsatellite polymorphism in the heme oxygenase-1 gene promoter is associated with susceptibility to emphysema. Am J Hum Genet 2000; 66:187-195.
40. Shimajiri S, Arima N, Tanimoto A et al. Shortened microsatellite d(CA)21 sequence down-regulates promoter activity of matrix metalloproteinase 9 gene. FEBS Lett 1999; 455:70-74.
41. Okladnova O, Syagailo YV, Tranitz M et al. A promoter-associated polymorphic repeat modulates PAX-6 expression in human brain. Biochem Biophys Res Commun 1998; 248:402-405.
42. Okladnova O, Syagailo YV, Tranitz M et al. Functional characterization of the human PAX3 gene regulatory region. Genomics 1999; 57:110-119.
43. Hough C, Cameron CL, Notley CR et al. Influence of a GT repeat element on shear stress responsiveness of the VWF gene promoter. J Thromb Haemost 2008; 6:1183-1190.
44. Wang B, Ren J, Ooi LL et al. Dinucleotide repeats negatively modulate the promoter activity of Cyr61 and is unstable in hepatocellular carcinoma patients. Oncogene 2005; 24:3999-4008.
45. Roberts RL, Gearry RB, Bland MV et al. Trinucleotide repeat variants in the promoter of the thiopurine S-methyltransferase gene of patients exhibiting ultra-high enzyme activity. Pharmacogenet Genomics 2008; 18:434-438.
46. Vedrine SM, Vourc'h P, Tabagh R et al. A functional tetranucleotide (AAAT) polymorphism in an Alu element in the NF1 gene is associated with mental retardation. Neurosci Lett 2011; 491:118-121.
47. Valverde P, Koren G. Purification and preliminary characterization of a cardiac Kv1.5 repressor element binding factor. Circ Res 1999; 84:937-944.
48. Gebhardt F, Zanker KS, Brandt B. Modulation of epidermal growth factor receptor gene transcription by a polymorphic dinucleotide repeat in intron 1. J Biol Chem 1999; 274:13176-13180.
49. Funke-Kaiser H, Thomas A, Bremer J et al. Regulation of the major isoform of human endothelin-converting enzyme-1 by a strong housekeeping promoter modulated by polymorphic microsatellites. J Hypertens 2003; 21:2111-2124.
50. Stuehr DJ. Mammalian nitric oxide synthases. Biochim Biophys Acta 1999; 1411:217-230.
51. Reif A, Kiive E, Kurrikof T et al. A functional pax6 promoter polymorphism interacts with adverse environment on functional and dysfunctional impulsivity. Psychopharmacology (Berl.) 2011; 214:239-248.
52. Retz W, Reif A, Freitag CM et al. Association of a functional variant of neuronal nitric oxide synthase gene with self-reported impulsiveness, venturesomeness and empathy in male offenders. J Neural Transm 2010; 117:321-324.
53. Laas K, Reif A, Herterich S et al. The effect of a functional pax6 promoter polymorphism on impulsivity is moderated by platelet MAO activity. Psychopharmacology (Berl.) 2010; 209:255-261.
54. Nelson RJ, Demas GE, Huang PL et al. Behavioural abnormalities in male mice lacking neuronal nitric oxide synthase. Nature 1995; 378:383-386.
55. Kelkar YD, Tyekucheva S, Chiaromonte F et al. The genome-wide determinants of human and chimpanzee microsatellite evolution. Genome Res 2008; 18:30-38.
56. Grindley JC, Davidson DR, Hill RE. The role of Pax-6 in eye and nasal development. Development 1995; 121:1433-1442.
57. Larsen KB, Lutterodt MC, Laursen H et al. Spatiotemporal distribution of pax6 and MEIS2 expression and total cell numbers in the ganglionic eminence in the early developing human forebrain. Dev Neurosci 2010; 32:149-162.
58. Stoykova A, Treichel D, Hallonet M et al. Pax6 modulates the dorsoventral patterning of the mammalian telencephalon. J Neurosci 2000; 20:8042-8050.
59. Chapouton P, Gartner A, Gotz M. The role of Pax6 in restricting cell migration between developing cortex and basal ganglia. Development 1999; 126:5569-5579.
60. Okladnova O, Syagailo YV, Mossner R et al. Regulation of PAX-6 gene transcription: alternate promoter usage in human brain. Brain Res Mol Brain Res 1998; 60:177-192.
61. Schedl A, Ross A, Lee M et al. Influence of pax6 gene dosage on development: overexpression causes severe eye abnormalities. Cell 1996; 86:71-82.
62. Han W, Leung KH, Fung WY et al. Association of pax6 polymorphisms with high myopia in han chinese nuclear families. Investigative Ophthalmology and Visual Science 2009; 50:47-56. doi:10.1167/iovs.07-0813.
63. Jiang B, Yap MKH, Leung KH et al. pax6 haplotypes are associated with high myopia in han chinese. PLoS ONE 2011; 6:e19587. doi:10.1371/journal.pone.0019587.

64. Lawson MJ, Zhang L. Housekeeping and tissue-specific genes differ in simple sequence repeats in the 5'-UTR region. Gene 2008; 407:54-62.
65. Warren WC, Hillier LW, Marshall Graves JA et al. Genome analysis of the platypus reveals unique signatures of evolution. Nature 2008; 453:175-183.
66. Jaworski A, Higgins NP, Wells RD et al. Topoisomerase mutants and physiological conditions control super coiling and Z-DNA formation in vivo. J Biol Chem 1991; 266:2576-2581.
67. Wolfl S, Martinez C, Rich A et al. Transcription of the human corticotropin-releasing hormone gene in NPLC cells is correlated with Z-DNA formation. Proc Natl Acad Sci U S A 1996; 93:3664-3668.
68. Wittig B, Wolfl S, Dorbic T et al. Transcription of human c-myc in permeabilized nuclei is associated with formation of Z-DNA in three discrete regions of the gene. EMBO J 1992; 11:4653-4663.
69. Li H, Xiao J, Li J et al. Human genomic Z-DNA segments probed by the Z alpha domain of ADAR1. Nucleic Acids Res 2009; 37:2737-2746.
70. Schroth GP, Chou PJ, Ho PS. Mapping Z-DNA in the human genome. Computer-aided mapping reveals a nonrandom distribution of potential Z-DNA-forming sequences in human genes. J Biol Chem 1992; 267:11846-11855.
71. Wong B, Chen S, Kwon JA et al. Characterization of Z-DNA as a nucleosome-boundary element in yeast Saccharomyces cerevisiae. Proc Natl Acad Sci U S A 2007; 104:2229-2234.
72. Liu H, Mulholland N, Fu H et al. Cooperative activity of BRG1 and Z-DNA formation in chromatin remodeling. Molecular Cell Biology 2006; 26:2550-2559.
73. Zhang J, Ohta T, Maruyama A et al. BRG1 interacts with Nrf2 to selectively mediate HO-1 induction in response to oxidative stress. Mol Cell Biol 2006; 26:7942-7952.
74. Xu YZ, Thuraisingam T, Marino R et al. Recruitment of SWI/SNF complex is required for transcriptional activation of SLC11A1 gene during macrophage differentiation of HL-60 cells. J Biol Chem 2011.
75. Peck LJ, Wang JC. Transcriptional block caused by a negative supercoiling induced structural change in an alternating CG sequence. Cell 1985; 40:129-137.
76. Wittig B, Dorbic T, Rich A. Transcription is associated with Z-DNA formation in metabolically active permeabilized mammalian cell nuclei. Proc Natl Acad Sci USA 1991; 88:2259-2263.
77. Tae HJ, Luo X, Kim KH. Roles of CCAAT/enhancer-binding protein and its binding site on repression and derepression of acetyl-CoA carboxylase gene. J Biol Chem 1994; 269:10475-10484.
78. Wu T, Ikezono T, Angus CW et al. Characterization of the promoter for the human 85 kDa cytosolic phospholipase A2 gene. Nucleic Acids Res 1994; 22:5093-5098.
79. Kulish VV, Heng L, Droge P. Z-dna-induced super-transport of energy within genomes. Physica A: Statistical Mechanics and its Applications 2007; 384:733 738. doi:DOI:10.1016/j.physa.2007.06.023.
80. Johnston BH. The S1-sensitive form of d(C-T)n.d(A-G)n: chemical evidence for a three-stranded structure in plasmids. Science 1988; 241:1800-1804.
81. Kohwi Y, Kohwi-Shigematsu T. Altered gene expression correlates with DNA structure. Genes Dev 1991; 5:2547-2554.
82. Maiti AK, Brahmachari SK. Poly purine.pyrimidine sequences upstream of the beta-galactosidase gene affect gene expression in Saccharomyces cerevisiae. BMC Mol Biol 2001; 2:11.
83. Amiri H, Nekhotiaeva N, Sun JS et al. Benzoquinoquinoxaline derivatives stabilize and cleave H-DNA and repress transcription downstream of a triplex-forming sequence. J Mol Biol 2005; 351:776-783.
84. Motallebipour M, Rada-Iglesias A, Westin G et al. Two polypyrimidine tracts in the nitric oxide synthase 2 gene: similar regulatory sequences with different properties. Mol Biol Rep 2010; 37:2021-2030.
85. Lu Q, Wallrath LL, Granok H et al. (CT)n (GA)n repeats and heat shock elements have distinct roles in chromatin structure and transcriptional activation of the Drosophila hsp26 gene. Mol Cell Biol 1993; 13:2802-2814.
86. Lu Q, Teare JM, Granok H et al. The capacity to form H-DNA cannot substitute for GAGA factor binding to a (CT)n*(GA)n regulatory site. Nucleic Acids Res 2003; 31:2483-2494.
87. Matharu NK, Hussain T, Sankaranarayanan R et al. Vertebrate homologue of Drosophila GAGA factor. J Mol Biol 2010; 400:434-447.
88. Beaulieu M, Barbeau B, Rassart E. Triplex-forming oligonucleotides with unexpected affinity for a nontargeted GA repeat sequence. Antisense Nucleic Acid Drug Dev 1997; 7:125-130.
89. Rustighi A, Tessari MA, Vascotto F et al. A polypyrimidine/polypurine tract within the Hmga2 minimal promoter: a common feature of many growth-related genes. Biochemistry 2002; 41:1229-1240.
90. Han YJ, de Lanerolle P. Naturally extended CT. AG repeats increase H-DNA structures and promoter activity in the smooth muscle myosin light chain kinase gene. Mol Cell Biol 2008; 28:863-872.
91. Nowling TK, Fulton JD, Chike-Harris K et al. Ets factors and a newly identified polymorphism regulate Fli1 promoter activity in lymphocytes. Mol Immunol 2008; 45:1-12.
92. Morris EE, Amria MY, Kistner-Grifn E et al. A GA microsatellite in the Fli1 promoter modulates gene expression and is associated with systemic lupus erythematosus patients without nephritis. Arthritis Res Ther 2010; 12:R212.

93. Zhang L, Eddy A, Teng YT et al. An immunological renal disease in transgenic mice that overexpress Fli-1, a member of the ets family of transcription factor genes. Mol Cell Biol 1995; 15:6961-6970.
94. Borrmann L, Seebeck B, Rogalla P et al. Human hmga2 promoter is coregulated by a polymorphic dinucleotide (TC)-repeat. Oncogene 2003; 22:756-760.
95. Zhou X, Benson KF, Ashar HR et al. Mutation responsible for the mouse pygmy phenotype in the develop mentally regulated factor HMGI-C. Nature 1995; 376:771-774.
96. Hodge JC, T Cuenco K, Huyck KL et al. Uterine leiomyomata and decreased height: a common hmga2 predisposition allele. Hum Genet 2009; 125:257-263.
97. Lanktree MB, Guo Y, Murtaza M et al. Meta-analysis of dense genecentric association studies reveals common and uncommon variants associated with height. Am J Hum Genet 2011; 88:6-18.
98. Suter B, Schnappauf G, Thoma F. Poly(dA.dT) sequences exist as rigid DNA structures in nucleosome-free yeast promoters in vivo. Nucleic Acids Res 2000; 28:4083-4089.
99. Shimizu M, Mori T, Sakurai T et al. Destabilization of nucleosomes by an unusual DNA conformation adopted by poly(dA) small middle poly(dT) tracts in vivo. EMBO J 2000; 19:3358-3365.
100. Bao Y, White CL, Luger K. Nucleosome core particles containing a poly(dA.dT) sequence element exhibit a locally distorted DNA structure. J Mol Biol 2006; 361:617-624.
101. Segal E, Widom J. Poly(dA:dT) tracts: major determinants of nucleosome organization. Curr Opin Struct Biol 2009; 19:65-71.
102. Struhl K. Naturally occurring poly(dA-dT) sequences are upstream promoter elements for constitutive transcription in yeast. Proc Natl Acad Sci USA 1985; 82:8419-8423.
103. Wu R, Li H. Positioned and G/C-capped poly(dA:dT) tracts associate with the centers of nucleosome-free regions in yeast promoters. Genome Res 2010; 20:473-484.
104. Uhlemann AC, Szlezak NA, Vonthein R et al. DNA phasing by TA dinucleotide microsatellite length determines in vitro and in vivo expression of the gp91phox subunit of NADPH oxidase and mediates protection against severe malaria. J Infect Dis 2004; 189:2227-2234.
105. Spek CA, Bertina RM, Reitsma PH. Unique distance and DNA-turn-dependent interactions in the human protein C gene promoter confer submaximal transcriptional activity. Biochem J 1999; 340: 513-518.
106. Vardhanabhuti S, Wang J, Hannenhalli S. Position and distance specificity are important determinants of cis-regulatory motifs in addition to evolutionary conservation. Nucleic Acids Res 2007; 35:3203-3213.
107. Levine AJ. p53, the cellular gatekeeper for growth and division. Cell 1997; 88:323-331.
108. Ebihara M, Ohba H, Hattori E et al. Transcriptional activities of cholecystokinin promoter haplotypes and their relevance to panic disorder susceptibility. Am J Med Genet B Neuropsychiatr. Genet 2003; 118B:32-35.
109. Loesch D, Hagerman R. Unstable mutations in the FMR1 gene and the phenotypes. In: Hannan AJ, ed. Tandem Repeat Polymorphisms: Genetic Plasticity, Neural Diversity and Disease. Austin/New York: Landes Bioscience/Springer Science+Business Media, 2012:78-114.
110. Darlow JM, Leach DR. Secondary structures in d(CGG) and d(CCG) repeat tracts. J Mol Biol 1998; 275:3-16.
111. Qin Y, Hurley LH. Structures, folding patterns, and functions of intramolecular DNA G-quadruplexes found in eukaryotic promoter regions. Biochimie 2008; 90:1149-1171.
112. Shklover J, Weisman-Shomer P, Yafe A et al. Quadruplex structures of muscle gene promoter sequences enhance in vivo MyoD-dependent gene expression. Nucleic Acids Res 2010; 38:2369-2377.
113. Matsugami A, Ouhashi K, Kanagawa M et al. An intramolecular quadruplex of (GGA)(4) triplet repeat DNA with a G:G:G:G tetrad and a G(:A):G(:A):G(:A):G heptad and its dimeric interaction. J Mol Biol 2001; 313:255-269.
114. Palumbo SL, Memmott RM, Uribe DJ et al. A novel G-quadruplex-forming GGA repeat region in the c-myb promoter is a critical regulator of promoter activity. Nucleic Acids Res 2008; 36:1755-1769.
115. Hafner M, Zimmermann K, Pottgiesser J et al. A purine-rich sequence in the human BM-40 gene promoter region is a prerequisite for maximum transcription. Matrix Biol 1995; 14:733-741.
116. Chamboredon S, Briggs J, Vial E et al. v-Jun downregulates the SPARC target gene by binding to the proximal promoter indirectly through Sp1/3. Oncogene 2003; 22:4047-4061.
117. Edwards S, Sirito M, Krahe R et al. A Z-DNA sequence reduces slipped-strand structure formation in the myotonic dystrophy type 2 (CCTG) x (CAGG) repeat. Proc Natl Acad Sci USA 2009; 106:3270-3275.
118. Kashi Y, King DG. Simple Sequence repeats as advantageous mutators in evolution. TRENDS in Genetics 2006; 22:253-259.
119. King DG. Evolution of simple sequence repeats as mutable sites. In: Hannan AJ, ed. Tandem Repeat Polymorphisms: Genetic Plasticity, Neural Diversity and Disease. Austin/New York: Landes Bioscience/Springer Science+Business Media, 2012:10-25.
120. Buschiazzo E, Gemmell NJ. Conservation of human microsatellites across 450 million years of evolution. Genome Biol Evol 2010; 2:153-165.

CHAPTER 5

DYNAMIC MUTATIONS
Where Are They Now?

Clare L. van Eyk* and Robert I. Richards

Discipline of Genetics, School of Molecular and Biomedical Sciences, The University of Adelaide, Adelaide, South Australia, Australia
Corresponding Author: Clare L. van Eyk—Email: clare.vaneyk@adelaide.edu.au

Abstract: Dynamic mutations are those caused by the expansion of existing polymorphic DNA repeat sequences beyond a copy number threshold. These genetic mutations can give rise to dominant, recessive or X-linked disorders, dependent upon the location of the repeat sequence with respect to the genes that are affected by the expansion. The distinguishing feature of these mutations is their instability, which is a function of the copy number of repeats and can occur in either meiosis or mitosis. For some of the resultant disorders there is a relationship between repeat copy number and age-at-onset and/or severity of symptoms of the disease. For this reason much effort is now focused on identifying the pathogenic pathways from the mutation to the disease symptoms in the hope of finding means of delaying onset, slowing progression or even preventing symptoms of the disease. The growing list of neurodegenerative and neuromuscular diseases caused by dynamic mutations includes Huntington's disease (HD), spinobulbar muscular atrophy (SBMA), dentatorubral-pallidoluysian atrophy (DRPLA), a number of spinocerebellar ataxias (SCAs), oculopharyngeal muscular dystrophy (OPMD), myotonic dystrophy Type 1 and 2 (DM1 and 2), Huntington's disease-like 2 (HDL-2), Friedrich's ataxia (FRDA), Fragile X associated tremor ataxia syndrome (FXTAS), Fragile XE (FRAXE) and Fragile XA (FRAXA). This chapter aims to give a brief overview of what is currently known about each disease and the mechanisms underlying pathogenesis.

INTRODUCTION

Since the identification of the first diseases resulting from the expansion of simple repeat sequences 20 years ago,[1,2] there has been a growing list of diseases attributed to

Tandem Repeat Polymorphisms: Genetic Plasticity, Neural Diversity and Disease,
edited by Anthony J. Hannan ©2012 Landes Bioscience and Springer Science+Business Media.

this sort of mutation. Termed "dynamic mutations," repeat expansions are defined by several unique features: (1) the mutation involves a change in copy number of the repeat sequence (usually an increase) which can occur both in the germ line and in somatic cells, with the rate of change dependent upon the initial size of the repeat; (2) some alleles (usually those consisting of uninterrupted repeats) have a greater likelihood of changes in copy number; (3) the copy number of the repeat sequence has a relationship to the severity and/or age-at-onset of the disease. These properties account for the unusual inheritance pattern (called "anticipation") involving increasing severity/incidence and/ or decreasing age-at-onset which is seen in families affected by many of the diseases associated with dynamic mutations.

MECHANISMS OF MUTATION

The processes involved in microsatellite repeat expansion are complex and vary between tissues and developmental stages. Both *cis* (repeat sequence, starting size, flanking sequence, methylation status and purity) and *trans* (DNA metabolism) factors influence the rate and incidence of change in copy number in any particular cell. Some of the processes involved in mutation are summarized below.

Replication

DNA replication has been shown to play a key role in repeat expansion in several models.[3-7] It is thought that the formation of unusual secondary structures in single-stranded DNA during replication as well as DNA polymerase slippage during replication of the lagging strand are responsible for changes in repeat size during replication. In *Eschericia coli*, a bimodal pattern of expansion of CAG repeats has been shown: expansions of repeat tracts which are smaller than an Okazaki fragment occurs incrementally, while expansions of large repeat tracts (> 80 repeats) tend to occur in a punctuated manner.[8] A similar bimodal correlation between expansion rate and Okazaki fragment size is also seen in several diseases (e.g., in FRDA[9]). This correlation is though to be the result of secondary structures formed by expanded repeat sequences which prevent the binding and cleavage of fragments by the Okazaki fragment endonuclease (FEN1) in a length-dependent manner[10,11] and therefore disrupt the removal of misplaced Okazaki fragments, resulting in changes in copy number of the repeat tract. However, since large somatic repeat expansions have also been observed in non-proliferative cells, for example in the striatum of a HD mouse models[12,13] and affected individuals,[14] DNA replication is likely not to be the only process involved. Indeed, mechanisms of expansion may also differ between tissues[15] and between different repeat sequences.[16]

Repair

The observation that increasing lengths of microsatellite repeats correlate with increasing chromosomal fragility suggests that DNA repair pathways may also play a role in changes in repeat size. Unusual hairpin DNA secondary structures formed by repeat expansions have been shown to be breakage 'hot spots' which are frequently repaired inefficiently by the base excision repair enzyme 7,8-dihydro-8-oxoguanine glycosylase (OGG1),[17,18] resulting in changes in repeat copy number. This suggests a link between oxidative stress, which induces mutations, and increases in repeat number observed in somatic cells as the individual ages. At least 2 components of the DNA mismatch repair complex, msh2 and msh3, are required for repeat expansion in several mouse models,[19-22] supporting a key role for inefficient repair of single stranded breaks in repeat regions in expansion.

Recombination

Recent evidence also supports a role for double stranded break repair pathways, such as those activated during recombination, in expansions and contractions of CAG/CTG repeats in yeast[23] and contractions of GAA repeats in an *E. coli* model of FRDA.[24] Double stranded breaks are known to be induced by replication fork stalling in *E. coli*[25] and therefore faulty replication may be linked to repeat expansions and contractions resulting from recombination. The repair of double stranded breaks by recombination can result in changes in repeat number by a number of different mechanisms, including out-of-frame re-annealing of newly synthesized DNA or unequal crossing over resulting in the expansion of one allele and the contraction of the other.[26]

MECHANISMS OF PATHOGENESIS

Disease-causing microsatellite repeat expansions can be classified into two categories: those which fall in translated regions of a gene and those which fall in an untranslated region of the genome (Fig. 1).[27,28] Expansions in translated regions encoding polyglutamine

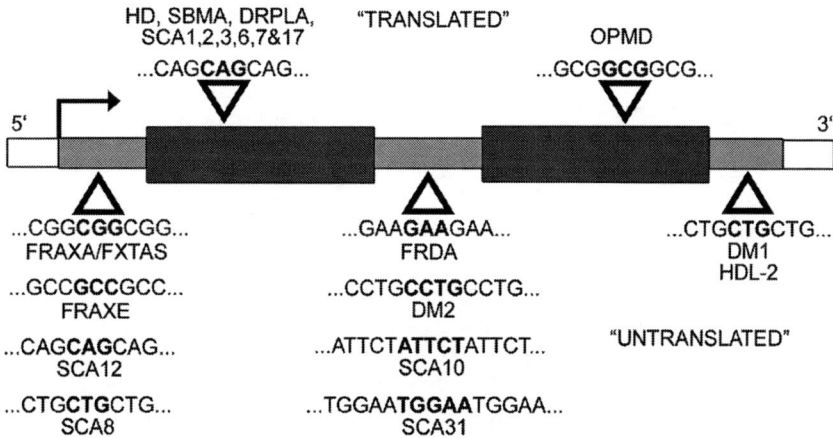

Figure 1. Schematic demonstrating locations and sequences of disease-causing expanded repeat sequences in humans. Expanded repeat sequences can fall either within translated (above) or untranslated (below) regions of genes. Diseases caused by translated repeats are the result of tri-nucleotide repeat expansions and include the polyglutamine diseases and the polyalanine diseases. Untranslated repeat expansions can be tri-, tetra-, or penta-nucleotide sequences and can fall either in 5′ untranslated or 3′ untranslated regions or within an intron. Expanded repeat diseases include a number of Spinocerebellar ataxias (SCAs), Huntington's disease (HD), spinobulbar muscular atrophy (SBMA), dentatorubral-pallidoluysian atrophy (DRPLA), oculopharyngeal muscular dystrophy (OPMD), Fragile X syndrome (FRAXA), Fragile X tremor ataxia syndrome (FXTAS), Fragile XE mental retardation syndrome (FRAXE), Friedrich's ataxia (FRDA), Myotonic dystrophy Types 1 and 2 (DM1 and 2) and Huntington's disease like-2 (HDL-2).

and polyalanine have been shown to cause disease through a mechanism which at least partly involves dominant toxicity of the expanded repeat-containing proteins themselves. Various tri-, tetra- and penta-nucleotide repeat sequences located in either 5′ or 3′ untranslated regions or introns have also been identified as disease-causing mutations. In a number of cases, these untranslated repeats have been suggested to exert toxicity through a mechanism involving RNA gain-of-function, although loss-of-function mechanisms also exist. Therefore, the location of the repeat expansion appears, to some extent, to determine the mechanism of pathogenesis (Fig. 2). However, there is significant clinical overlap between the translated and untranslated repeat diseases which has led to recent proposals of alternative, or additional, mechanisms of pathogenesis.

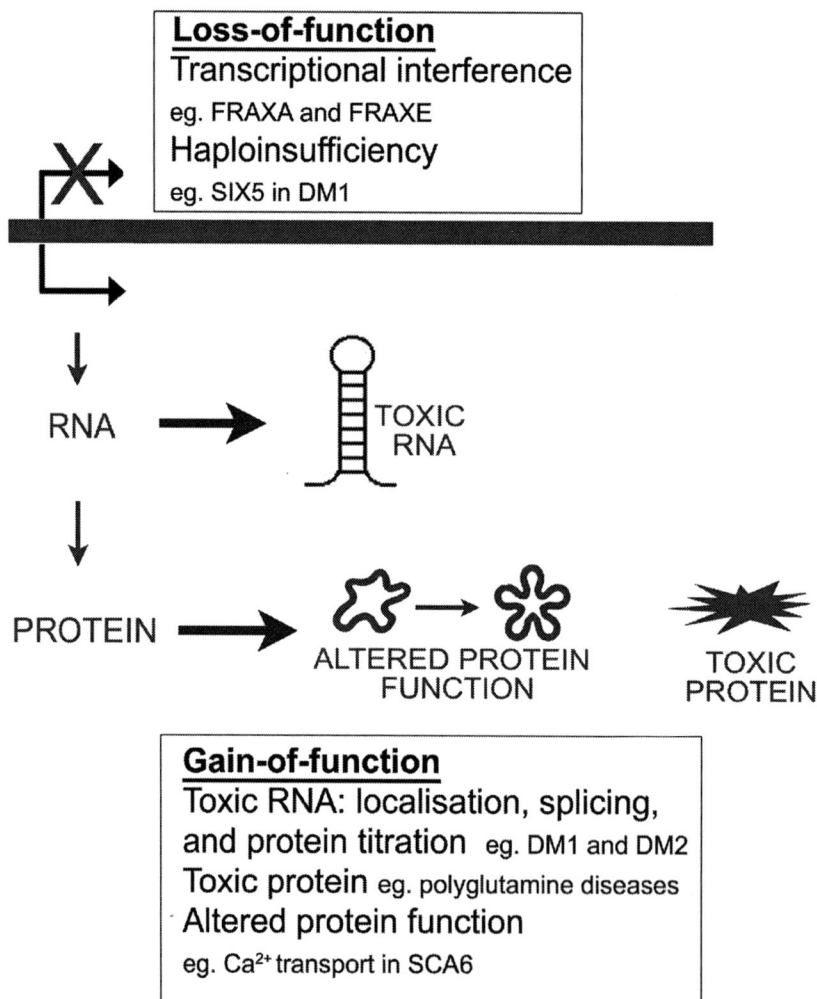

Figure 2. Proposed mechanisms of pathogenesis in the expanded repeat diseases. Multiple pathogenic pathways have been described for expanded repeat diseases, including both loss-of-function and gain-of-function. A complex interplay of these mechanisms is likely to play a role in pathogenesis in many cases.

Translated Repeat Diseases

Polyglutamine (polyQ) Diseases

To date there are 9 diseases known to be caused by expansion of a CAG repeat encoding glutamine (Table 1). These diseases are Huntington's disease (HD), spinobulbar muscular atrophy (SBMA), dentatorubral-pallidoluysian atrophy (DRPLA) and a number of spinocerebellar ataxias (SCA1,2,3,6,7&17). Despite their presence in entirely unrelated proteins, overlapping features of the polyglutamine diseases suggest that there are dominant toxic properties of expanded polyglutamine. These features include progressive neurological degeneration which is generally late-onset, repeat number-dependent age-of-onset and the formation of aggregates by the polyglutamine-containing proteins.[29] Differences in the specificity of affected cells are thought to be a result of the gene in which the repeat falls, however it is unclear why mutant forms of ubiquitously expressed proteins should elicit effects on such a limited population of cells. Some features of these diseases are summarized in Table 1 and below.

Huntington's Disease

Huntington's disease (HD), the most common of the polyglutamine diseases, is caused by expansion of a CAG repeat resulting in an expanded N-terminal polyglutamine tract

Table 1. Diseases caused by expansion of CAG encoding polyglutamine

Disease	Polyglutamine-Containing Protein	Proposed Function	CAG Repeat Number	
			Disease	Normal
HD	Huntington	Organelle trafficking and axonal transport	37–265	9–36
SBMA	Androgen Receptor	Ligand-activated Transcription Factor	38–62	10–36
DRPLA	Atrophin-1	Nuclear receptor Co-repressor	49–88	7–23
SCA1	Ataxin-1	RNA processing	39–82	6–39
SCA2	Ataxin-2	RNA metabolism	36–63	15–31
SCA3/MJD	Ataxin-3	Protein turnover/ DNA Repair	52–86	12–44
SCA6	CACNA1A	Voltage-gated calcium channel sub-unit	19–33	4–18
SCA7	Ataxin-7	Transcriptional regulation	37–306	4–35
SCA17	TATA box-binding protein	Transcriptional initiation	47–63	25–44

HD—Huntington's disease; SBMA—spinobulbar muscular atrophy; DRPLA—dentatorubral-pallidoluysian atrophy; SCA—spinocerebellar ataxia; MJD—Machado Joseph disease. Larger repeats within the normal range are normally interrupted by CAT for SCA1 and CAA for SCA2 while disease-causing alleles consist of pure CAG repeats.

in the Huntingtin (HTT) protein. The disease presents as progressive neurodegeneration resulting in loss of motor and cognitive function. Neurodegeneration preferentially affects the medium spiny GABA neurons in the striatum.[29] The function of the 350kDa HTT protein is not known, although its association with microtubules and synaptic vesicles suggests a role in organelle trafficking and axonal transport.[30] A direct association of mutant HTT with clathrin-coated membranes has also been demonstrated, suggesting that perturbation of endocytic pathways may play a role in HD pathogenesis.[31] Recent studies have demonstrated axonal transport defects in both mammalian neurons and a *Drosophila* model expressing the mutant HTT protein, as well as in HD brains, characterized by aggregation of vesicles, mis-localization of mitochondria and apoptosis.[32-34]

HTT protein shows ubiquitous expression in neurons and is also expressed at low levels throughout the body.[35] A number of neuronal cell types which express HTT at all times survive in Huntington's patients—for example the striatal cholinergic interneurons—while the striatal spiny neurons most affected in Huntington's disease do not consistently express HTT[36] and it is therefore unclear how selective neurodegeneration is elicited. Furthermore, the mutant protein is expressed at similar or even slightly reduced levels in comparison to the normal protein in affected regions of the brain[29] and therefore there does not appear to be a simple relationship between expression of the mutant protein and vulnerability.

Spinalbulbar Muscular Atrophy (SBMA)

SBMA or Kennedy's disease is caused by a CAG repeat expansion in the *androgen receptor (AR)* gene on the X chromosome and is the only one of the polyglutamine diseases not autosomally dominantly inherited.[37] The AR is a ligand activated transcription factor which is translocated to the nucleus in response to testosterone, a process which is perturbed by the expansion of the glutamine tract in the AR.[38] It has been suggested that higher circulating androgen and testosterone levels could explain why only males are affected by SBMA.[39] Disease is not simply a result of loss-of-function of the *AR* gene, since mutations other than expansion of the CAG repeat do not result in SBMA and SBMA individuals show only limited androgen insensitivity, however recent reports suggest that loss-of-function may contribute to disease progression, since the AR is essential for neuronal health in a YAC mouse model.[40] The disease predominantly affects spinal and bulbar motor neurons resulting in muscular atrophy and weakness.[39]

Dentatorubral-Pallidoluysian Atrophy (DRPLA)

DRPLA individuals commonly show seizures, involuntary muscle movement and chorea as well as dementia[41] as a result of generalized neurodegeneration in the cortex, globus pallidus, striatum and cerebellum.[42] The polyglutamine tract in DRPLA is located in Atrophin-1 (43–44) which has been characterized as a nuclear receptor co-repressor and is predicted to interact with a number of transcription factors and members of the histone de-acetylase (HDAC) family.[45] Expression of expanded polyglutamine-containing Atrophin-1 is sufficient to induce symptoms of DRPLA through a process which involves proteolytic processing and aggregation of an 120 kDa fragment containing the expanded polyglutamine tract.[46] Pathogenesis does not appear to be mediated through a functional interaction of the mutant Atrophin-1 protein with the wild-type protein, since homozygous deletion of the C-terminal region of the wild-type protein in a mouse model does not induce neurodegeneration and also fails to modify polyglutamine-induced phenotypes.[47]

The Spinal Cerebellar Ataxias (SCAs)

The SCAs are a group of diseases characterized by progressive degeneration of the cerebellum resulting in late-onset ataxia and lack of coordination. 70% of SCA patients also show degeneration of the peripheral nervous system involving both axonal and primary neuropathy.[48] There are currently 28 autosomal dominant SCAs recognized, 17 of which are caused by a known mutation and 6 of which are recognized as polyglutamine diseases[49] as listed in Table 1.

SCA1 is caused by a CAG expansion in the *Ataxin-1* gene which encodes a protein containing an RNA binding motif, with a proposed role in RNA metabolism.[50] The RNA binding capacity of Ataxin-1 has been shown to be dependent on the length of the CAG repeat and therefore it has been suggested that loss of Ataxin-1 protein function, perhaps resulting in aberrant RNA metabolism, may play a role in disease progression.[50] The expanded polyglutamine-containing protein is found in nuclear aggregates in disease models[51] however, while nuclear localization of the polyglutamine-containing protein is required for disease progression, nuclear aggregation is not[51] and it is therefore unclear whether aggregates are a component of pathogenesis. While disease-associated *Ataxin-1* alleles consist of pure CAG repeats, normal alleles of *Ataxin-1* contain CAT interruptions, encoding histidine, in the CAG repeat which are thought to reduce aggregation of the protein and may prevent expansion of the repeat tract.[52]

SCA2 is caused by a CAG expansion within the *Ataxin-2* gene. It is characterized by degeneration in the cerebellum and brainstem, although it is unclear how specificity is elicited since the Ataxin-2 protein is widely expressed in the brain.[53] The function of the Ataxin-2 protein is unknown although it has been implicated in RNA metabolism, a role which is supported by its interaction with cytoplasmic poly-A-binding protein[54,55] and poly-ribosomes,[54] and a *Drosophila* ortholog has been characterized as a regulator of actin filament formation.[56] Both mutant and wild-type Ataxin-2 are found exclusively in the cytoplasm and mutant Ataxin-2 does not form nuclear inclusions such as those thought to be pathogenic in a number of the other polyglutamine diseases.[57] Normal alleles of *Ataxin-2* are frequently interrupted with CAA repeats which, while still encoding glutamine, are thought to alter the structure at the DNA and RNA level.[53]

SCA3 or Machado Joseph disease (MJD) is the result of a CAG expansion in the gene encoding Ataxin-3, *MJD1*. The phenotype involves progressive ataxia as well as peripheral neuropathy which affects both motor and sensory neurons.[58] The function of Ataxin-3 is unknown, however interactions with DNA repair proteins have been reported. HHR23 proteins, which are important in nucleotide excision repair, have been shown to be localized to nuclear inclusions in MJD individuals via their interaction with the mutant protein.[59] Interactions with components of the proteasome and ubiquitin-binding factors, which are also present in nuclear inclusions in MJD, suggest a role for Ataxin-3 in protein turnover.[60] More recent studies also implicate Ataxin-3 in Ca^{2+} signaling via an interaction of the mutant protein with the Type 1 inositol 1,4,5-trisphosphate receptor, an intracellular calcium channel.[61] Interestingly, nuclear inclusions in MJD have also been demonstrated to contain wild-type Ataxin-3 protein which is recruited in a polyglutamine length-dependent manner, suggesting that the polyglutamine expansion may affect the normal function of the protein.[62]

SCA6 is the result of a C-terminal polyglutamine expansion in the α (1A) subunit of the neuronal P/Q-type voltage-gated calcium channel encoded by the *CACNA1A* gene.[63] Neurodegeneration in SCA6 is primarily localized to the Purkinje cells of the cerebellum, which coincides with high expression of the *CACNA1A* gene.[64] There are two known disorders caused by missense mutations in the *CACNA1A* gene—episodic ataxia Type 2 (EA2) and familial hemiplegic migraine—both of which give phenotypes similar to SCA6. It is therefore unclear whether there is toxic gain-of-function in SCA6, or whether alteration of the kinetic properties of the encoded channel is sufficient to explain the neurodegeneration observed.[63] This evidence, along with the observation that the pathogenic threshold for CAG repeats is much lower in SCA6 than the other polyglutamine diseases, led to questions about the classification of SCA6 as a polyglutamine disease.[63] More recently, aggregation of the mutant CACNA1A calcium channel has been demonstrated in a knock-in mouse model of SCA6, coinciding with age-dependent neurodegeneration with no concurrent change in electrophysiological function of neurons observed.[65] This evidence suggests that a simple change to calcium channel function is unlikely to explain SCA6 pathology.

The polyglutamine tract in **SCA7** is located within Ataxin-7. SCA7 individuals show neuronal loss in the cerebellum, brainstem and spinal cord. There is also retinal degeneration which typically leads to blindness.[66] The regions where neurodegeneration is observed coincide with regions where Ataxin-7 is expressed. Ataxin-7 is thought to play a role in transcriptional regulation since it forms complexes with other transcriptional regulators including histone acetyl-transferases.[67,68] Transcriptional dysregulation has been observed in the absence of ataxia in mice expressing expanded *Ataxin-7,* perhaps suggesting that this is an early component of pathology.[69]

SCA17 is caused by an expanded CAG repeat in the gene encoding the TATA box-binding protein (TBP) which is a transcription initiation factor. TBP is ubiquitously expressed, yet pathogenesis in SCA17 is limited to the cerebellum and particularly affects the Purkinje cells.[70] Expression of TBP containing an expanded polyglutamine tract in both cellular and mouse models has been demonstrated to result in decreased expression of the nerve growth factor (NGF) receptor, TrkA, which is required for Purkinje neuron survival.[71] This relationship may go some way to explaining specificity of neurodegeneration in SCA17.

Despite the presence of the expanded polyglutamine tract in unrelated genes and the degeneration of a specific sub-set of neurons in each disease, there are a number of features of the polyglutamine diseases which suggest that common pathogenic mechanisms may be involved. While a number of pathways have been demonstrated to be perturbed in models of polyglutamine pathogenesis—including axonal transport, Ca^{2+} homeostasis, transcription, protein turn-over, RNA metabolism and mitochondrial function–the primary cause of dysfunction remains unclear.

Pathogenesis and Aggregate Formation

Aggregates of polyglutamine-proteins are a feature of all of the polyglutamine diseases, however their role as a protective or pathogenic agent is contentious. They are present in neurons which undergo degeneration in each disease, and are also commonly found in neurons which do not undergo degeneration. Both cytoplasmic and intranuclear inclusions are seen in HD,[72] while only intranuclear inclusions are seen in brains of SBMA, DRPLA and SCA1,3,6,7 and 17 patients. In SCA6 and some cases of SCA2 only cytoplasmic aggregates have been reported.[73]

The role of the nuclear and cytoplasmic aggregates formed by mutant HTT in neurodegeneration has been much debated. It has been reported that the N-terminal fragment of the mutant HTT protein forms aggregates selectively in striatal neurons and is predictive of cell death.[74] However, striatal cells transfected with mutant HTT in conditions where inclusions cannot form actually show an increase in associated cell death, suggesting that formation of nuclear inclusions can actually play a protective role, perhaps by sequestering soluble forms of the mutant protein.[75] In support of this, several studies suggest that intermediates of aggregation, including soluble monomeric conformers and insoluble oligomers, may actually be more toxic than the aggregates themselves.[76-78] There is some debate over the validity of cell culture models for neurodegeneration and it has been suggested that animal models, which typically show increased incidence of cell death correlating with formation of aggregates, are more representative of the situation in HD individuals.[79] Other expanded repeat diseases tell a different story: there is no formation of nuclear inclusions in SCA2[57] yet many clinical manifestations of the disease overlap with the other SCAs where nuclear inclusions are observed and in the case of SCA1, are even necessary for pathogenesis.[51] This observation further supports the idea that aggregates may not actually play a causative role in neurodegeneration, however the role of nuclear and cytoplasmic aggregates in pathogenesis remains unclear.

Polyalanine Diseases

There are a further nine diseases caused by the expansion of repeats encoding polyalanine. Polyalanine tracts are found in 494 human proteins and are highly polymorphic. They are thought to play a role in transcriptional regulation and are known to be regions of DNA binding in many proteins,[80] however expansion of these repeat regions beyond a threshold is pathogenic. Eight of the nine known polyalanine expansions occur in transcription factors that play important roles during development and the resulting phenotypes are congenital malformation syndromes.[81] In these cases, the same phenotypes can arise from different mutations in the affected gene, suggesting that they are due to loss of function of the encoded transcription factor.[82] However the final disease, oculopharyngeal muscular dystrophy (OPMD), is not congenital and instead shares some clinical similarities with the other expanded repeat diseases.

OPMD is a late-onset, progressive disease characterized by muscle weakness. The phenotype is caused by expansion of a polyalanine-encoding GCG repeat in nuclear poly(A) binding protein 1 (PABPN1) to 8–13 copies.[83] PABPN1 is important in regulating the length of poly(A) tails after mRNA processing through stimulating poly(A) polymerase. In OPMD, aggregates of mutant PABPN1 are observed in muscle cells, the major site of pathology. Like polyglutamine aggregates, aggregates in OPMD contain ubiquitin and components of the proteasome.[84]

While many polyalanine tracts are encoded by GCC or GCG codons, a –1 frameshift of a CAG or CUG repeat would also result in translation of alanine and a –2 frameshift in translation of serine and therefore a role for polyalanine in pathogenesis of the polyglutamine diseases has been proposed. It has been shown that the expanded CAG repeat tract in the *Ataxin-3* gene is prone to –1 frameshifts, resulting in hybrid proteins containing alanine and glutamine tracts.[85] Frameshifts are thought to be elicited through the formation of secondary structures, particularly hairpins, in RNA containing long repeat tracts. In a cellular model of SCA3 treatment with anisomycin, a ribosome-interacting drug which reduces −1 frameshift, results in a reduction of the cellular toxicity of CAG repeats, suggesting a role for these polyalanine tracts in toxicity of polyglutamine diseases.[85] Very low levels of +1 and +2 frameshifts which would also result in polyalanine or polyserine tracts have also been demonstrated in mutant HTT.[86] Polyalanine and polyserine containing proteins are also among identified modifiers for mutant HTT, further supporting a role for frameshift in the pathogenic pathway of HD.[87]

There is some evidence to suggest that polyalanine tracts are unlikely to play a major role in intrinsic toxicity of polyglutamine. Both CAG and CAA-encoded polyglutamine tracts have been demonstrated to show a similar range of pathogenic phenotypes in a *Drosophila* model,[88] and therefore the ability of the repeat tract to undergo a frameshift to encode polyalanine does not appear to play a major role in toxicity in all cases. It is not clear whether this result indicates specific properties of polyglutamine tracts in *Drosophila* and therefore the possibility that frameshifts are an important component of toxicity in mammalian cells cannot be ruled out.

Untranslated Repeat Expansions

Loss-of-Function

There are currently 11 diseases attributed to the presence of expanded untranslated repeats in the genome. Three of these, Fragile X, XE and Friedrich's ataxia, are the result of loss-of-function of the gene in which the repeat falls and therefore have a recessive inheritance pattern.

Fragile X syndrome (FRAXA) is an X-linked disorder and the most common inherited form of mental retardation. Individuals with FRAXA display mild to severe cognitive impairment, as well as macro-orchidism, distinct facial dysmorphisms and a range of anxiety disorders. The mutation causing FRAXA is an expansion of a CGG repeat located within the 5′ untranslated region of the *Fragile X mental retardation* gene, *FMR1*.[1] Expansion of this sequence beyond approximately 200 repeats results in hyper-methylation of a proximal CpG island in the promoter region of *FMR1*, causing transcriptional silencing.[89] This silencing produces a deficiency or loss of the FMR protein (FMRP), an RNA binding protein which plays important roles in shuttling of RNAs and regulation of local protein translation.[90]

The second X-linked expanded repeat disease, **Fragile XE mental retardation syndrome (FRAXE)**, is the result of 5′ untranslated GCC repeat expansion which also results in methylation and transcriptional silencing, but in this case the mutation falls in the *FMR2* gene.[91,92] Interestingly, FMR2 protein has also been shown to be an RNA binding protein, one splice-target of which is FMR1 mRNA.[93]

Friedrich's Ataxia is unique in being the only autosomal recessive expanded repeat disease. In 98% of cases, the causative mutation is a GAA repeat expansion within intron 1 of the *Frataxin (FXN)* gene, with the remaining 2% of cases involving point mutations in the *FXN* gene.[94] Reduced transcription and translation of the FXN gene product as a result of alterations to methylation and histone marks in the repeat-flanking region has been demonstrated.[95] The FXN protein is proposed to be a mitochondrial protein with roles in cellular iron metabolism. Loss of FXN may result in iron overload in mitochondria, leading to oxidative stress and eventual neurodegeneration.[96]

Gain-of-Function

Dominant diseases caused by expanded untranslated repeats are thought to result from RNA gain-of-function. Unlike the translated-repeat diseases there is no toxic protein encoded by the repeat, yet these diseases show dominance and similar neurodegenerative phenotypes to the polyglutamine diseases (Table 2).

Myotonic dystrophy Type 1 and 2 (DM1 and 2) are the best characterized of the untranslated repeat diseases. DM1 was the earliest known example of a dominant disorder caused by repeat expansion in a non-coding region of a gene. It results from a CTG expansion above around 50 repeats in the 3'UTR of *Dystrophia myotonica-protein kinase (DMPK)* gene. The disease manifests as myopathy and progressive cardiac defects, often coupled with the formation of characteristic posterior iridescent cataracts and insulin resistant diabetes.[97] There is also evidence of some CNS pathology, including damage to cortical regions of the brain,[98] resulting in variable and progressive cognitive impairment in individuals with DM1.[99] It is clear that DM1 is not simply the result of loss-of-function of *DMPK* since mouse *DMPK* knockout models do not recapitulate all aspects of the disease and show much milder cardiac and muscular defects even when both alleles are knocked-out.[100] A role for loss-of-function of the nearby *SIX5* gene has also been suggested however, while knockout models of *SIX5* do have cataracts, they are not the posterior iridescent sort seen in DM1[101,102] and therefore even a reduction in expression of both genes cannot explain all aspects of pathology.

Table 2. Untranslated Repeat Diseases

Disease	Repeat	Gene and Region	Mechanism	Repeat Number Disease	Normal
FRAXA	CGG	*FMR1*, 5'UTR	Protein L-O-F	>200- >2000	6-52
FRAXE	GCC	*FMR2*, 5'UTR	Protein L-O-F	>200	6-25
FRDA	GAA	*FXN*, intron	Protein L-O-F	70- >1000	5-30
FXTAS	CGG	*FMR1*, 5'UTR	RNA G-O-F	55-200	6-52
DM1	CTG	*DMPK*, 3'UTR	RNA G-O-F	50-3000	5-37
DM2	CCTG	*ZNF9*, intron	RNA G-O-F	75-11000	10-26
SCA8	CTG/ CAG	*KLHL1*, non-coding *Ataxin-8*, polyQ	RNA G-O-F/ Protein G-O-F	107-127	16-37
SCA10	ATTCT	*Ataxin-10*, intron	Unknown	800-4500	10-29
SCA12	CAG	*PPP2R2B*, 5'UTR	Unknown	55-78	9-28
HDL2	CTG	*Junctophilin-3*, depends on splicing	Unknown	51-57	14-19
SCA31	TGGAA	*TK2/BEAN*, intron	RNA G-O-F	≥100	0

FRAXA—Fragile X; FRAXE—Fragile XE; FRD—Friedrich's Ataxia; FXTAS—Fragile X-associated tremor-ataxia syndrome; DM—Myotonic dystrophy; SCA—spinal cerebellar ataxia; HDL2—Huntington's disease-like 2; UTR—untranslated region; G-O-F—gain-of-function; L-O-F—loss-of-function. SCA31 involves the insertion of a large fragment containing a TGGAA repeat and therefore unaffected individuals do not have a repeat number.

The most significant evidence for an RNA gain-of-function mechanism in DM1 is that expression of a CUG repeat in a completely unrelated mRNA has been shown to result in myopathy and myotonia in a mouse transgenic model.[103] Transgenic mice expressing 45 kb of the human *DM1* region containing at least 300 repeats also display symptoms of DM1, including CNS pathology.[100] It has been demonstrated by in situ analysis that the expanded repeat in DM1 is transcribed and mRNAs containing the repeat are spliced in the normal manner but retained in the nucleus in foci associated with nuclear components, including the splicing factor Muscleblind (MBNL).[104] A number of targets of MBNL are aberrantly spliced in DM1 individuals including cardiac troponin T (TNNT2),[105] insulin receptor (IR),[104,105] chloride channel-1 (CLCN-1),[104] tau,[106] myotubularin-related protein-1 (MTMR1), fast skeletal troponin T (TNNT3), N-methyl-D-aspartate (NMDAR1),[106] amyloid precursor protein (APP),[106] ryanodine receptor (RyR)[107] and sarcoplasmic/endoplasmic reticulum Ca^{2+} ATPase1&2 (SERCA1&2).[107] These targets normally show developmental or tissue-specific regulation of splicing which is lost in DM1 individuals such that the adult splice-form or correct tissue isoform fails to be expressed. Interestingly, splicing mis-regulation can be separated from the co-localization of MBNL, since MBNL has also been found in association with CAG-repeat containing foci but is not coupled with mis-splicing of MBNL targets in this case.[105] Some pathologies associated with DM1 are directly related to mis-spliced MBNL targets; for example, it has been demonstrated that a reduction in CLCN-1 expression equivalent to the loss of expression of the adult isoform observed in DM1 results in myotonia[108] and failure to express the correct muscular isoform of IR results in failure of skeletal muscle cells to respond to insulin.[109]

DM2 results from the expansion of a CCTG repeat within an intron of the *ZNF9* gene. In some cases, this expansion can result in a repeat tract which is tens of thousands of repeats long.[110] Despite the presence of repeats in completely unrelated genes, DM1 and DM2 have very similar pathologies. MBNL has also been shown to co-localize to CCUG repeat-containing foci,[111] resulting in splicing alterations including changes to CLCN-1, TNNT3 and IR splicing.[112,113] This observation strongly supports a common gain-of-function mechanism of pathogenesis, however there are also a number of differences between the diseases: proximal muscles are most affected in DM2 and distal in DM1, DM1 shows CNS symptoms while DM2 does not and there is no congenital form of DM2.[114] DM1 is also generally associated with more severe pathogenesis.[114] These disease features may be related to the function and expression pattern of the gene in which the expanded repeat resides.

A second splicing factor CUG-binding protein-1 (CUG-BP1) has also been suggested to play a role in DM1 and DM2 pathogenesis. While CUG-BP1 does not associate with repeat containing foci, MBNL and CUG-BP1 have antagonistic roles in regulating splicing such that the sequestration of MBNL results in an increase in the levels of CUG-BP1 splice-forms, perhaps through an increase in CUG-BP1 activity.[115] Several lines of evidence support this hypothesis: expression of human CUG-BP1 in a mouse model is sufficient to induce splicing changes in skeletal muscle and the heart, as well as muscular defects reminiscent of those observed in DM1.[116,117] In a *Drosophila* model, neurodegeneration in the eye caused by expression of a large CUG repeat RNA can be suppressed by human MBNL overexpression and is enhanced when human CUG-BP1 is overexpressed.[118] This result provides further evidence that the expanded repeat RNA itself can explain a large proportion of the pathology associated with DM1 and 2, since CUG repeat RNA alone is sufficient to cause neurodegeneration which can be altered by MBNL and CUG-BP1 protein levels.

Fragile X tremor-ataxia syndrome (FXTAS) is caused by a CGG expansion in the 5′UTR of the *Fragile X mental retardation 1 (FMR1)* gene within the pre-mutation range (55–200 repeats) for Fragile X syndrome. Unlike Fragile X, which gives characteristic mental retardation and anxiety disorders, FXTAS does not result from loss-of-function of the *FMR1* gene. Clinical manifestations of FXTAS (including late-onset neurodegeneration presenting as gait instability, cognitive decline and tremors) cannot be explained simply by disruption of the *FMR1* gene; indeed levels of *FMR1* transcripts are reported to be elevated by up to 8 times normal levels in FXTAS individuals and protein levels are reported to be normal.[119,120]

Individuals with FXTAS show ubiquitin-positive inclusions in neurons and astrocytes throughout the cerebrum and brain-stem which also contain CGG repeat RNAs.[121] The composition of these inclusions has been investigated and a number of proteins identified including the splicing factor MBNL, previously implicated in the myotonic dystrophies, several intermediate filament proteins including Lamins A/C and Internexin and Heterogeneous nuclear ribonucleoprotein A2 (hnRNP A2).[121] The FMR1 protein is not present in the inclusions. These observations support a similar pathogenic pathway in FXTAS to that described for DM1 and DM2, where sequestration of proteins into inclusions with the expanded repeat-containing RNA results in loss of the normal function of those proteins which is responsible for disease pathogenesis. Expression of ectopic CGG repeat RNA in a mouse model has been shown to be sufficient to induce nuclear inclusions and death in Purkinje cells which strongly supports this hypothesis.[122]

SCA8 is characterized by the slow progressive cerebellar ataxia typical of the SCAs,[123] but with quite variable degrees of pathology in individual families.[124] The mechanism of pathogenesis in SCA8 is a source of some debate since the expanded repeat region has recently been discovered to be transcribed in a bi-directional manner, resulting in the production of both a CUG repeat within the 5′UTR of the *KLHL1* gene and a nearly pure polyglutamine tract encoded by the CAG repeat transcribed from the opposite strand.[125] The KLHL1 protein has been shown to be involved in regulation of neurite outgrowth via an actin-binding domain and a role in calcium influx regulation through P/Q-type calcium channels has also been demonstrated.[126] It has been suggested that the transcription of the CAG repeat from the opposite strand to *KLHL1* may be involved in regulation of mRNA levels of KLHL1 since the two strands both show expression in cells which are functionally involved in processes affected by SCA8.[127] Furthermore, deletion of *KLHL1* in a mouse model has also been shown to result in gait abnormalities and loss of motor control, an effect which was also reproduced by targeted deletion in Purkinje cells alone,[128] suggesting that a reduction in KLHL1 function may play a major role in SCA8 pathology. Polyglutamine-containing inclusions have also been detected in both a mouse model of SCA8 and patient tissue, raising the possibility that a mixture of RNA-mediated and polyglutamine-mediated toxicity may be at play in the disease situation.[129]

Expression of the expanded CUG repeat-containing 5′UTR of *KLHL1* alone in the *Drosophila* eye has been demonstrated to elicit a neurodegenerative phenotype.[130] A number of modifiers of a SCA8 model in the *Drosophila* eye have been identified, with the majority of these being RNA splicing factors, RNA-binding proteins, RNA helicases, translational regulators and transcription factors.[130] Loss-of-function of the Drosophila homolog of MBNL enhances neurodegeneration in this model, as does loss-of-function of the double-stranded RNA-binding protein Staufen which is recruited to the repeat RNA.[130] These results support a role for RNA toxicity in SCA8 pathogenesis.

SCA10 is caused by a pentanucleotide repeat—ATTCT—of which there can be up to 4500 repeats within intron 9 of the *Ataxin-10* gene.[131] The disease manifests as cerebellar dysfunction often involving seizures, with cognitive and neuro-psychiatric impairment.[132] AUUCU repeat-containing RNA has been shown to form foci when overexpressed in cell culture[132] and the repeat tract itself has been demonstrated to have the potential to form a hairpin secondary structure under physiological conditions.[133] Ataxin-10 protein has been characterized as essential for the survival of cerebellar neurons, however since the disease shows dominant inheritance and *Ataxin-10* heterozygous mutant mice do not recapitulate features of SCA10, a loss-of-function mechanism cannot explain disease pathology.[134]

SCA12 is caused by a CAG expansion at the 5′ end of the gene encoding the brain-specific regulatory subunit of the protein phosphatase PP2A holoenzyme (PPP2R2B). PP2A has been shown to be involved in the DNA repair checkpoint[135] and to play a role in induction of neuronal apoptosis via translocation to mitochondria.[136] Since the repeat can fall either in the 5′UTR or an upstream promoter of *PP2R2B* depending upon alternative splicing, it has been suggested that the expansion may cause upregulation of PPP2R2B resulting in altered regulation of the PP2A enzyme and therefore altered phosphorylation of down-stream targets.[137]

Huntington's disease-like-2 **(HDL-2)** is one of the clearest examples of phenotypic overlap between the translated and untranslated repeat diseases. It is caused by a CTG/CAG expansion in a variably spliced exon of *Junctophilin-3 (JPH3)* which results in a disease which is commonly misdiagnosed as HD.[138] Characteristics of HDL-2 include striatal and cortical neurodegeneration coupled with formation of nuclear inclusions such as those typical of HD.[139] Alternative transcripts contain the repeat either in the 3′UTR or translated as a polyalanine or polyleucine tract. While a mouse model expressing an HDL-2 BAC has been reported to show bi-directional transcription of the locus resulting in production of a transcript encoding an expanded polyglutamine tract from the opposite strand to *JPH3*, no transcripts encoding polyglutamine have been detected in patient tissue to date and therefore the relevance of these findings to human disease remains unclear.[140] RNA foci have been detected in frontal cortex from HDL-2 brains and expression of an untranslated CUG repeat-containing form of JPH3 in HEK293 and HT22 cells also resulted in formation of RNA foci which co-localized with MBNL, supporting a RNA pathogenesis model for HDL-2.[141]

SCA31 is an unusual spinocerebellar ataxia in that, instead of being caused by the expansion of a repeat sequence from a normal copy number, it is the result of an insertion of a large fragment including a long penta-nucleotide TGGAA repeat which falls in the introns of two genes transcribed from opposite strands: *Thymidine Kinase 2 (TK2)* and *Brain expressed associated with NEDD4 (BEAN)*.[142] Very few individuals have been identified who are not affected and carry the insertion without the TGGAA repeat expansion (0.23% of controls) and it is therefore thought that either the increase in size of the insertion with the expansion or the expanded TGGAA repeat itself is responsible for the pathogenic effect in SCA31 individuals.[142] Nuclear RNA foci containing the expanded TGGAA repeat sequence have been identified in Purkinje cells of patients, supporting the possibility of an RNA gain of function mechanism.[142]

RNA as a Pathogenic Agent

There are a number of characteristics of the polyglutamine diseases which cannot be explained by loss-of-function of the gene in which the repeat falls. The striking overlap in phenotypes associated with the polyglutamine diseases, which has led to misdiagnosis of SCA17 and DRPLA as HD, indicates that there are pathogenic mechanisms involved which are not gene-specific.[143] The most parsimonious explanation for such a phenotypic overlap appears to be a common pathogenic pathway for the polyglutamine diseases. Several groups have demonstrated that expanded polyglutamine peptides are intrinsically toxic both in Drosophila models and transfected cells[88,144-146] and there is evidence to suggest that in HD, DRPLA, SBMA and SCA3 at least, caspase cleavage can release the polyglutamine tract from the disease protein.[147] However, it is likely that the functional properties of the expanded polyglutamine-containing proteins also contribute to the pathology, since there are unique clinical features associated with each disease.

The overlap between the polyglutamine and untranslated repeat phenotypes is also striking. Not least, the fact that there are SCAs caused by both polyglutamine and untranslated repeats which show the typical set of phenotypes for this group of diseases, cerebellar neurodegeneration and progressive ataxia, suggests some common causal link for the two sets of diseases. Either there is a common pathogenic agent between the two sets of diseases or an entirely separate pathogenic pathway is involved in each case, resulting in largely overlapping phenotypes. Since the presence of repeat-containing RNA is a common factor between polyglutamine and untranslated diseases, RNA has been suggested as a common pathogenic agent in the two groups of diseases.

A common feature of all disease-associated expanded repeat RNAs is their predicted ability to form stable hairpin secondary structures which increase in stability the longer the repeat region grows. This property has been demonstrated for CUG,[148] CAG,[148] CGG,[148-150] CCTG[148] and AUUCU[133] repeats and appears to be integral to pathogenicity in the untranslated repeat diseases[151] (Fig. 3). Since the expanded CAG repeats which code for polyglutamine in the polyglutamine diseases can also form a stable hairpin structure at the RNA level with very similar stability to CUG repeats in vitro,[148] it seems likely that these repeats may be similarly toxic to cells. General acceptance that the polyglutamine proteins themselves are the pathogenic agent in the polyglutamine diseases has meant that other pathogenic mechanisms have not been thoroughly investigated. There is a large pool of evidence for polyglutamine being a pathogenic agent in the translated repeat diseases, however there is equally strong evidence of a role for RNA gain-of-function pathogenesis in DM1 and 2[111] and FXTAS[122] which demonstrates the ability of RNA to act as a pathogenic agent.

Recent evidence also supports a role for expanded repeat RNA in the translated repeat diseases: upregulating expression of *Drosophila* Mbl or overexpressing the human ortholog MBNL1 results in the enhancement of a neurodegenerative eye phenotype, as

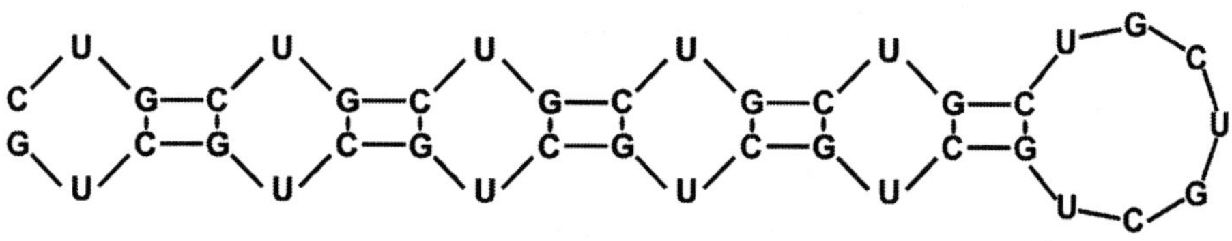

Figure 3. Predicted secondary structure formed by CUG repeat RNA. CUG repeat RNA forms a hairpin structure with a mismatch every third base along the stem of the hairpin. In myotonic dystrophy Type 1 (DM1), the secondary structure formed by expanded CUG repeats in the dystrophia myotonica-protein kinase (DMPK) transcript results in sequestration of RNA binding proteins, including muscleblind-like (MBNL). A similar mechanism may also play a role in other expanded repeat diseases.

well as a decrease in life-span associated with expression of human SCA3 containing a CAG-encoded polyglutamine tract.[152] This effect was not seen when the same experiment was performed using a polyglutamine tract encoded by a mixed CAG/CAA repeat, which is unable to form a hairpin in the same manner as a pure CAG repeat tract, suggesting that the interaction is occurring at the RNA level and is sequence-dependent.[152] This result supports the hypothesis that RNA toxicity may at least be one component of the pathogenic mechanism in the translated repeat diseases.

Emerging evidence suggests an additional role for expanded repeat RNA as an intermediary in toxicity. Recently published data demonstrates that both CAG and CUG expanded repeat sequences are able to be translated even in the absence of an AUG initiation codon through a mechanism known as repeat-associated non-AUG translation (RAN translation).[153] RAN translation is thought to result from hairpin structures formed by expanded repeat RNAs acting as Internal Ribosome Entry Sites (IRES) and could potentially result in production of homopolymeric peptides in all three reading frames of each repeat sequence. As a consequence of these findings, homopolymeric amino acid sequences have emerged as potential contributors to RNA-mediated toxicity in the 'untranslated' repeat diseases. Furthermore, a contribution of repeat peptides encoded by alternate frames must also be considered in the 'translated' repeat diseases.

Bidirectional Transcription

Another common feature of microsatellite repeat sequences which has recently become apparent is their tendency to be transcribed in a bi-directional manner. Examinations so far have revealed anti-sense transcripts produced from every repeat locus tested.[129,154-156] Production of antisense transcripts has been suggested to be an important component of gene regulation throughout the genome which may act through various mechanisms including collisions between RNA polymerases on opposite strands, changes to alternative splice site selection and alterations to histone modifications and DNA methylation by recruitment of chromatin remodeling complexes.[157-159] For example, expansion of the CGG repeat sequence in *FMR1* has been demonstrated to result in an alteration to the relative ratios of the sense and antisense transcripts present,[156] suggesting another mechanism by which repeat expansions might alter regulation of the genes in which they reside.

Bi-directional transcription may also allow the production of multiple toxic agents from some expanded repeat sequences. For example, in SCA8 both expanded CAG and CUG repeat-containing transcripts have been detected, resulting in the expression of expanded polyglutamine as well as toxic RNA.[129] Therefore, further investigation of expression patterns of the regions surrounding expanded repeats is warranted. Bi-directional transcription has also been shown to greatly enhance instability of CAG*CUG repeat sequences compared with transcription of only the sense or anti-sense[160] and therefore may play a key role in somatic instability observed in the expanded repeat diseases.

CONCLUSION: WHERE NEXT FOR DYNAMIC MUTATIONS?

Despite identification of dynamic mutations as a cause of disease 20 years ago, the mechanisms leading from mutation to cell death have remained surprisingly enigmatic. Recent advances in understanding these mechanisms, which have frequently come from modeling pathogenesis in animals, have revealed increasing levels of complexity. For example, while there is clearly a role for the expanded polyglutamine-containing proteins in pathogenesis of the polyglutamine diseases, other mechanisms such as a role for RNA gain-of-function, RAN translation of expanded repeat containing RNAs to produce alternative homopolymeric repeat tracts and and bi-directional transcription of the repeat-containing genes may also be important and are therefore key areas under current investigation using animal models.[161-165]

Furthermore, while there are clearly common features of the expanded repeat diseases, including aspects of the pathogenic mechanism, the cellular specificity of toxicity in each disease suggests that there are also important differences. How repeat expansions in unrelated genes which are generally widely expressed can result in diseases with such cell-specificity remains an intriguing problem. Another major question remaining in the field surrounds the relative contribution of each of the proposed pathogenic agents to toxicity in the different diseases. This information will be vital for determining effective therapeutic targets in attempts to alleviate symptoms or halt progression in affected individuals.

ACKNOWLEDGMENTS

Sections of this chapter, including Table 1 and 2, have been adapted from: van Eyk CL. Investigation of RNA-mediated pathogenic pathways in a Drosophila model of expanded repeat disease. PhD Thesis, Discipline of Genetics, University of Adelaide, Adelaide, Australia, 2010.[166] Small excerpts of text in this chapter have been adapted from. van Eyk CL, McLeod CJ, O'Keefe LV, and Richards RI. Comparative toxicity of polyglutamine, polyalanine and polyleucine tracts in drosophila models of expanded repeat disease. Hum Mol Genet 2011; 21: Epub ahead of print.[162]

REFERENCES

1. Kremer EJ, Pritchard M, Lynch M et al. Mapping of DNA instability at the fragile X to a trinucleotide repeat sequence p(CCG)n. Science 1991; 252:1711-4.
2. La Spada AR, Wilson EM, Lubahn DB et al. Androgen receptor gene mutations in X-linked spinal and bulbar muscular atrophy. Nature 1991; 352:77-9.
3. Martorell L, Monckton DG, Gamez J et al. Progression of somatic CTG repeat length heterogeneity in the blood cells of myotonic dystrophy patients. Hum Mol Genet 1998; 7:307-12.
4. Cleary JD, Nichol K, Wang YH, Pearson CE. Evidence of cis-acting factors in replication-mediated trinucleotide repeat instability in primate cells. Nat Genet 2002; 31:37-46.
5. Yang Z, Lau R, Marcadier JL et al. Replication inhibitors modulate instability of an expanded trinucleotide repeat at the myotonic dystrophy type 1 disease locus in human cells. Am J Hum Genet 2003; 73:1092-105.
6. Liu G, Chen X, Bissler JJ et al. Replication-dependent instability at (CTG)•(CAG) repeat hairpins in human cells. Nat Chem Biol 2010; 6:652-9.
7. Freudenreich CH, Lahiri M. Structure-forming CAG/CTG repeat sequences are sensitive to breakage in the absence of Mrc1 checkpoint function and S-phase checkpoint signaling: implications for trinucleotide repeat expansion diseases. Cell Cycle 2004; 3:1370-4.
8. Sarkar PS, Chang HC, Boudi FB, Reddy S. CTG repeats show bimodal amplification in E. coli. Cell 1998; 95:531-40.
9. Pollard LM, Sharma R, Gomez M et al. Replication-mediated instability of the GAA triplet repeat mutation in Friedreich ataxia. Nucleic Acids Res 2004; 32:5962-71.
10. Spiro C, Pelletier R, Rolfsmeier ML et al. Inhibition of FEN-1 processing by DNA secondary structure at trinucleotide repeats. Mol Cell 1999; 4:1079-85.
11. Yang J, Freudenreich CH. Haploinsufficiency of yeast FEN1 causes instability of expanded CAG/CTG tracts in a length-dependent manner. Gene 2007; 393:110-5.

12. Møllersen L, Rowe AD, Larsen E et al. Continuous and periodic expansion of CAG repeats in Huntington's disease R6/1 mice. PLoS Genet 2010; 6:e1001242.
13. Kennedy L, Shelbourne PF. Dramatic mutation instability in HD mouse striatum: does polyglutamine load contribute to cell-specific vulnerability in Huntington's disease? Hum Mol Genet 2000; 9:2539-44.
14. Kennedy L, Evans E, Chen CM et al. Dramatic tissue-specific mutation length increases are an early molecular event in Huntington disease pathogenesis. Hum Mol Genet 2003; 12:3359-67.
15. Cleary JD, Tome S, Lopez Castel A et al. Tissue- and age-specific DNA replication patterns at the CTG/CAG-expanded human myotonic dystrophy type 1 locus. Nat Struct Mol Biol 2010; 17:1079-87.
16. Edwards SF, Hashem VI, Klysik EA, Sinden RR. Genetic instabilities of (CCTG).(CAGG) and (ATTCT). (AGAAT) disease-associated repeats reveal multiple pathways for repeat deletion. Mol Carcinog 2009; 48:336-49.
17. Jarem DA, Wilson NR, Delaney S. Structure-dependent DNA damage and repair in a trinucleotide repeat sequence. Biochemistry 2009; 48:6655-63.
18. Kovtun IV, Liu Y, Bjoras M et al. OGG1 initiates age-dependent CAG trinucleotide expansion in somatic cells. Nature 2007; 447:447-52.
19. van den Broek WJ, Nelen MR, Wansink DG et al. Somatic expansion behaviour of the (CTG)n repeat in myotonic dystrophy knock-in mice is differentially affected by Msh3 and Msh6 mismatch-repair proteins. Hum Mol Genet 2002; 11:191-8.
20. Tome S, Holt I, Edelmann W et al. MSH2 ATPase domain mutation affects CTG*CAG repeat instability in transgenic mice. PLoS Genet 2009; 5(5): e1000482.
21. Foiry L, Dong L, Savouret C et al. Msh3 is a limiting factor in the formation of intergenerational CTG expansions in DM1 transgenic mice. Hum Genet 2006; 119:520-6.
22. Wheeler VC, Lebel LA, Vrbanac V et al. Mismatch repair gene Msh2 modifies the timing of early disease in Hdh(Q111) striatum. Hum Mol Genet 2003; 12:273-81.
23. Sundararajan R, Gellon L, Zunder RM, Freudenreich CH. Double-strand break repair pathways protect against CAG/CTG repeat expansions, contractions and repeat-mediated chromosomal fragility in Saccharomyces cerevisiae. Genetics 2010; 184:65-77.
24. Pollard LM, Bourn RL, Bidichandani SI. Repair of DNA double-strand breaks within the (GAA*TTC)n sequence results in frequent deletion of the triplet-repeat sequence. Nucleic Acids Res 2008; 36:489-500.
25. Michel B, Ehrlich SD, Uzest M. DNA double-strand breaks caused by replication arrest. EMBO J 1997; 16:430-8.
26. Richard GF, Paques F. Mini- and microsatellite expansions: the recombination connection. EMBO Rep 2000; 1:122-6.
27. Tsuji S. Molecular genetics of triplet repeats: unstable expansion of triplet repeats as a new mechanism for neurodegenerative diseases. Intern Med 1997; 36:3-8.
28. Richards RI. Dynamic mutations: a decade of unstable expanded repeats in human genetic disease. Hum Mol Genet 2001; 10:2187-94.
29. Sieradzan KA, Mann DMA. The selective vulnerability of nerve cells in Huntington's disease. Neuropathol Appl Neurobiol 2001; 27:1-21.
30. Gutekunst CA, Levey A, Heilman C et al. Identification and localization of Huntingtin in brain and human lymphoblastoid cell lines with anti-fusion protein antibodies. Proc Natl Acad Sci USA 1995; 92:8710-4.
31. Velier J, Kim M, Schwarz C et al. Wild-type and mutant Huntingtins function in vesicle trafficking in the secretory and endocytic pathways. Exp Neurol 1998; 152:34-40.
32. Trushina E, Dyer RB, Badger JD II et al. Mutant Huntingtin impairs axonal trafficking in mammalian neurons in vivo and in vitro. Mol Cell Biol 2004; 24:8195-209.
33. Sinadinos C, Burbidge-King T, Soh D et al. Live axonal transport disruption by mutant huntingtin fragments in Drosophila motor neuron axons. Neurobiol Dis 2009; 34:389-95.
34. Gunawardena S, Her LS, Brusch RG et al. Disruption of axonal transport by loss of huntingtin or expression of pathogenic polyQ proteins in Drosophila. Neuron 2003; 40:25-40.
35. Landwehrmeyer GB, McNeil SM, Dure LSt et al. Huntington's disease gene: regional and cellular expression in brain of normal and affected individuals. Ann Neurol 1995; 37:218-30.
36. Fusco FR, Chen Q, Lamoreaux WJ et al. Cellular localization of Huntingtin in striatal and cortical neurons in rats: lack of correlation with neuronal vulnerability in Huntington's disease. J Neurosci 1999; 19:1189-202.
37. Hardy J, Orr H. The genetics of neurodegenerative diseases. J Neurochem 2006; 97:1690-9.
38. Chamberlain NL, Driver ED, Miesfeld RL. The length and location of CAG trinucleotide repeats in the androgen receptor N-terminal domain affect transactivation function. Nucleic Acids Res 1994; 22:3181-6.
39. MacLean HE, Warne GL, Zajac JD. Spinal and bulbar muscular atrophy: androgen receptor dysfunction caused by a trinucleotide repeat expansion. J Neurol Sci 1996; 135:149-57.
40. Thomas PS Jr., Fraley GS, Damien V et al. Loss of endogenous androgen receptor protein accelerates motor neuron degeneration and accentuates androgen insensitivity in a mouse model of X-linked spinal and bulbar muscular atrophy. Hum Mol Genet 2006; 15:2225-38.

41. Uyama E, Kondo I, Uchino M et al. Dentatorubral-pallidoluysian atrophy (DRPLA): clinical, genetic, and neuroradiologic studies in a family. J Neurol Sci 1995; 130:146-53.
42. Burke JR, Wingfield MS, Lewis KE et al. The Haw River syndrome: dentatorubropallidoluysian atrophy (DRPLA) in an African-American family. Nat Genet 1994; 7:521-4.
43. Nagafuchi S, Yanagisawa H, Ohsaki E et al. Structure and expression of the gene responsible for the triplet repeat disorder, dentatorubral and pallidoluysian atrophy (DRPLA). Nat Genet 1994; 8:177-82.
44. Margolis RL, Li S-H, Scott Young W et al. DRPLA gene (Atrophin-1) sequence and mRNA expression in human brain. Brain Res Mol Brain Res 1996; 36:219-26.
45. Wang L, Rajan H, Pitman JL et al. Histone deacetylase-associating Atrophin proteins are nuclear receptor corepressors. Genes Dev 2006; 20:525-30.
46. Schilling G, Wood JD, Duan K et al. Nuclear accumulation of truncated atrophin-1 fragments in a transgenic mouse model of DRPLA. Neuron 1999; 24:275-86.
47. Yu J, Ying M, Zhuang Y et al. C-terminal deletion of the atrophin-1 protein results in growth retardation but not neurodegeneration in mice. Dev Dyn 2009; 238:2471-8.
48. van de Warrenburg BPC, Notermans NC, Schelhaas HJ et al. Peripheral nerve involvement in spinocerebellar ataxias. Arch Neurol 2004; 61:257-61.
49. Carlson KM, Andresen JM, Orr HT. Emerging pathogenic pathways in the spinocerebellar ataxias. Curr Opin Genet Dev 2009; 19:247-53.
50. Yue S, Serra HG, Zoghbi HY, Orr HT. The spinocerebellar ataxia type 1 protein, ataxin-1, has RNA-binding activity that is inversely affected by the length of its polyglutamine tract. Hum Mol Genet 2001; 10:25-30.
51. Klement IA, Skinner PJ, Kaytor MD et al. Ataxin-1 nuclear localization and aggregation: role in polyglutamine-induced disease in SCA1 mice. Cell 1998; 95:41-53.
52. Sharma D, Sharma S, Pasha S, Brahmachari SK. Peptide models for inherited neurodegenerative disorders: conformation and aggregation properties of long polyglutamine peptides with and without interruptions. FEBS Lett 1999; 456:181-5.
53. Imbert G, Saudou F, Yvert G et al. Cloning of the gene for spinocerebellar ataxia 2 reveals a locus with high sensitivity to expanded CAG/glutamine repeats. Nat Genet 1996; 14:285-91.
54. Satterfield TF, Pallanck LJ. Ataxin-2 and its Drosophila homolog, ATX2, physically assemble with polyribosomes. Hum Mol Genet 2006; 15:2523-32.
55. Ralser M, Albrecht M, Nonhoff U et al. An integrative approach to gain insights into the cellular function of human ataxin-2. J Mol Biol 2005; 346:203-14.
56. Satterfield TF, Jackson SM, Pallanck LJ. A Drosophila homolog of the polyglutamine disease gene SCA2 is a dosage-sensitive regulator of actin filament formation. Genetics 2002; 162:1687-702;.
57. Huynh DP, Figueroa K, Hoang N, Pulst S-M. Nuclear localization or inclusion body formation of ataxin-2 are not necessary for SCA2 pathogenesis in mouse or human. Nat Genet 2000; 26:44-50.
58. Klockgether T, Schols L, Abele M et al. Age related axonal neuropathy in spinocerebellar ataxia type 3/Machado-Joseph disease (SCA3/MJD). J Neurol Neurosurg Psychiatry 1999; 66:222-4.
59. Wang G-h, Sawai N, Kotliarova S et al. Ataxin-3, the MJD1 gene product, interacts with the two human homologs of yeast DNA repair protein RAD23, HHR23A and HHR23B. Hum Mol Genet 2000; 9:1795-803.
60. Doss-Pepe EW, Stenroos ES, Johnson WG, Madura K. Ataxin-3 interactions with Rad23 and valosin-containing protein and its associations with ubiquitin chains and the proteasome are consistent with a role in ubiquitin-mediated proteolysis. Mol Cell Biol 2003; 23:6469-83.
61. Chen X, Tang TS, Tu H et al. Deranged calcium signaling and neurodegeneration in spinocerebellar ataxia type 3. J Neurosci 2008; 28:12713-24.
62. Jia NL, Fei EK, Ying Z et al. PolyQ-expanded ataxin-3 interacts with full-length ataxin-3 in a polyQ length-dependent manner. Neurosci Bull 2008; 24:201-8.
63. Frontali M. Spinocerebellar ataxia type 6: channelopathy or glutamine repeat disorder? Brain Res Bull 2001; 56:227-31.
64. Gazulla J, Tintore MA. The P/Q-type voltage-dependent calcium channel as pharmacological target in spinocerebellar ataxia type 6: gabapentin and pregabalin may be of therapeutic benefit. Med Hypotheses 2007; 68:131-6.
65. Watase K, Barrett CF, Miyazaki T et al. Spinocerebellar ataxia type 6 knockin mice develop a progressive neuronal dysfunction with age-dependent accumulation of mutant CaV2.1 channels. Proc Natl Acad Sci USA 2008.
66. Michalik A, Martin JJ, Van Broeckhoven C. Spinocerebellar ataxia type 7 associated with pigmentary retinal dystrophy. Eur J Hum Genet 2004; 12:2-15.
67. Ström A-L, Forsgren L, Holmberg M. A role for both wild-type and expanded ataxin-7 in transcriptional regulation. Neurobiol Dis 2005; 20:646-55.
68. Palhan VB, Chen S, Peng G-H et al. Polyglutamine-expanded ataxin-7 inhibits STAGA histone acetyltransferase activity to produce retinal degeneration. Proc Natl Acad Sci USA 2005; 102:8472-7.

69. Chou AH, Chen CY, Chen SY et al. Polyglutamine-expanded ataxin-7 causes cerebellar dysfunction by inducing transcriptional dysregulation. Neurochem Int 2010; 56:329-39.
70. van Roon-Mom WMC, Reid SJ, Faull RLM, Snell RG. TATA-binding protein in neurodegenerative disease. Neuroscience 2005; 133:863-72.
71. Shah AG, Friedman MJ, Huang S et al. Transcriptional dysregulation of TrkA associates with neurodegeneration in spinocerebellar ataxia type 17. Hum Mol Genet 2009; 18:4141-52.
72. DiFiglia M, Sapp E, Chase KO et al. Aggregation of Huntingtin in neuronal intranuclear inclusions and dystrophic neurites in brain. Science 1997; 277:1990-3.
73. Wells RD, Ashizawa T, eds. Genetic Instabilities and Neurological Diseases. San Diego: Academic Press, 2006.
74. Li H, Li S-H, Johnston H, Shelbourne PF, Li X-J. Amino-terminal fragments of mutant huntingtin show selective accumulation in striatal neurons and synaptic toxicity. Nat Genet 2000; 25:385-9.
75. Saudou F, Finkbeiner S, Devys D, Greenberg ME. Huntingtin Acts in the Nucleus to Induce Apoptosis but Death Does Not Correlate with the Formation of Intranuclear Inclusions. Cell 1998; 95:55-66.
76. Nagai Y, Inui T, Popiel HA et al. A toxic monomeric conformer of the polyglutamine protein. Nat Struct Mol Biol 2007; 14:332-340.
77. Takahashi T, Kikuchi S, Katada S et al. Soluble polyglutamine oligomers formed prior to inclusion body formation are cytotoxic. Hum Mol Genet 2008; 17:345-356.
78. Lajoie P, Snapp EL. Formation and toxicity of soluble polyglutamine oligomers in living cells. PLoS One 2010; 5:e15245.
79. Michalik A, Van Broeckhoven C. Pathogenesis of polyglutamine disorders: aggregation revisited. Hum Mol Genet 2003; 12:R173-86.
80. Lavoie H, Debeane F, Trinh Q-D et al. Polymorphism, shared functions and convergent evolution of genes with sequences coding for polyalanine domains. Hum Mol Genet 2003; 12:2967-79.
81. Albrecht A, Mundlos S. The other trinucleotide repeat: polyalanine expansion disorders. Curr Opin Genet Dev 2005; 15:285-93.
82. Brown LY, Brown SA. Alanine tracts: the expanding story of human illness and trinucleotide repeats. Trends Genet 2004; 20:51-8.
83. Brais B, Bouchard JP, Xie YG et al. Short GCG expansions in the PABP2 gene cause oculopharyngeal muscular dystrophy. Nat Genet 1998; 18:164-7.
84. Calado A, Tome FM, Brais B et al. Nuclear inclusions in oculopharyngeal muscular dystrophy consist of poly(A) binding protein 2 aggregates which sequester poly(A) RNA. Hum Mol Genet 2000; 9:2321-8;.
85. Toulouse A, Au-Yeung F, Gaspar C et al. Ribosomal frameshifting on MJD-1 transcripts with long CAG tracts. Hum Mol Genet 2005; 14:2649-60.
86. Davies JE, Rubinsztein DC. Polyalanine and polyserine frameshift products in Huntington's disease. J Med Genet 2006; 43:893-6.
87. Berger Z, Davies JE, Luo S et al. Deleterious and protective properties of an aggregate-prone protein with a polyalanine expansion. Hum Mol Genet 2006; 15:453-65.
88. McLeod CJ, O'Keefe LV, Richards RI. The pathogenic agent in Drosophila models of 'polyglutamine' diseases. Hum Mol Genet 2005; 14:1041-8.
89. Pieretti M, Zhang FP, Fu YH et al. Absence of expression of the FMR-1 gene in fragile X syndrome. Cell 1991; 66:817-22.
90. Garber K, Smith KT, Reines D, Warren ST. Transcription, translation and fragile X syndrome. Curr Opin Genet Dev 2006; 16:270-5.
91. Gu Y, Shen Y, Gibbs RA, Nelson DL. Identification of FMR2, a novel gene associated with the FRAXE CCG repeat and CpG island. Nat Genet 1996; 13:109-13.
92. Gecz J, Gedeon AK, Sutherland GR, Mulley JC. Identification of the gene FMR2, associated with FRAXE mental retardation. Nat Genet 1996; 13:105-8.
93. Bensaid M, Melko M, Bechara EG et al. FRAXE-associated mental retardation protein (FMR2) is an RNA-binding protein with high affinity for G-quartet RNA forming structure. Nucleic Acids Res 2009; 37:1269-79.
94. Delatycki MB, Knight M, Koenig M et al. G130V, a common FRDA point mutation, appears to have arisen from a common founder. Hum Genet 1999; 105:343-6.
95. Kumari D, Biacsi RE, Usdin K. Repeat expansion affects both transcription initiation and elongation in friedreich ataxia cells. J Biol Chem 2011; 286:4209;.
96. Puccio H, Koenig M. Recent advances in the molecular pathogenesis of Friedreich ataxia. Hum Mol Genet 2000; 9:887-92.
97. Ranum LP, Day JW. Myotonic dystrophy: RNA pathogenesis comes into focus. Am J Hum Genet 2004; 74:793-804.
98. Giorgio A, Dotti MT, Battaglini M et al. Cortical damage in brains of patients with adult-form of myotonic dystrophy type 1 and no or minimal MRI abnormalities. J Neurol 2006; 253:1471-7.

99. Sistiaga A, Urreta I, Jodar M et al. Cognitive/personality pattern and triplet expansion size in adult myotonic dystrophy type 1 (DM1): CTG repeats, cognition and personality in DM1. Psychol Med 2010; 40:487-95.
100. Seznec H, Agbulut O, Sergeant N et al. Mice transgenic for the human myotonic dystrophy region with expanded CTG repeats display muscular and brain abnormalities. Hum Mol Genet 2001; 10:2717-26.
101. Personius KE, Nautiyal J, Reddy S. Myotonia and muscle contractile properties in mice with SIX5 deficiency. Muscle Nerve 2005; 31:503-5.
102. Klesert TR, Cho DH, Clark JI et al. Mice deficient in Six5 develop cataracts: implications for myotonic dystrophy. Nat Genet 2000; 25:105-9.
103. Mankodi A, Logigian E, Callahan L et al. Myotonic dystrophy in transgenic mice expressing an expanded CUG repeat. Science 2000; 289:1769-73.
104. Ranum LPW, Cooper TA. RNA-mediated neuromuscular disorders. Annu Rev Neurosci 2006; 29:259-77.
105. Ho TH, Savkur RS, Poulos MG et al. Colocalization of muscleblind with RNA foci is separable from mis-regulation of alternative splicing in myotonic dystrophy. J Cell Sci 2005; 118:2923-33.
106. Jiang H, Mankodi A, Swanson MS et al. Myotonic dystrophy type 1 is associated with nuclear foci of mutant RNA, sequestration of muscleblind proteins and deregulated alternative splicing in neurons. Hum Mol Genet 2004; 13:3079-88.
107. Kimura T, Nakamori M, Lueck JD et al. Altered mRNA splicing of the skeletal muscle ryanodine receptor and sarcoplasmic/endoplasmic reticulum Ca2+-ATPase in myotonic dystrophy type 1. Hum Mol Genet 2005; 14:2189-200.
108. Mankodi A, Takahashi MP, Jiang H et al. Expanded CUG Repeats Trigger Aberrant Splicing of ClC-1 Chloride Channel Pre-mRNA and Hyperexcitability of Skeletal Muscle in Myotonic Dystrophy. Mol Cell 2002; 10:35-44.
109. Savkur RS, Philips AV, Cooper TA et al. Insulin receptor splicing alteration in myotonic dystrophy type 2. Am J Hum Genet 2004; 74:1309-13.
110. Finsterer J. Myotonic dystrophy type 2. Eur J Neurol 2002; 9:441-7.
111. Mankodi A, Urbinati CR, Yuan Q-P et al. Muscleblind localizes to nuclear foci of aberrant RNA in myotonic dystrophy types 1 and 2. Hum Mol Genet 2001; 10:2165-70.
112. Salvatori S, Furlan S, Fanin M et al. Comparative transcriptional and biochemical studies in muscle of myotonic dystrophies (DM1 and DM2). Neurol Sci 2009; 30:185-92.
113. Botta A, Vallo L, Rinaldi F et al. Gene expression analysis in myotonic dystrophy: indications for a common molecular pathogenic pathway in DM1 and DM2. Gene Expr 2007; 13:339-51.
114. Mankodi A, Teng-Umnuay P, Krym M et al. Ribonuclear inclusions in skeletal muscle in myotonic dystrophy types 1 and 2. Ann Neurol 2003; 54:760-8.
115. Ho TH, Bundman D, Armstrong DL, Cooper TA. Transgenic mice expressing CUG-BP1 reproduce splicing mis-regulation observed in myotonic dystrophy. Hum Mol Genet 2005; 14:1539-47.
116. Ho TH, Bundman D, Armstrong DL, Cooper TA. Transgenic mice expressing CUG-BP1 reproduce splicing mis-regulation observed in myotonic dystrophy. Hum Mol Genet 2005; 14:1539-47.
117. Ladd AN, Taffet G, Hartley C et al. Cardiac tissue-specific repression of CELF activity disrupts alternative splicing and causes cardiomyopathy. Mol Cell Biol 2005; 25:6267-78.
118. de Haro M, Al-Ramahi I, De Gouyon B et al. MBNL1 and CUGBP1 modify expanded CUG-induced toxicity in a Drosophila model of myotonic dystrophy type 1. Hum Mol Genet 2006; 15:2138-45.
119. Hagerman RJ. Lessons from fragile X regarding neurobiology, autism, and neurodegeneration. J Dev Behav Pediatr 2006; 27:63-74.
120. Hagerman RJ, Ono MY, Hagerman PJ. Recent advances in fragile X: a model for autism and neurodegeneration. Curr Opin Psychiatry 2005; 18:490-6.
121. Iwahashi CK, Yasui DH, An H-J et al. Protein composition of the intranuclear inclusions of FXTAS. Brain 2006; 129:256-71.
122. Hashem V, Galloway JN, Mori M et al. Ectopic expression of CGG containing mRNA is neurotoxic in mammals. Hum Mol Genet 2009; 18:2443-51.
123. Day JW, Schut LJ, Moseley ML et al. Spinocerebellar ataxia type 8: Clinical features in a large family. Neurology 2000; 55:649-57.
124. Gupta A, Jankovic J. Spinocerebellar ataxia 8: variable phenotype and unique pathogenesis. Parkinsonism Relat Disord 2009; 15:621-6.
125. Nemes JP, Benzow KA, Moseley ML et al. The SCA8 transcript is an antisense RNA to a brain-specific transcript encoding a novel actin-binding protein (KLHL1). Hum Mol Genet 2000; 9:1543-51.
126. Aromolaran KA, Benzow KA, Cribbs LL et al. T-type current modulation by the actin-binding protein Kelch-like 1 (KLHL1). Am J Physiol Cell Physiol 2010; 298(6):C1353-62.
127. Chen WL, Lin JW, Huang HJ et al. SCA8 mRNA expression suggests an antisense regulation of KLHL1 and correlates to SCA8 pathology. Brain Res 2008; 1233:176-84.

128. He Y, Zu T, Benzow KA et al. Targeted deletion of a single Sca8 ataxia locus allele in mice causes abnormal gait, progressive loss of motor coordination, and Purkinje cell dendritic deficits. J Neurosci 2006; 26:9975-82.

129. Moseley ML, Zu T, Ikeda Y et al. Bidirectional expression of CUG and CAG expansion transcripts and intranuclear polyglutamine inclusions in spinocerebellar ataxia type 8. Nat Genet 2006; 38:758-69.

130. Mutsuddi M, Marshall CM, Benzow KA et al. The spinocerebellar ataxia 8 noncoding RNA causes neurodegeneration and associates with staufen in Drosophila. Curr Biol 2004; 14:302-8;.

131. März P, Probst A, Lang S et al. Ataxin-10, the spinocerebellar ataxia type 10 neurodegenerative disorder protein, is essential for survival of cerebellar neurons. J Biol Chem 2004; 279:35542-50.

132. Lin X, Ashizawa T. Recent progress in spinocerebellar ataxia type 10 (SCA10). Cerebellum 2005; 4:37-42.

133. Handa V, Yeh HJ, McPhie P, Usdin K. The AUUCU repeats responsible for spinocerebellar ataxia type 10 form unusual RNA hairpins. J Biol Chem 2005; 280:29340-5.

134. Keren B, Jacquette A, Depienne C et al. Evidence against haploinsuffiency of human ataxin 10 as a cause of spinocerebellar ataxia type 10. Neurogenetics 2010; 11:273;.

135. Petersen P, Chou DM, You Z et al. Protein phosphatase 2A antagonizes ATM and ATR in a Cdk2- and Cdc7-independent DNA damage checkpoint. Mol Cell Biol 2006; 26:1997-2011.

136. Dagda RK, Merrill RA, Cribbs JT et al. The spinocerebellar ataxia 12 gene product and protein phosphatase 2A regulatory subunit Bbeta2 antagonizes neuronal survival by promoting mitochondrial fission. J Biol Chem 2008; 283: 36241-36248.

137. Holmes SE, Hearn EOr, Ross CA, Margolis RL. SCA12: an unusual mutation leads to an unusual spinocerebellar ataxia. Brain Res Bull 2001; 56:397-403.

138. Holmes SE, O'Hearn E, Rosenblatt A et al. A repeat expansion in the gene encoding junctophilin-3 is associated with Huntington disease-like 2. Nat Genet 2001; 29:377-8.

139. Margolis RL, O'Hearn E, Rosenblatt A et al. A disorder similar to Huntington's disease is associated with a novel CAG repeat expansion. Ann Neurol 2001; 50:373-80.

140. Wilburn B, Rudnicki DD, Zhao J et al. An antisense CAG repeat transcript at JPH3 locus mediates expanded polyglutamine protein toxicity in Huntington's disease-like 2 mice. Neuron 2011; 70:427-440.

141. Rudnicki DD, Holmes SE, Lin MW et al. Huntington's disease–like 2 is associated with CUG repeat-containing RNA foci. Ann Neurol 2007; 61:272-82.

142. Sato N, Amino T, Kobayashi K et al. Spinocerebellar ataxia type 31 is associated with "inserted" penta-nucleotide repeats containing (TGGAA)n. Am J Hum Genet 2009; 85:544-57.

143. Stevanin G, Fujigasaki H, Lebre A-S et al. Huntington's disease-like phenotype due to trinucleotide repeat expansions in the TBP and JPH3 genes. Brain 2003; 126:1599-603.

144. Marsh JL, Walker H, Theisen H et al. Expanded polyglutamine peptides alone are intrinsically cytotoxic and cause neurodegeneration in Drosophila. Hum Mol Genet 2000; 9:13-25.

145. Raspe M, Gillis J, Krol H et al. Mimicking proteasomal release of polyglutamine peptides initiates aggregation and toxicity. J Cell Sci 2009; 122:3262-71.

146. Nakayama H, Hamada M, Fujikake N et al. ER stress is the initial response to polyglutamine toxicity in PC12 cells. Biochem Biophys Res Commun 2008; 377:550-5.

147. Wellington CL, Ellerby LM, Hackam AS et al. Caspase cleavage of gene products associated with triplet expansion disorders generates truncated fragments containing the polyglutamine tract. J Biol Chem 1998; 273:9158-67.

148. Sobczak K, de Mezer M, Michlewski G et al. RNA structure of trinucleotide repeats associated with human neurological diseases. Nucleic Acids Res 2003; 31:5469-82.

149. Amrane S, Mergny JL. Length and pH-dependent energetics of (CCG)n and (CGG)n trinucleotide repeats. Biochimie 2006; 88:1125-34.

150. Zumwalt M, Ludwig A, Hagerman PJ, Dieckmann T. Secondary structure and dynamics of the r(CGG) repeat in the mRNA of the fragile X mental retardation 1 (FMR1) gene. RNA Biol 2007; 4:93-100.

151. Yuan Y, Compton SA, Sobczak K et al. Muscleblind-like 1 interacts with RNA hairpins in splicing target and pathogenic RNAs. Nucleic Acids Res 2007; 35:5474-86.

152. Li LB, Yu Z, Teng X, Bonini NM. RNA toxicity is a component of ataxin-3 degeneration in Drosophila. Nature 2008; 453:1107-11.

153. Zu T, Gibbens B, Doty NS et al. Non-ATG-initiated translation directed by microsatellite expansions. Proc Natl Acad Sci USA 2011; 108:260-265.

154. Batra R, Charizanis K, Swanson MS. Partners in crime: bidirectional transcription in unstable microsatellite disease. Hum Mol Genet 2010; 19:R77-82.

155. Cho DH, Thienes CP, Mahoney SE et al. Antisense transcription and heterochromatin at the DM1 CTG repeats are constrained by CTCF. Mol Cell 2005; 20:483-9.

156. Ladd PD, Smith LE, Rabaia NA et al. An antisense transcript spanning the CGG repeat region of FMR1 is upregulated in premutation carriers but silenced in full mutation individuals. Hum Mol Genet 2007; 16:3174-87.

157. Lavorgna G, Dahary D, Lehner B et al. In search of antisense. Trends Biochem Sci 2004; 29:88-94.
158. Lapidot M, Pilpel Y. Genome-wide natural antisense transcription: coupling its regulation to its different regulatory mechanisms. EMBO Rep 2006; 7:1216-22.
159. Morris KV, Santoso S, Turner AM et al. Bidirectional transcription directs both transcriptional gene activation and suppression in human cells. PLoS Genet 2008; 4: e1000258.
160. Nakamori M, Pearson CE, Thornton CA. Bidirectional transcription stimulates expansion and contraction of expanded (CTG)*(CAG) repeats. Hum Mol Genet 2011; 20:580-8.
161. Lawlor KT, O'Keefe LV, Samaraweera SE et al. Double-stranded RNA is pathogenic in Drosophila models of expanded repeat neurodegenerative diseases. Hum Mol Genet 2011; 20:3757-3768.
162. van Eyk CL, McLeod CJ, O'Keefe LV et al. Comparative toxicity of polyglutamine, polyalanine and polyleucine tracts in Drosophila models of expanded repeat disease. Hum Mol Genet 2011; 21: Epub ahead of print.
163. van Eyk CL, O'Keefe LV, Lawlor KT et al. Perturbation of the Akt/Gsk3-beta signalling pathway is common to Drosophila expressing expanded untranslated CAG, CUG and AUUCU repeat RNAs. Hum Mol Genet 2011; 20:2783-2794.
164. Li LB, Yu Z, Teng X et al. RNA toxicity is a component of ataxin-3 degeneration in Drosophila. Nature 2008; 453:1107-1111.
165. Yu Z, Teng X, Bonini NM. Triplet repeat-derived siRNAs enhance RNA-mediated toxicity in a Drosophila model for myotonic dystrophy. PLoS Genet 2011; 7: e1001340.
166. van Eyk CL. Investigation of RNA-mediated pathogenic pathways in a Drosophila model of expanded repeat disease. PhD Thesis, Discipline of Genetics, University of Adelaide, Adelaide, Australia, 2010.

CHAPTER 6

UNSTABLE MUTATIONS IN THE *FMR1* GENE AND THE PHENOTYPES

Danuta Loesch*,[1] and Randi Hagerman[2]

[1]Department of Psychology, LaTrobe University, Melbourne, Victoria, Australia; [2]Department of Pediatrics, MIND Institute, University of California at Davis Medical Center, Sacramento, California, USA
*Corresponding Author: Danuta Loesch—Email: d.loesch@latrobe.edu.au

Abstract: Fragile X syndrome (FXS), a severe neurodevelopmental anomaly, and one of the earliest disorders linked to an unstable ('dynamic') mutation, is caused by the large (>200) CGG repeat expansions in the noncoding portion of the *FMR1* (Fragile X Mental Retardation-1) gene. These expansions, termed full mutations, normally silence this gene's promoter through methylation, leading to a gross deficit of the Fragile X Mental Retardation Protein (FMRP) that is essential for normal brain development. Rare individuals with the expansion but with an unmethylated promoter (and thus, FMRP production), present a much less severe form of FXS.

However, a unique feature of the relationship between the different sizes of CGG expanded tract and phenotypic changes is that smaller expansions (<200) generate a series of different clinical manifestations and/or neuropsychological changes. The major part of this chapter is devoted to those *FMR1* alleles with small (55-200) CGG expansions, termed 'premutations', which have the potential for generating the full mutation alleles on mother-offspring transmission, on the one hand, and are associated with some phenotypic changes, on the other. Thus, the role of several factors known to determine the rate of CGG expansion in the premutation alleles is discussed first. Then, an account of various neurodevelopmental, cognitive, behavioural and physical changes reported in carriers of these small expansions is given, and possible association of these conditions with a toxicity of the elevated *FMR1* gene's transcript (mRNA) is discussed.

The next two sections are devoted to major and well defined clinical conditions associated with the premutation alleles. The first one is the late onset neurodegenerative disorder termed fragile X-associated tremor ataxia syndrome (FXTAS). The wide range of clinical and neuropsychological manifestations of this syndrome, and their relevance to elevated levels of the *FMR1* mRNA, are described. Another distinct disorder linked to the CGG repeat expansions within the premutation range is fragile X-associated primary ovarian insufficiency (FXPOI) in females, and an account

---

Tandem Repeat Polymorphisms: Genetic Plasticity, Neural Diversity and Disease,
edited by Anthony J. Hannan ©2012 Landes Bioscience and Springer Science+Business Media.

of the spectrum of manifestations of this disorder, together with the latest findings suggesting an early onset of the ovarian changes, is given.

In the following section, the most recent findings concerning the possible contribution of *FMR1* 'grey zone' alleles (those with the smallest repeat expansions overlapping with the normal range i.e., 41-54 CGGs), to the psychological and clinical manifestations, already associated with premutation alleles, are discussed. Special emphasis has been placed on the possibility that the modest elevation of 'toxic' *FMR1* mRNA in the carriers of grey zone alleles may present an additional risk for some neurodegenerative diseases, such as those associated with parkinsonism, by synergizing with either other susceptibility genes or environmental poisons.

The present status of the treatment of fragile X-related disorders, especially FXS, is presented in the last section of this chapter. Pharmacological interventions in this syndrome have recently extended beyond stimulants and antipsychotic medications, and the latest trials involving a group of GluR5 antagonists aim to ascertain if these substances have the potential to reverse some of the neurobiological abnormalities of FXS.

INTRODUCTION

The trinucleotide expansion of CGG repeats in the 5' untranslated region (5'-UTR) of the fragile X mental retardation 1 gene (*FMR1*) located at Xq27.3 was sequenced in 1991.[1] Although the phenotype of fragile X syndrome (FXS) was first described by Lubs,[2] it was not until 1991 that those with FXS were found to have >200 CGG repeats described as a full mutation in *FMR1*. Individuals who were carriers of smaller expansions between 55 to 200 CGG repeats (premutation) were originally thought to be unaffected clinically. However, in 1991 an elevated rate of premature ovarian failure (POF) was documented in carriers compared to controls[3] and later confirmed by many other groups (reviewed in refs. 4,5). POF has been renamed the fragile X- associated primary ovarian insufficiency (FXPOI) to emphasize the association with the premutation and also the occasional ability of women to reproduce such that the ovary has not completely failed.[6] Subsequently in 2001, the fragile X-associated tremor ataxia syndrome (FXTAS) was discovered in aging carriers[7,8] and it includes not only tremor and ataxia but also neuropathy, autonomic dysfunction, neuropsychiatric problems and cognitive decline sometimes leading to dementia.[9,10] This chapter delineates the history and development of the spectrum of involvement in these fragile X-associated disorders, with special emphasis on premutation (PM) carriers.

As the role of elevated *FMR1* mRNA in carriers of the PM alleles[11] has been researched and the concept of RNA toxicity leading to FXPOI or FXTAS has been developed[12,13] a variety of additional phenotypes has been described in those carriers. They include developmental problems in a subgroup of young male carriers including autism, autism spectrum disorder (ASD), attention deficit hyperactivity disorder (ADHD), shyness, anxiety and seizures.[10,14-18] In many adults with the premutation including both males and females, psychopathology is common including anxiety and depression compared to controls.[19-24] Cognitive changes, particularly executive function deficits, can begin before the onset of FXTAS in carriers[25] and there is evidence of early white matter disease reflected in diffusion tensor imaging changes before the onset of FXTAS.[26] Most recently autoimmune dysfunction including fibromyalgia and hypothyroidism has been found to be more common in carriers compared to controls.[22,27] Therefore a large spectrum of involvement associated with the premutation constitutes a newly emerging group of disorders leading to new avenues in research and clinical management.

LARGE EXPANSIONS OF CGG REPEAT—THE CAUSE OF FRAGILE X SYNDROME

Fragile X syndrome (FXS) was one of the earliest conditions found to be linked to the unstable mutation. It is the most common inherited form of intellectual disability caused by the large expansion in the *FMR1* gene, where cognitive and behavioural impairments are associated with minor physical features.[28] These expansions defined as 'full mutation' are usually associated with the methylation-coupled inactivation of *FMR1* gene leading to gross deficit of the protein gene product, FMRP.[29,30] Abnormalities of FXS are primarily caused by the depletion of FMRP,[31] and experimental evidence indicates that FMRP is essential for normal brain development.[32-35] More specifically, studies based on rat brains have shown that FMRP is involved in synaptogenesis, especially in the cerebral cortex, cerebellum and hippocampus,[36] and in modifying synaptic structure in response to environmental stimulation.[37-40] FMRP is an RNA binding protein and it carries mRNAs to the synapse where it typically inhibits translation until stimulation occurs and then it allows translation.[41] In the absence of FMRP there is dramatic up-regulation of protein production in the hippocampus.[42] FMRP regulates the translation of hundreds of proteins and they include many of the proteins also associated with autism when they are mutated including Neuroligins 3 and 4, Neurorexins, SHANK3, amyloid precursor protein (APP), Arc, MAPKinase, CYFIP 1 and 2 and many more.[43-48] There is also upregulation of mTOR in animals and humans with FXS[49,50] similar to other forms of autism such as Tuberous Sclerosis.[51] The molecular overlap between FXS and autism is likely the cause of the high rate of autism (30%) and also autism spectrum disorder (30%; ASD) in FXS.[16,52-54] The behavioral phenotype also includes a high rate of anxiety disorders[55] and ADHD.[56-58] Significant mood instability leading to aggression or severe tantrums occurs in about 30 to 40% of adolescents or adults.[28] This problem frequently leads to out of home placement and the use of psychotropic medication.

The physical phenotype of FXS includes a long face and prominent ears but approximately 30% of children will not have these features, although hyperextensible finger joints are seen in the majority along with double jointed thumbs.[28] Large testicles (macroorchidism) begin to occur at about 8 years of age and they reach their maximal size, typically at 2 to 3 times normal (60 ml) at about 15 to 16 years.[59]

Approximately 20% to 40% of individuals with FXS have mosaicism, meaning some cells with the premutation and some cells with the full mutation (size mosaicism) or a mixture of some cells with a lack of methylation and some cells with the full methylation (methylation mosaicism). Occasional individuals have a complete lack of methylation but a CGG repeat number >200. Those with a lack of methylation or those with a majority of their cells with the premutation typically have the highest levels of FMRP and they are often high functioning with an IQ > 70.[31,60] Also, individuals with FXS and FMRP levels >30% to 50% of normal are likely to maintain an IQ above 70 as adults and demonstrate fewer physical features of FXS.[31,61] Although these individuals seem to be better off in childhood and early adulthood, they may also have an elevated level of *FMR1* mRNA because of an increased rate of translation compared to those with a full mutation that is fully methylated.[62] The higher level of *FMR1* mRNA may put them at risk for neurological disease with aging similar to those with the premutation.[63] Hall et al[63] reported a case of a male with a high level of mosaicism who experienced significant neurodegeneration in his 60s, although his diagnosis was consistent with Parkinson's disease (PD). Although there is concern for aging problems in those with mosaicism, a recent report of more than

60 individuals with FXS who were aging demonstrated no difference in the molecular studies in those who developed PD and those who did not.[64] So far there has never been an individual with FXS who developed FXTAS, although PD is more common in aging in FXS than what is seen in the general population.[64]

Females with FXS are typically less affected than males because they have a second X chromosome. Their level of FMRP correlates with their activation ratio (the proportion of cells that have the normal X as the active X).[30,31] Approximately 25 to 30 percent of females have an IQ below 70 but the majority of females have an IQ in the borderline or low normal range.[65] The majority of girls or women with the full mutation experience some anxiety and common diagnoses are selective mutism, social phobia and specific phobia.[55] ADHD is seen in at least 30% of women and executive function deficits are common even when ADHD is not apparent.[66] Even women with a normal IQ and the full mutation can experience attentional problems and anxiety difficulties.[67] Usually girls respond well to stimulant medication to treat ADHD and selective serotonin reuptake inhibitors (SSRIs) to treat anxiety.[68] Counseling with a psychologist is also helpful for these problems especially if the women are under additional stress from having children with FXS and they need support and behavioral management guidance.[69]

SMALL CGG EXPANSION ALLELES AS THE SOURCE OF THE FULL MUTATION ALLELES

Apart from the flurry of studies on psychological and clinical manifestations of FXS, a major interest was in the origin of the large expansions leading to this severe developmental abnormality. According to the earliest reports, the mothers of the affected offspring did not manifest any obvious developmental anomalies reminiscent of those seen in FXS, but had a potential to generate offspring with this syndrome. Hence the term premutation (PM) was coined to reflect this potential, as well as the apparently unaffected status of the carriers of these PM alleles.[70]

The reason for, and the pattern of, these intergenerational differences were explained by the molecular data from large pedigrees which showed that these high risk parents carried small-size expansions of CGG repeat in the *FMR1* gene, which did not cause gene silencing, but further expanded if transmitted from mothers to offspring into the full mutation range, with the rates of expansion increasing proportionally to the size of repeat.[71-77] The detailed risk, as up-dated by Nolin and colleagues[78] in a large multicentre study of female carriers (with correction for ascertainment) were as follows. For PM carriers: 3.7% for the repeat range of 55-59, 5.3% for the range of 60-69, 31.1% for the range of 70-79, 57.8% for the range of 80-89, 80.1% for the range of 90-99, between 94.4 and 100% for the range of 100-139 repeats, and 100% for repeat sizes >139 repeats. It was established that the PM alleles carried by the fathers were usually stable, and on occasions they might even be reduced in size on transmission to their daughters.

The introduction of PCR techniques[79] allowed for the recognition of the potential for expansions of the alleles below the PM range (~35 to 54), and the risks for expansion of these small size (intermediate size, or GZ) alleles of 49-54 range were established.[78,80-87] Instability in maternal transmission of GZ alleles over one generation has been reported as 2%-4% on average, with the possibility of expanding to the full mutation within two generations. The lowest small expansion allele reported to expand to the full mutation had 56 CGG repeats.[88]

Unlike for the premutation (PM) alleles, grey zone (GZ) alleles have a much higher instability rate (~16%) if transmitted through the male than the female carriers.[89] It is the potential for expansion that determined the latest guidelines for an overall classification of *FMR1* alleles.[90] The upper bound of normal alleles (that is, showing no meiotic or mitotic instability) was set at ~44 CGG repeats; the range of intermediate size or GZ alleles (showing only minor increase or decrease in repeat number, but never expanding to full mutation) was set between 45 and 54 CGG repeats; and the alleles in the PM range (set between 55 and 200 repeats), if carried by females, showed a gradually increasing risk of expansion to the full mutation. The relevance of this classification in the light of the most recent molecular and clinical findings in carriers of small expansions will be discussed in the next sections.

The above figures represent an average risk for different categories of *FMR1* alleles, but the individual risks may vary widely between different carriers, and between different families. In some families small expansions may be transmitted unchanged throughout several generations, and in others the risk of expansion from PM in the mother to full mutation in the offspring may be high or very high. This variation prompted a search for those molecular and other factors which might predispose to, or protect from, further expansions of the small CGG repeat size alleles. The major modifying factor is the sex of a parent, where, as described above, the risk of expansion of PM alleles is limited to the mothers but for the GZ alleles, it is relatively higher for the fathers than from the mothers.[89] In addition, the data from large pedigrees suggested that the sex of offspring might be another modifying factor, with male offspring manifesting higher rate of large expansions from the parent than female offspring;[91,92] however this concept still remains controversial.[78,93]

Some other molecular factors associated with *FMR1* gene have been found to be influential in expansion of small size CGG repeats. Earlier hypotheses implicated a role for the number and position of AGG triplets. It has been argued that the AGG triplets interspersed within the CGG repeat tract stabilize this repeat, and in their absence increased length of 'pure' CGG are generated, which are more prone to replication slippage.[80,94-98] However, more recent data suggested that the role of AGG is a late event in progressive CGG expansion[83] although definitely related to the risk for expansion in the next generation.[99]

The role of molecular markers flanking the CGG repeat in the expansion process has been well established. In earlier studies, these markers comprised dinucleotide (CA) microsatellites which, in combination, generated polymorphic haplotypes.[94,100] The frequencies of some of these haplotypes differed significantly between normal as opposed to fragile X both full mutation and PM alleles in all populations tested.[97,101,102] The fact that the haplotype associations in the smallest expansion (GZ) alleles showed a pattern similar to this seen in fragile X alleles indicated that these alleles evolved from among GZ alleles.[94,103,104]

Currently, single nucleotide polymorphisms, SNPs, are more widely used and are now recognized as important markers in studies of fragile X, because of greater discriminatory power of large number of SNPs that can be tested to identify haplotypes that segregate with different CGG repeat size allele categories.[104,105] The finding that the haplotypes combining the two SNPs, ATL1-alleles A and G[104,106] and FMRb-alleles A and G[103,106] exhibit strong associations with PM/full mutation, as well as with GZ alleles provided further evidence for the role of GZ alleles along the pathway leading from the normal to the fragile X full mutation alleles.

PHENOTYPIC FEATURES IN PM CARRIERS

The effect of small expansions on the phenotypes is an important problem considering their high population prevalence. The PM (using 55 repeats as the lower bound) is relatively common in the general population affecting 1 in 113 to 259 females and 1 in 260 to 813 males.[83,107-111] Population prevalence of intermediate (GZ) size alleles is spectacularly higher reaching 1 in ~ 30 males;[81,83,84] and may be at least twice as common in females.[112,113] Therefore, any evidence for clinical and neuropsychological involvement in carriers of these alleles would have a significant bearing on the causes of health problems in a previously unrecognized but sizeable group of individuals.

Phenotypic effects of small expansions has taken a long time to be appreciated since the term 'PM' was first coined by Pembrey and colleagues in 1985[70] who stated that it 'causes no harm other then predisposing to the final event', with 'final event' being the full mutation associated with the FXS. It took more than a decade to change the belief in a clear-cut division of carriers of fragile X into two distinct categories: 'affected' with FXS, as opposed to unaffected 'carrier females' and 'transmitting males'. Other than the finding of FXPOI in 1991,[3] the earliest studies of small samples of male PM carriers,[114,115] and both male and female carriers[116] demonstrated minor yet significant physical and intellectual impairments. Physical problems found in approximately 25% of PM carriers included prominent ears and hyperextensible finger joints.[117,118] More stringent analysis of cognitive status used genetic statistical models based on pedigree data, which allowed for adjusting the measures obtained from the carriers for these measures in other relatives. This analysis provided evidence for a significance effect of PM on standard cognitive (Wechsler) measures, including performance IQ, Block Design, Perceptual Organization, and Object Assembly, in the adult male carriers and Symbol Search, in the adult female carriers,[119] as illustrated in Figure 1.

A significant effect of PM irrespective of age was also established, using pedigree models, on some FSIQ-adjusted executive function test scores.[120] These included impairment of the motor planning, inhibition and working memory measured by the Behavioural Dyscontrol Scale (BDS),[121] in the male carriers, and of visuospatial memory and visuospatial constructional ability, assessed by Rey Complex Figure Design and Recognition Test (RCFT),[122] in the female carriers. Similar findings of significant impairments on tests of executive function (Verbal Fluency, Trail Making Test and Tower of London) and memory were reported in a sample of 20 unrelated male PM carriers.[123] This data may indicate that the CGG expansion alleles within the PM range may affect specific neuronal circuits manifesting as relevant neuropsychological deficits. These deficits may also include, specifically for the female sex, impairment in arithmetics as reported in 39 PM women based on performance on the Wide Range Achievement Test-3,[124] visual pathway deficits involving low social/high temporal frequency contrast sensitivity and frequency-doubling veneer,[125] and motion perception deficits.[126] That the PM alleles affect brain functioning has been also shown in CGG knock-in (KI) female mice where, apart from deficits in processing special information, poor performance on tests of temporal order of presenting two objects, specifically related to the upper-end of PM range (150-200 CGGs) was encountered.[127] This latter finding is reminiscent of the results of a recent study based on 60 female PM carriers, where a (significant) decrease in IQ (Wechsler Scales) verbal and performance scores was more evident for the females in the upper PM range (CGG>100), and with a 10% decrement of FMRP expression.[128]

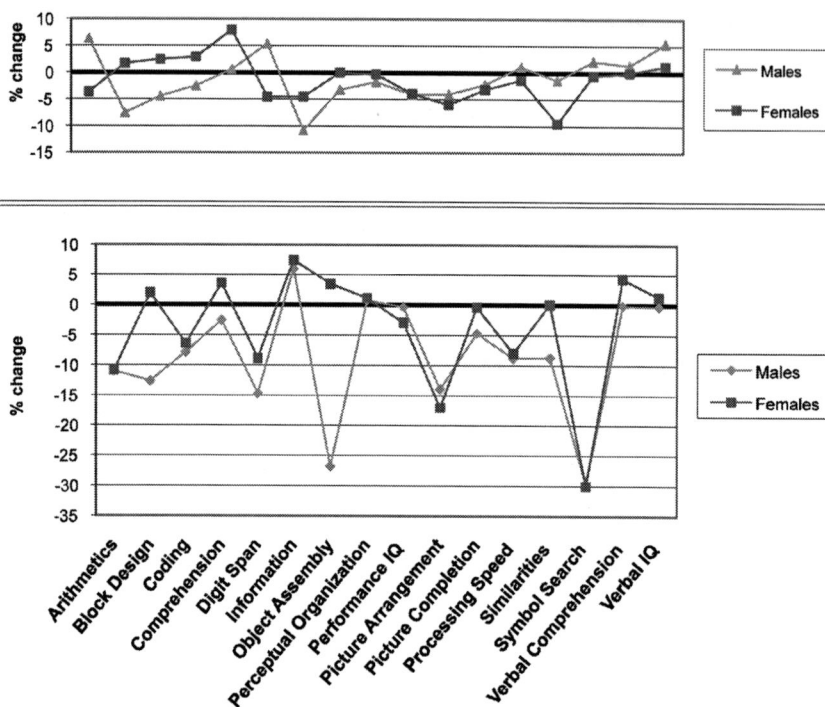

Figure 1. Standardized percent deviation of FSIQ-adjusted summary and subtest (Wechsler) scores from the normal mean due to the effect of premutation (upper figure), and full mutation (lower figure) separately for males and females, based on pedigree models. The baseline represents the FSIQ of a 100.[119] Reproduced from Loesch DZ et al. Am J Med Genet A 2003; 122A(1):13-23.[119]

Although these findings have been largely based on adult males and females, the overlap of the observed cognitive deficits with some of those seen in FXS and their relevance to minor FMRP deficits and the upper end of PM range, indicated a significant neurodevelopmental component in the origin of these changes. The presence of the associated behavioural deficits, including autism spectrum, and attention deficit hyperactivity disorder (ADHD), in boys carrying the PM allele, support this notion.[14,16,129] Another supportive piece of evidence has been provided by demonstration of several minor physical anomalies in unrelated PM male and female carriers, which are typical of the FXS phenotype,[117,118] and in families,[130] as well as of the distinct abnormalities in brain structure on MR images.[131-134] Notably, reduced hippocampal volume was found to be associated with memory deficits in both male and female carriers,[131] or with anxiety-related psychological symptoms in female carriers.[134] In another study, the wide range of structural brain anomalies, including reduced volume of the whole brain, caudate and thalamic nuclei bilaterally were found to be associated with changes in metabolic rates in several brain areas in female carriers.[132] The results of the study, which used fMRI to measure brain responses to fearful faces and fearful social images, demonstrated diminished activation of amygdala and several other brain areas that mediate social cognition in the group of PM adult male carriers compared with age and IQ-matched controls.[135] This reduction was also significantly associated with self-report of psychological symptoms on the Symptom Checklist-90-revised (SCL-90-R).

A significant association between reduced activity and volume of the hippocampus during a memory recall task, and psychological symptom severity, was also reported in two other studies based on male[136] and female[134] PM carriers.

It is clear from these and other reports (to be discussed below) that it is difficult to draw the clear demarcation line between the phenotypic abnormalities which can be related to neurodevelopmental changes, and those determined by the late onset progressive brain pathology. While the first category of changes appears to represent a mild version of physical and neuropsychological anomalies linked to the full mutation that may be related mainly to subtle deficits of FMRP at the upper end of the PM range,[31,137] evidence for the late onset age-dependent changes suggest alternative or additional major pathogenic effects, related to the elevated levels of expanded *FMR1* transcript, which may account for the late neurodegenerative changes, or for both neurodegenerative and neurodevelopmental abnormalities (as detailed in the later sections of this chapter).

The finding of age-dependent deficit in the ability to inhibit prepotent responses,[138] and in working memory tasks, particularly executive control of memory,[139,140] in a sample of 40 adult PM male carriers, may additionally illustrate an overlap between these 2 dimensions. The progress in severity of cognitive decline was shown to be more striking in the older age group of carrier males aged >50 years, with 33.3% penetrance of dementia for the average age of 63.4 years, which represented sixfold increase compared with noncarriers, and further increasing with age and allele size.[141] However, a critical review of neuropsychological status of male and female PM carriers[142] recommends caution in the interpretation of these data and the use of larger samples obtained from several sources.

Although no psychiatric pathology has been found in the earliest study of female carriers,[143] it has been widely reported among numerous subsequent studies of adult carrier males and females. In a large comprehensive study by Franke et al in 1998[144] an overall rate of a lifetime diagnosis of affective disorders was found as high as 55.7%; of anxiety disorders, 41%; of panic disorder, 11.5%; and 18% was a lifetime risk of social phobia, which was independent on the effect of having children with FXS. Although distinct schizoaffective disorders have only been reported in rare individuals,[145] schizotypal features and avoidant personality disorder occurred in respectively 8.2% and 4.9% of carriers.[144] Later studies provided confirmatory evidence for the high prevalence of mood, especially depressive disorders, in female PM carriers. The Spanish study based on a small sample of PM carriers and mothers of FXS children showed higher scores for depression (based on SCL-90-R and Beck Depression Inventory) compared with noncarrier mothers of mentally retarded children.[146] In the much larger study by Roberts et al in 2009,[147] 93 PM females were compared to 2,159 controls from the National Comorbidity Survey Replication (NCS-R) US dataset. The results based on Structured Clinical Interview for the DSM-IV (SCID-1) showed a significant increase in the risk of lifetime major depressive disorder (43% in carriers compared with 31.9% controls), which is close to the earlier estimate of nearly 60% risk of all affective disorders by Franke et al.[144] The risk of lifetime panic disorder and current agoraphobia were not high (8.6% and 3.2%, respectively), but they were significantly elevated compared with control results. Interestingly, the risks were higher in the lower end of PM range, which suggested that the link of the affective disorders as seen in the carriers with minor FMRP depletion is unlikely.

Significantly higher levels of obsessive-compulsive disorder were reported in the later study of 122 female and 26 male carriers of PM alleles.[20] Increase in depression was reported in 34 PM carrier mothers of FXS children who were compared, using SCL-90 and Beck Depression Inventory, with 39 noncarrier mothers of mentally retarded child and 39 mothers from the general (Spanish) population.[146] Notably, the results of a survey of 199 female and 57 male PM carriers from North Carolina enrolling in a national survey of families of children with FXS showed that the range of problems such as anxiety, depression or learning difficulties in females, and attention problems, aggression and anxiety in males, tend to cluster in the same individuals.[148] These confounding effects may be additional reasons for difficulties in the phenotypic risk estimates for the carriers, and thus for conflicting results from different studies, which may have already been affected by small sample sizes, inadequate psychometric measures, and biases related to preselection of the tested or control samples. In the large study by Hunter and colleagues in 2008,[149] the effect of small CGG expansions on mood and anxiety was investigated using regression models (adjusting for potential confounders) in a sample of 119 male and 446 female PM carriers aged 18-50 ascertained from either fragile X families or the general population. The results have shown that repeat length was (marginally) associated with depression and negative affect in males, and with negative affect in females, which suggested that, although PM carriers may be at risk of emotional morbidity, the effect of repeat size appears subtle. However, considering the later finding by Roberts et al[19] that the lower end of the range of PM alleles may have the largest effect on the risk of mood disorder, the linear model[149] analysis may not have been the most appropriate one for probing this kind of a relationship.

TOXICITY OF THE ELEVATED *FMR1* TRANSCRIPT CONTRIBUTES TO ABNORMAL PHENOTYPES

Search for possible pathophysiological mechanisms involved in abnormal phenotypes or specific disorders related to PM alleles resulted in a discovery of an elevated level of *FMR1* transcript (mRNA) in peripheral blood lymphocytes of PM carriers,[11] which was up to 10-fold higher than normal in carriers of the higher end (>100) alleles, and 2-4 times higher than normal in carriers of the lower end (55-100) alleles. The original finding was confirmed and expanded in several later studies.[150-155] This elevation results from increased transcriptional activity rather than an increased mRNA stability,[11,151,156] and direct evidence for an increased transcription rate has later been obtained using a nuclear run-on experiment.[157] The elevated *FMR1* transcript levels were found to be associated with the size of CGG repeat expansion within the PM-size range.[11,150,152,153,155] The data obtained from 63 male and 61 female carriers (with the effect of inactivation accounted for in the females) showed that this relationship was linear and consistent in both sexes.[150] However, in the more recent study of 90 PM females using piecewise regression model, the relationship between mRNA levels and CGG repeat was shown to be much stronger above the threshold of 90-100 repeats after correction for the activation ratio (AR); whereas without the correction, a significant regression was only limited to the lower portion (<100) of PM range.[158] These findings have also provided evidence for skewed inactivation in females carrying PM alleles with >100 repeats, and these data emphasize the need for applying the relevant correction in correlation analyses involving molecular and phenotypic changes. A recently proposed statistical method of covariate adjusted correlation analysis between CGG repeat number and *FMR1* mRNA, accommodating both additive and multiplicative effects of activation ratio nonparametrically,[159] will allow the more accurate evaluation of the effect of this ratio on molecular measures in females.

FRAGILE X-ASSOCIATED TREMOR/ATAXIA SYNDROME (FXTAS)

Clinical Manifestations and Prevalence in Males

The fragile X-associated tremor/ataxia syndrome (FXTAS), a recently identified form of late-onset (>50 years) neurodegenerative disorder, was first described in a sample of five older PM carrier males who were the grandfathers of children with FXS.[7] The major clinical features of this progressive disorder comprise intention or postural/action tremor, cerebellar gait and limb ataxia, and parkinsonism.[7-9,160-164] Although severe rigidity typical of Parkinson's disease is uncommon, mild rigidity in the upper limbs was reported in 71% of FXTAS patients, 40% out of this group manifested mild resting tremor, and 57% showed bradykinesia.[165] The three scales recognized as the measures of tremor (Clinical Rating Scale for Tremor, CRST), ataxia (International Cerebellar Ataxia Scale, ICARS), and parkinsonism (Unified Parkinson's Disease Rating Scale, UPDRS) were used in several different studies of male PM carriers, and these scores were found significantly higher compared with age and sex-matched controls.[162,163,165,166] Gait ataxia usually starts from balance problems manifesting as difficulty with tandem walking and progresses till the person is unable to walk without support or is wheel-chair bound. The tremor most often involves upper extremities, with postural/action tremor initially, and resting tremor at the later stages. Autonomic dysfunction and neuropathy have also been reported,[7,160,165,167] and the latter has been confirmed by the results of a tibial nerve conduction study showing significantly slower velocity and prolonged F-wave latency in typical FXTAS patients compared with controls and unaffected PM carriers.[168] The analysis of the progression of the major motor signs conducted in 55 patients showed that tremor was usually the first sign of the disorder occurring at the median age of 60 years; median delay of onset of ataxia was 2 years; onset of falls, 6 years; dependence on walking aid, 15 years; and death, 21 years.[163] Approximately 60% of male PM carriers manifesting tremor/ataxia also show increased T2 signal intensity in white matter of the MCP on MR images.[169] Other MR imaging findings include widespread cerebral, cerebellar and brainstem atrophy, as well as patchy or confluent areas of hyperintensity on T2-weighted images in periventricular and deep white matter of the cerebral hemispheres and corpus callosum.[8,165,166,170] Patchy or confluent T2 hyperintensities were recently reported in the basis pontis of a female affected with FXTAS,[171] and our own observations in the affected male carriers suggest that these changes may have been underreported.[172] The pattern of white matter pathology as seen on MR images and histological preparations in FXTAS is distinct from that associated with hypertensive vascular disease,[173] with the changes being in the form of various degrees of spongiosis and loss of axons and myelin in the areas shown as having T2 hyperintensities. An example of the typical white matter and other changes as recorded on MR images of a PM carrier affected with FXTAS is given in Figure 2.

The frequencies of these characteristic abnormalities are significantly higher in the affected carriers than in age and sex matched controls.[169,170,174] Confirmatory evidence for the relevance of these changes to the effect of PM alleles have been provided by the MRI volumetric studies, showing significant volume loss of the cerebellum, brainstem, thalamus, hippocampus and cerebral cortex in the affected carriers compared with age-matched normal controls.[133,169,173,174]

The characteristic neurohistological changes of FXTAS are ubiquitin-positive intranuclear inclusions abundant in both neuronal and astrocytic nuclei throughout the

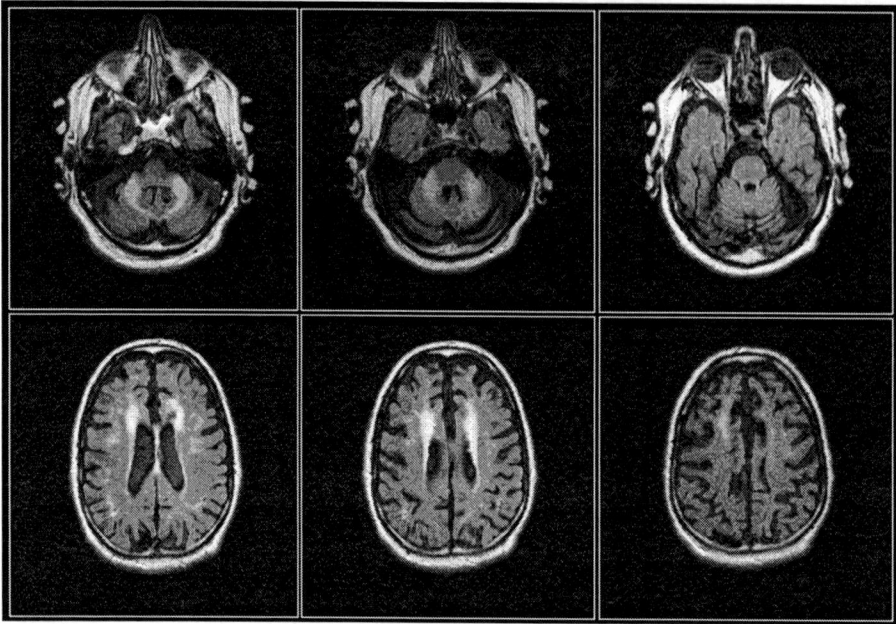

Figure 2. Axial FLAIR images of the brain of the 64 year old male carrier of 80 CGG repeats with typical manifestations of FXTAS comprising marked cerebellar ataxia, intention tremor, autonomic failure (originally diagnosed with multiple system atrophy, MSA). Marked bilateral T2 hyperintense signals within the middle cerebellar peduncles (MCP sign), combined with symmetrical T2 hyperintense signals within the corona radiate and considerable cerebral and cerebellar atrophy are evident from the images.[166] Reproduced from Loesch DZ et al. Clinical Genetics 2005; 67(5):412-417.

cerebrum, cerebellum (apart from Purkinje cells) and brainstem, being most numerous in the hippocampal formation (~50%, contrasting with only 5%-10% in cerebral cortex); the inclusions were also found in autonomic neurons and astrocytes of the spinal cord.[176] Although they contain about 20 different proteins, they are negative for PAS, tau, silver, polyglutamine and alpha synuclein,[173,176] which distinguishes them from the intranuclear inclusions of the CAG repeat (polyglutamine) disorders, and obviously from cytoplasmic inclusions associated with multiple system atrophy or Lewy body dementia (Fig. 3). The same inclusions were also found in neuronal nuclei of a KI mice carrying expanded (~100 CGG) repeat in the mouse gene.[177] Although the inclusions were not originally seen in the astrocytes in the KI mouse they have subsequently been documented in astrocytes in addition to neurons as in PM carriers.[178]

Cognitive Deficits and Psychiatric Symptoms

The earlier studies of cognitive functioning of men with FXTAS showed marked impairment of executive cognitive abilities, with only slightly lowered performance on standard cognitive IQ scores.[179,180] Moreover, significantly higher scores on the Neuropsychiatric Inventory (NPI) scales (total, aggression, depression, apathy and irritability) in the smaller sample of FXTAS males compared with normal controls were reported and interpreted as indicative of fronto-subcortical dementia.[181] In another

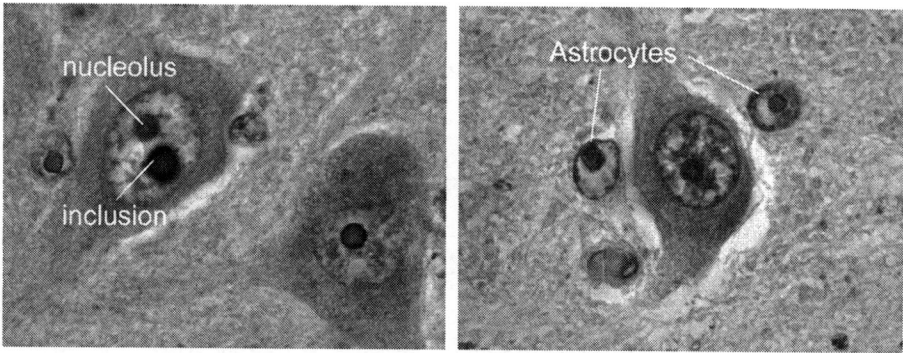

Figure 3. Eosinophilic inclusions in the nucleus of both neurons (left panel) and astrocytes (right panel) from a patient with FXTAS. Picture courtesy of Claudia Greco MD and Paul Hagerman MD, PhD.

study of 15 patients with FXTAS, twelve were diagnosed with dementia, and seven others with mood and anxiety disorders.[182] A larger study of 68 adults with FXTAS, including 50 males and 18 females, found dementia in approximately 50% of men that was similar in severity to Alzheimer's dementia.[183] However, none of the 18 females with FXTAS demonstrated dementia or significant cognitive impairment.[183] These data, considered in context with disease duration, indicated that deficits in executive cognitive functioning (ECF), working memory and speed, as well as capacity of information processing, combined with obvious personality changes, typically represent the early stages of FXTAS. Amongst the most affected aspects of executive functioning is the initiation of purposeful goal directed activity, and the inhibition of inappropriate or irrelevant behaviour. Standard cognitive measures using Wechsler scale usually reveals deficits of working memory represented by Arithmetic and Digit Span scores, whereas other verbal IQ subtests, especially language and verbal comprehension, are relatively unaffected. The Folstein Mini Mental State Examination (MMSE) usually remains within normal limits until the late stages of a disorder because it does not include the assessment of executive functioning. In the latest stages of the disorder, however, severity of dementia is fast progressing as the result of diffuse deficits in other cognitive functions. That these deficits are related to FXTAS has been shown by the results of a study based on 109 men divided into FXTAS+ carriers, FXTAS- carriers, and noncarriers. This data showed that, while FXTAS- carriers performed worse than controls on some aspects of ECF, FXTAS+ carriers performed worse than controls on working memory, recall of information, declarative learning and memory, information processing speed and temporal sequencing, as well as on ECF tests.[25] On the other hand, there is obvious overlap, especially in ECF impairments, between FXTAS and asymptomatic carriers, with the former showing more severe and widespread deficits. This overlap may be further illustrated by the latest study based on analysis of functional magnetic resonance imaging (fMRI) BOLD signals obtained during the performance of verbal working memory from 15 FXTAS+, 15 FXTAS– and age matched normal control individuals.[26] The results have shown that activation of prefrontal cortex, which may underlie executive and memory deficits, was reduced in both carrier samples compared to normal controls; in addition, FXTAS+ group showed reduced activation of the right ventral inferior frontal cortex.

A large scale comprehensive longitudinal study is still required to establish the sequence of different aspects of cognitive and neuropsychiatric changes in carriers.[9] This is because their relevance to FXTAS has not yet been clearly established, considering that, although neuropsychiatric manifestations often follow the establishment of motor disability, some, such as anxiety or depression, also occur in carriers not affected with FXTAS, or prior to the onset of neurological changes.

Prevalence and Atypical Manifestations

The initial clinical and genetic descriptions of male PM carriers preselected for tremor and ataxia was shortly followed by estimates of prevalence of this syndrome by screening of unselected population of PM carriers over 50 years of age ascertained through fragile X families with at least one child diagnosed with the FXS. The results of two parallel studies conducted in the US[165] and Australia[166] provided compelling evidence for the association between the PM carrier status and neurological involvement. The US study showed age-related increase in prevalence of tremor/ataxia, ranging from 17% to 75% between the age of 51 and 85, with the overall prevalence of 37% for this age range. A similar study based on Australian families generated an overall prevalence of 41.7% for the comparable age range. In both studies, an excess of manifestations of tremor/ataxia in the sample of PM carriers was significant compared to the sample of noncarriers matched for age. In a more recent study of 44 male PM carriers aged >50 years and ascertained through 151 fragile X Spanish families from Barcelona 45.5% of respondents surveyed over the phone reported tremor, balance problems, falls and/or gait ataxia.[22] However, the extensive variability of neurological manifestations needs to be considered while interpreting the prevalence results, since these estimates were biased towards considering syndromic manifestations rather than any less specific clinical changes indicating a significant neurological involvement. Shortly after the first description of FXTAS stringent diagnostic criteria were recommended, which included intention tremor or gait ataxia, and MCP sign ('definite' FXTAS), with less stringent criteria including problems of executive function deficits and brain atrophy incorporated into possible or probable FXTAS.[8] Thus, the combination of intention tremor and gait ataxia, and parkinsonism, was classified as 'probable' FXTAS and the combination of intention tremor or gait ataxia and cerebral white matter lesion and atrophy, as 'possible' FXTAS. Modifications of the FXTAS diagnostic criteria was made with addition of the FXTAS inclusions in the definite criteria.[12]

However, it appears that neurological involvement in PM carriers does not always meet those criteria, as illustrated by a number of published cases,[166,184-186] and those known through personal communications or our unpublished observations. Some carriers manifested typical MCP change with minimal or no clinical signs of tremor/ataxia, and others showed cerebral white matter lesions without the MCP sign, combined with some atypical neurological manifestations, psychiatric manifestations or autoimmune disease such as fibromyalgia and/or hypothyroidism[22,27] or multiple sclerosis.[187,188] Severe and fast progressing dementia was a major feature in a 62-year-old PM carrier, who at the later stages of his condition manifested typical FXTAS signs including gait ataxia and white matter degeneration including MCP sign.[189] A more recent study in France demonstrated a high rate of cognitive changes even before the onset of tremor or ataxia in carriers compared to controls.[141]

Another atypical form reported in a 58-year-old female PM carrier presented with dementia and parkinsonism, and typical ubiquitin positive intranuclear inclusions in neurons and astrocytes, but no MCP sign or ataxia.[190] The occurrence of low symptomatic or atypical manifestations determined by the same major etiological factor as a typically diagnosed case of FXTAS warrants the term 'FXTAS spectrum', which better reflects the fact that neurological, cognitive and psychiatric involvements associated with neurodegenerative changes in PM carriers extend beyond the classical definition of FXTAS. This new definition also covers the cases of co-occurrence of some of the clinical features of FXTAS with other neurodegenerative disorders, such as Parkinson's disease with Lewy bodies and Alzheimers disease, which will cause a more rapid decline in function than is seen in FXTAS without other neurodegenerative disorders. However, there are still clinical problems associated with the PM that may not be on the FXTAS spectrum, such as the autoimmune disease, but do seem to be related to the RNA toxicity and as the molecular mechanisms are clarified, a deeper understanding of these effects will be known.

The observed variability of manifestations is predictable considering complex mechanisms which might be involved in the origin of late onset neurological involvement, where the toxicity of the expanded *FMRI* transcript may interact with the number of other genetic or environmental effects throughout the lifespan (as detailed in Pathology section below), all contributing to the late-onset neurodegenerative changes, with FXTAS representing the most severe form of involvement. The occurrence of features of both FXTAS and Alzheimer's disease in a carrier of PM[189] is a good illustration of such combined effects towards more severe and complex manifestations; on the other hand, some other factors may be protective in nature thus determining less severe or vestigial manifestations, as well as the fact that at least 30% of individuals carrying PM alleles do not show any neurological involvement, even at advanced age. FXTAS appears to be clustered in families and a family without any neurological manifestations appears to have a lower incidence of involvement in subsequent members. Although the rates of FXTAS are low in daughters of men who have had FXTAS, other symptoms such as sleep disturbances, psychopathology and intermittent tremor are increased in these women as they age compared to women who do not have a father with FXTAS.[191]

Genotype-Phenotype Relationships

Confirmatory evidence for the link between the effect of *FMRI* small expansion alleles and specific neurodegerative processes has been provided by the results of genotype-phenotype correlations. The earliest data showed that *FMRI* CGG repeat length was significantly associated with the total brain volume loss in a sample of male PM carriers aged >50 years.[175] This association was also reported in a sample of male carriers affected with FXTAS.[169] A strong significant association was found between CGG repeat number and the percentage of neurons and astrocytes with intranuclear inclusions in the cortical grey matter and hippocampus;[173] a similar association was found between the number of repeats and the proportion of inclusion-bearing neurones in dentate gyrus and inferior colliculus in the expanded (CGG)n mice.[192] Some important relationships have also been reported between the number of CGG repeats and clinical outcomes. A negative correlation was found between CGG repeat number and the age of death[173] and age of onset of motor signs in FXTAS.[193] In a sample of male PM carriers who were not preselected for FXTAS diagnosis, CGG repeat size was found associated with the degree of motor impairment represented by all three motor rating scales,[194] and both neuropathic features[195] and abnormal nerve conduction.[168] Only a few of those relationships were also significant in female carriers.

Relationships between the phenotypic changes and elevated *FMR1* mRNA levels in PM carriers are less obvious than those with the size of CGG repeat expansion. This is not unpredictable since, while there is close correspondence between the size of CGG expansion between blood and brain cells,[196] there are blood-brain discrepancies between the levels of mRNA transcript.[154] Nevertheless, significant relationships between the elevated *FMR1* mRNA and psychological symptoms measured by SCL-90-R Global Severity Index was reported in 68 male and 144 female PM carriers both with and without a diagnosis of FXTAS,[20] with the relationships predominantly concerning obsessive-compulsive disorder and psychoticism. Moreover, in another study, a (negative) correlation was found between the levels of *FMR1* mRNA and the right ventral inferior frontal activity on fMRI during performance on working verbal memory test in 30 female PM carriers (15 with and 15 without FXTAS) compared with 12 matched healthy controls.[26] However, since the studies concerned with such relationships have been scarce, global interpretation of the findings so far is limited.

Since the discovery of FXTAS, there have been several studies from different locations assessing the prevalence of this syndrome in neurological disorders including essential tremor, cerebellar ataxia, Parkinson's disease, and multiple system atrophy (MSA).[197-210] Overall, the prevalence of PM alleles amongst groups of patients with late onset movement disorders has been lower than expected.[211] The diagnostic groups with the highest prevalence of PM carriers were males over 50 with cerebellar ataxia, with the overall prevalence of 17/872, 2%, ranging from 0%-7%[197-199,201,208] and individuals with multiple system atrophy-cerebellar type (MSA-C), with an overall prevalence of 2% with a range of 0% to 3.95%.[204,208] Surveys of male patients originally diagnosed with atypical,[200] or typical[203,205,212] Parkinson's disease (PD) have failed to show a statistically significant excess of PM carriers. Three carriers were found among 776 patients with idiopathic PD[213] and, although this was at least a three-fold greater rate than the normal population prevalence, it was not statistically significant. However, a significant excess of PM male carriers in the sample comprising typical and atypical idiopathic PD was reported in the latest screening study.[214]

Clinical Manifestations and Prevalence of FXTAS in Females

FXTAS has also been reported in female PM carriers.[22,206,215-220] However, it occurs at a much lower rate (~8% to 16.5%) than in male carriers, and clinical manifestations and cognitive decline tend to be milder and occur at a later age than in men; this is largely because of the protective role of the second X chromosome.[211,216-218] In another study[22] based on a sample of 85 Spanish females aged >50 years, FXTAS symptoms were encountered in 14 (16.5%) individuals using less stringent diagnostic criteria.

Brain changes as seen on MR images are also less severe in FXTAS females than in their male counterparts. Although a significant reduction in brain volumes and increased severity of white matter disease were found in a FXTAS group of 15 females compared with 20 age-matched controls,[169] these changes were less pronounced than in a FXTAS group of 36 males; the MCP sign occurred in only 13% of affected females compared with 58%, in males, and there was no significant association between the volumetric measures and either the severity of FXTAS manifestations, or CGG repeat expansion.

However, females presenting with typical symptoms of FXTAS had greater medical comorbidity than males.[27] In this comprehensive study of 146 PM female carriers aged 20-75 years and 69 age-matched controls, 18 carriers (12.3%), who met the criteria of FXTAS diagnosis, had significantly higher prevalence by history of thyroid disorders (50%), hypertension (61.1%), fibromyalgia (43.8%), diagnosed peripheral neuropathy (52.9%), and persistent muscular pain (76.5%) compared to controls. In non-FXTAS group of carriers, the frequencies of chronic muscle pain, paraesthesias in extremities and history of tremor, were significantly elevated compared with normal controls. The size of CGG repeat and the levels of mRNA were both significantly higher in the FXTAS than in non-FXTAS groups, but the ARs were surprisingly not different. The more recent similar study based on a larger sample of 280 female PM carriers from Spain (with 195 between 4-50 years of age and 85 older than 50 years) found lower cormorbidity than reported in the American sample with thyroid disease (15.9%), and muscle pain (24.4%).[22] However females, just as their male counterparts, may present with atypical clinical manifestations of FXTAS, such as in the case of a 58-year-old PM female recently described,[190] who developed progressive dementia and parkinsonian signs, but neither action-intention tremor nor ataxia or MCP sign. Magnetic resonance imaging, however, showed extensive subcortical *wmd,* consistent with the distribution of *wmd* on post-morten examination, which showed white matter spongiosis in this region, as well as widespread ubiquitin-positive intranuclear inclusions in both neurons and astrocytes.

To date there has been no systematic study of cognitive status of females affected with FXTAS. But the analysis of psychological symptoms obtained from the SCL-90-R showed a significantly elevated somatization, obsessive-compulsive symptoms, depression, and overall symptom severity in a sample of 22 women affected with FXTAS.[20]

Pathogenesis of FXTAS

The associations of FXTAS and of the other related neurological manifestations with CGG repeat expansion within the PM range, the elevation of the expanded *FMR1* mRNA, as well as the presence of this mRNA in the intranuclear inclusions, gave rise to the hypothesis of a toxic 'gain-of-function' effect of excessive *FMR1* mRNA on brain tissue.[7,151,173,176] This concept originated from pathogenesis of myotonic dystrophy,[221] where another (CUG) repeat expansion in the gene's promoter leads to sequestration of Muscleblind-like 1 (MBNL1) protein found in the intranuclear inclusions, leading to dysregulation and splicing of several other mRNAs and thus clinical involvement through deficit of the relevant proteins.[222,223] According to this model hypothesised for small CGG expansions in *FMR1* mRNA, the 'toxic' effect of elevated transcript leads to dysregulation of specific candidate proteins, such as purine-rich element binding protein (Pur α), or nuclear ribonucleoprotein A2/B1 (hnRNP A2/B1), which are sequestrated from their normal function and deposited in cells' nuclei causing their premature death. In support of this claim, the lack of Pur α protein in knock-out (KO) mice resulted in severe loss of neurones in the hippocampus and cerebellum, and development of tremor and seizures at two weeks of age,[224] whereas over-expression of this protein alleviated manifestations of neurodegeneration in a fly FXTAS model.[225] However, it has also been speculated that the elevated toxic RNA may not be directly involved in the sequestration of these proteins, but may act as a trigger of stress responses leading to over-expression of candidate proteins involved in neuroprotection.[13,178,225] Increased stress responses to toxic gain of function of *FMR1* mRNA with expanded CGG repeat have most recently been shown in cultured human fibroblasts and CNS tissue obtained from PM carriers.[226] These data provide evidence for a somewhat modified model of a toxicity of the expanded mRNA, in which increased expression of this transcript results in altered expression and disruption of the nuclear Lamin A/C architecture and induction of the expression of stress response genes. Since this type of cellular pathology may be induced within a very short time frame, the authors suggest that *FMR1* mRNA-induced abnormal skeletal organization in neural cells may take place in early development, thus underlying both neurodevelopmental and late-onset clinical conditions relevant to the PM carrier status. The risk and the type of manifestations of the latter may be determined by the presence and the nature of additional genetic effects varying between individuals, as well as some environmental factors such as pollutants.[227]

Identification of the contents of intranuclear inclusions has been an important step towards better understanding of the mechanisms leading to FXTAS.[151,228,229] Unlike in many other disorders associated with unstable repeat expansions, there was lack of a principal protein, and ubiquinated proteins represented only a minor component in the inclusions' deposits, which indicated that their formation is unlikely to result from accretion of specific abnormal proteins. The minor proteins deposited included lamin A/C, two RNA directly binding proteins: Pur α, hnRNP A2/B1 and MBL1, along with the *FMR1* mRNA. Recent evidence of the sequestration of Sam 68, an important RNA splicing protein,[230] and the recent evidence of DROSHA sequestration in the inclusions suggests that dysregulation of miRNAs is involved in the toxicity of the premutation.[231]

Although precise mechanisms leading from the elevated RNA to neurodegeneration still remain speculative, sufficient evidence for direct RNA toxicity has been provided by the results of animal and cell based studies (reviewed in: Garcia-Arocena and Hagerman 2010[13]). The earliest findings in *Drosophila melanogaster* showed that human PM-size CGG repeat transcripts ectopically expressed in neural eye tissue caused neurodegeneration in a dosage- and repeat length-dependent manner.[232] Furthermore, the ubiquitin-positive intranuclear neuronal inclusions and neurodegeneration similar to that in human PM carriers affected with FXTAS were observed in the FMR1 KI mice;[177] these mice also manifested cognitive decline, neuromotor, and behavioural disturbances associated with the elevated mRNA.[233] Further evidence for the RNA with the expanded CGG repeats to cause neurodegeneration in FXTAS was provided by the findings showing that this RNA expressed in Purkinje neurons of experimental mice outside the context of *FMR1* mRNA resulted in neuronal pathology presenting as intranuclear inclusions, neuronal death and behavioural changes.[234] In the most recent study of PM KI mice the presence of ubiquitin-positive intranuclear inclusions were found in neurons throughout the brain in cortical and subcortical regions increasing in number and size with advanced age;[127] importantly, these inclusions were also present in protoplasmic astrocytes including Bergman glia in the cerebellum, which might have some relevance to a particularly severe white matter neurodegeneration in cerebellar peduncles.

The typical inclusions have also been found in the anterior and posterior pituitary gland in humans as well as in KI PM mice and throughout the limbic system including the amygdala[235-237] which imply the possibility of a dysfunction of hypothalamus-pituitary-adrenal gland axis, and may explain some of neuroendocrine and psychiatric problems seen in carriers with and without obvious FXTAS. The inclusions have also been located in the Leydig and mucoid cells in the testicles obtained from human PM carriers which may explain the low testosterone levels and frequent onset of impotence before the tremor and ataxia of FXTAS.[235]

Another important step towards understanding of pathophysiological mechanisms of broader neuropathology linked to the RNA with expanded CGG repeats has been made in the latest study based on female KI mice heterozygous for the PM allele, where cultured neurones manifested dendritic changes and lower cell viability.[238] This finding, combined with the recent report of abnormal cellular distribution of lamin A/C isoforms in embryonic fibroblasts of KI mice with the expanded CGG repeat in the murine *FMR1* gene[226] point to both neurodegenerative and neurodevelopmental effects of the expanded-repeat mRNA; notably, in the same study, altered lamin A/C expression/organization and increased stress response were also found in cultured skin fibroblasts and CNS tissue from male PM carriers either with or without FXTAS. These data provide strong evidence for biological processes underlying the postulated link between early onset or mid-life pathologies associated with elevated levels of *FMR1* transcript, such as behavioural and cognitive deficits, psychiatric symptoms or FXPOI (described later in this chapter), on the one hand, and late-onset neurodegenerative disorders, on the other.

Additional difficulty in understanding the individual steps leading from the toxic RNA to changes at the cellular and clinical level may be related to the likely involvement of other factors and processes interacting with, and modifying the effects of the expanded toxic RNA. After all, only certain proportions of PM carriers are affected with either neurodevelopmental or neurodegenerative conditions, or present the symptoms of both of them. Clearly, individual genetic make-up, including background genes of minor effects, contributes to the variability in penetrance at the individual as well as the family level. The role of environmental exposure, especially to neurotoxic agents, postulated in context with the severity of expression of FXTAS,[166] and recently demonstrated by several clinical examples[227] should not be underestimated. This aspect is especially relevant considering the latest evidence suggesting the involvement of mitochondrial dysfunction in pathogenesis of neurological involvement in PM carriers based on fibroblast and brain samples,[239] and on blood leucocytes from PM and GZ allele carriers.[240] Finally, the contribution of the recently identified antisense *FMR1* mRNA, which is also elevated in PM carriers[241,242] to neurotoxicity, needs to be further investigated. Our preliminary data[240] have shown that the elevated levels of this transcript may play a pivotal role in the severity of neurological involvement in the carriers of small expansion *FMR1* alleles.

FRAGILE X-ASSOCIATED PRIMARY OVARIAN INSUFFICIENCY

Another distinct disorder linked to the *FMR1* PM alleles is Primary Ovarian Insufficiency (POI). This term was proposed[243] to reflect, apart from the cessation of menstrual periods before age 40, a spectrum of diminished ovarian functions occurring in female carriers of the *FMR1* PM alleles. The more specific and currently used term of 'Fragile X-associated Primary Ovarian Insufficiency' (FXPOI) was introduced to describe any degree of POI attributable to the *FMR1* aetiology.[244] FXPOI affects approximately 20% of PM carriers, consistently in many population/ethnic groups,[245-249] representing 20-fold increase compared with the risk of 1% seen in the general population. Estimated frequencies of PM carriers among females reporting to infertility clinics with familial or sporadic FXPOI were: 11.5% with 95% CI: 5.4%-20.8%, for familiar sample; and 3.2% with 95% CI: 1.4%-6.2%, in sporadic sample, compared with the population frequency of these carriers of ~0.4%, as reviewed by De Caro et al.[244]

Notably, data obtained[4] from over 500 women with a wide range of CGG repeats from the normal range into the high end of the PM range demonstrated a significant association between CGG repeat number and prevalence of FXPOI that was nonlinear. For those with a repeat <40 the prevalence of POI was 0.9%, for those with 41-58 repeats it increased to 2.2%, for those with 59-79 repeats it was 5.9%, for those with 80-99 repeats it reached 18.6%, but for those with $\geq$100 repeats the prevalence decreased to 12.5%. These authors suggested that perhaps those with a high CGG repeat number may have some ovarian target cells with a full mutation, or some cells have early methylation at a lower CGG repeat number that protects them from the toxicity of the PM. An overall effect of the CGG repeat size on age of menopause has been reported,[4] with low end CGG repeats demonstrating menopause 2.5 years earlier than controls, and medium to high end PM carriers demonstrating menopause 4 years earlier than low- end carriers. Nonlinear association between the CGG repeat number and the risk of FXPOI and age of menopause was later confirmed in different samples.[250-252] In the 2007 study based on data from 948 women with the wide range of repeats sizes, the mean age of menopause for carriers of 59-79 repeats and >100 repeats was similar (48.5 and 47.5, respectively), but was ~3 years lower (44.9) in carriers with 80-100 repeats. Although the highest decrease in menopausal age was in carriers of CGG expansions with the 80-100 range, it was reduced across the total PM range compared with 52.3 + 0.5 years found in the sample of noncarriers.

Although the *FMR1* PM represents a common major gene affecting the age of menopause in particular, and the female reproductive system in general, other factors modifying this effect should be considered. Smoking has been shown to aggravate this major effect in carriers of PM by 1 year. The identity of other major genes possibly interacting with the effect of *FMR1* has not yet been established. However, the effect of the background genes (residual genetic variance) in the presence of the major effect of the *FMR1* PM, was recently estimated in a study based on 230 fragile X families and 219 families from the general population.[253] The analysis used a random effect version of the Cox proportional hazards model allowing for shared polygenic (additive) effects and the effects of confounders.[254] The results confirmed significant major effect of *FMR1* PM, with the greatest effect of mid-size group, and provided evidence of a substantial additive (background) genetic component even after adjusting for the major effect; estimated additive genetic variance ranged from 0.55 to 0.96 (depending on different model assumptions), with P-values ranging from 0.0002 to 0.0027. Examples from single families further illustrate a significance of shared genetic effect on the risk of menopausal anomalies. In one family[255] premature menopause occurred in all six sisters carrying PM; in another family[256] two sisters, one PM carrier and one noncarrier, presented with POI. The reported examples of menopausal problems co-existing with the other *FMR1* PM- associated conditions such as FXTAS or psychiatric involvement in one and the same carrier,[220] or between two generations of carriers[191] are also relevant to a significance of shared family effect, which should be considered in individual risk estimates and counselling.

The features reflecting pathology of the reproductive system in PM females extend beyond, though are not irrelevant to, the earlier menopausal age. The earliest and most comprehensive endocrine study of the hormonal profile of female PM carriers demonstrated that those carriers that were cycling normally had endocrine dysfunction.[243] These authors studied 11 normally ovulating PM carriers (ages 23 to 41 years) and demonstrated a significantly shortened cycle, elevated follicle stimulating hormone (FSH) throughout the cycle (91% with elevations >2 SD above the mean), elevated Inhibin -B in the follicular phase, and elevated Inhibin- A and progesterone in the luteal phase compared to controls. These findings suggested a decreased number of follicles and granulosa cell dysfunction or decreased cell number in the corpus luteum compared to controls. In addition, 45% (5 of 11) of these carriers had a history of infertility as defined by 1 year of unprotected intercourse without a pregnancy. This study demonstrated sub-clinical ovarian dysfunction in PM females who do not have POI. For women at the premenopausal stage there was a CGG repeat, and AR effect on the FSH level, but only when controls and carriers were included together. Consistent finding of an increase in FSH serum levels in PM carrier who were still cycling was reported in the other studies.[257,258] This effect remained after adjusting for smoking and use of contraceptives[257] or age and shared family effects.[258] However, significantly increased FSH levels were limited only to women aged 30-39, which suggested late onset of ovarian dysfunction.[4] But the results of the later study showed that this dysfunction may occur at an earlier age. Using another biomarker of ovarian reserve, mullerian-inhibiting substance (AMH), which, unlike FSH, is only expressed in growing follicles, a significant reduction of the AMH levels was shown for all age groups within the 18-50 years range in the higher (>70 repeats) versus lower (<70 repeats) risk PM carriers.[244] Some other abnormalities within the FXPOI spectrum include irregular menses,[245,252] shorter cycles,[243,252] and irregular length and skipped cycles in the mid-size repeat PM carriers;[252] in the same group of carriers these authors also found increased twinning. Moreover, the findings of significant decrease of mineral bone density[259] in PM carriers compared with normal controls, and a significant increase in osteoporosis, with the highest increase of 11.9% in the mid-range PM carriers based on large samples,[260] are of particular concern.

The mechanisms involved in ovarian insufficiency in small expansion carriers are still unknown. The fundamental question to be answered is what stage of individual growth/development is most relevant to the damaging effect of the processes related to these expansions. Do these events lead to a smaller ovarian endowment already present at birth? Do they lead to alteration of recruitment of follicles later in life? Or do they lead to follicular atresia and thus premature degeneration? It is also unknown if these processes directly influence the ovarian tissue, or act via an altered hypothalamo-pituitary-gonadal axis, especially considering that the typical intranuclear inclusion formation has been found in neurons of the anterior and posterior pituitary gland in a male PM carrier.[235] Although these changes have not yet been investigated in female carriers, it has been postulated that these degenerative changes in gonadotropic cells lead to dysfunction of hypothalamic-pituitary-gonadal axis and thus to abnormal follicular recruitment.[244] Alternatively, FXPOI may result from direct damage to the ovaries either by increased FMRP levels during specific stages of development, or by the 'gain of function' toxic effect of elevated *FMR1* mRNA leading to premature ovarian degeneration later in life.[244] The recent important finding of an overproduction of FMRP among carriers of 80-89 repeat alleles[261] may give some support to the hypothesis of this protein's related damage to ovaries, since most clinical manifestations of FXPOI have been most prominent in carriers of medium size repeat PM alleles.

The lack of convincing evidence for any anomalies in the age of menarche or around menarcheal age in PM carriers[259,262] suggest that the changes related to the effect of this PM occur at later stages of reproductive life. However, one study[252] found a small but significant difference in the age of menarche only in a subgroup of low repeat PM carriers (12.18 ± 1.38 in the carriers compared with 12.44 ± 1.51 in noncarriers), although overall the age of menarche was not different from controls. Considering scarcity and inconsistency of the data concerning adolescent period in female carriers, more studies are required to identify the onset and the progress of FXPOI, which might be related to the specific effect of *FMR1* PM. Evidence that it may extend over to the adolescence period has been provided by data showing that the onset of menopause was below the age 30 in 30% of PM carriers,[247,263] and in rare cases, it was at the age of 18,[256,264] and the lowest reported ages were between 13 and 17 years in three carriers.[263]

DO THE INTERMEDIATE SIZE (GZ) ALLELES CAUSE ABNORMAL PHENOTYPES?

Understanding of phenotypic effects of GZ alleles is important considering their bridging position between the common and fragile X (PM and full mutation) allele categories, as indicated by haplotype associations (discussed above). Most importantly, given their high population frequency of ~3-4 per 100 males,[83,84,97,112,265] even a small effect contributing to the developmental or late-onset conditions, if substantiated, is likely to be borne by a considerable number of individuals.

The GZ alleles, with the originally recommended range of 45-54 CGGs,[90] but varying between different studies (reviewed by Mitchell et al[84]) have, until recently, been included in the normal category with regard to phenotypic expression. However, the breakthrough has come with a demonstration of the elevated *FMR1* transcript in a sample of 32 Australian GZ male carriers identified in the SEN (special educational needs) population.[266]

This study also showed that, if the regression is applied across a wider range of *FMR1* alleles, including both GZ and PM allele categories, the sharp increase in mRNA levels that is proportional to the number of CGG repeats seen in the GZ range, slows down in the PM range (Fig. 4). The onset of this disturbance indicated by this study's data is at approximately 39 repeats; however more observations, especially at the upper end of the 'normal' alleles, are required to estimate this threshold more precisely. In the earliest study of small CGG expansions that included only two GZ carriers, increased transcription and a slight decrease in translation of the gene was also reported.[155]

Findings from another study, based on human cell lines transfected with the *FMR1* 5'-UTR containing CGG repeat lengths ranging in size from 0 to 99 repeats and a downstream reporter gene, confirmed a transcription defect in GZ class of alleles.[156] Although a different reporter gene (luciferase and not *FMR1*) gene was used, the results demonstrated an increase in transcription levels for constructs possessing either PM or GZ alleles, further supporting the later findings based on human subjects.[266] If, indeed, the elevated levels of *FMR1* mRNA cause neurodevelopmental and neurodegenerative changes in PM carriers, one can postulate that the elevated levels observed in carriers of GZ alleles, even if more moderate, could cause conditions reminiscent of those seen in a significant proportion of PM carriers.

However, in contrast with the PM alleles, establishing the pathogenic role of the GZ allele is not straightforward. *First,* these alleles have been relatively poorly defined with respect to repeat size, and the range of this category has differed between different studies (reviewed in refs. 83,84,97,112,267). *Second,* measurement of CGG repeat number is not consistently performed across all laboratories, and comparisons between

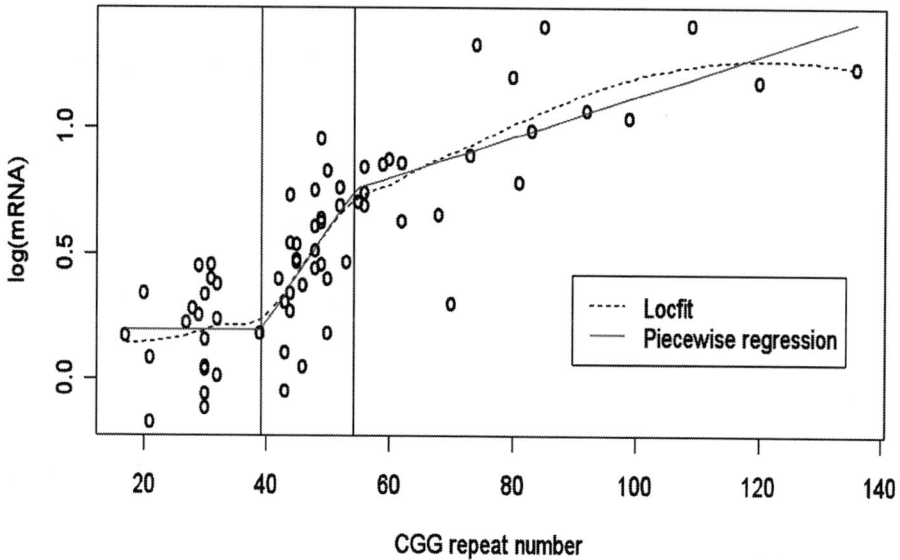

Figure 4. Nonparametric local fit (Locfit) and piecewise regression fit to the log-transformed mRNA levels versus CGG repeat numbers for data from all three allelic categories: Normal, GZ and PM.[266] Reproduced from Loesch DZ et al. J Med Genet 2007; 44(3):198-204.[266]

studies may not be valid. *Third*, unlike for PM alleles, GZ alleles cannot, except for rare cases, be identified through cascade testing of fragile X families, so that the prevalence and the type of phenotypic involvements cannot be determined in an unbiased manner. On the other hand, the results obtained from screening population groups with specific disorder or developmental delay would be biased and thus inconclusive. Therefore, the most stringent approach would be to test an entirely unbiased population, such as a series of consecutive newborns initially screened for *FMR1* CGG size, and their phenotypes subsequently assessed. Such an undertaking, while ideal, would be logistically daunting. The next best is to study a well defined population of affected individuals, and to compare the CGG_n profile with that of an appropriate control group.

The earliest systematic study based on screening of a well defined population preselected for significant learning problems (SEN) suggested phenotypic (neurodevelopmental) effects of GZ alleles.[267] The results provided evidence for an increased frequency of these alleles in this tested population compared with that in control X chromosomes. However, the studies have not been consistent, with some confirming this association,[82,89] and others failing to do so.[96,268-270] One study[82] found that 4.4% of 3,732 UK SEN males had GZ alleles, which represented a significant increase compared with non-SEN samples. Our own study conducted in the population of 1253 SEN male students from Tasmanian schools also found increased frequency of GZ alleles (3.45%) compared with 2.43% in 578 consecutive male births in the same state.[84] Although the frequency figures are similar to those obtained in the earlier studies, the SEN-control difference was insignificant. The possibility of an increased risk of learning or behavioural problems in carriers of GZ alleles has also been suggested by the fact these carriers are sometimes identified among children with learning or behavioural deficits attending genetic or psychiatric clinics. A summary of physical and neuropsychological test results from a sample of such children, combined with the carriers identified through comprehensive testing of fragile X families (the total of 10 boys aged 4-15 years, including 6 PM and 4 GZ carriers), was presented.[15] A majority of participants manifested psychiatric (autism spectrum and/or hyperactivity), psychological (expressive language and/or other minor cognitive impairments), and several physical characteristics (all normally associated with the full mutation). However, no relationship was found between clinical affectedness and either CGG repeat size or FMRP levels. Although descriptions of this kind are informative, such data are not amenable to statistical comparison with normal samples, because of ascertainment bias caused by preselection of participants for the traits examined in the study. Nevertheless, the results of a screening of a small but well defined sample of children diagnosed with autism spectrum disorders (ASDs) suggested a significant role of GZ alleles in the risk for this condition, and also suggested possible mechanisms involved in their effect.[271] It is clear from these results that the question whether or not GZ alleles are associated with neurodevelopmental problems is still unresolved and requires further population-based study, proper controls and more sensitive and specialized assessment tools.

The stronger evidence for phenotypic effects of GZ alleles has been provided by the data showing a significant association of these alleles with FXPOI spectrum of abnormalities.[272,273] Grouping the alleles into sizes of 35-54 and 55+ repeats, of the women having suffered POI, 14.2% (15 of 106) had an allele in the range of 35-54, whereas in the general population the figure was 6.5% (21 of 322).[272] The increase in the risks is significant, and the data speak for an effect of the CGG repeat length, whether GZ or PM, upon ovarian function.[249] In the later study of small samples (27 POI and 32 control women), six POI (22%) women and one control woman (3%) carrying *FMR1* alleles of >40 repeats, were identified, and this represented a significant increase in the POI sample.[274] An interesting new perspective on this issue has been provided by the latest finding, though based on a small samples of 23 tested and 11 control individuals that the fall in ovarian reserve represented by the FSH and AMH markers is already apparent in carriers of repeats >30.[275] In another study, the rise in serum concentration of FSH was associated with CGG repeat size within the normal range but exceeding 30 CGGs, both in 80 females with POI and 70 controls.[276]

Since FXTAS has been another late-onset condition found associated with a toxicity of elevated RNA in PM carriers, a similar, though less apparent, association may be expected for GZ carriers with elevated expanded *FMR1* mRNA. In rare studies, the small samples of patients with Parkinson's disease spectrum were screened for the GZ alleles and, although there was a trend towards the increased prevalence (reaching 7%), no significant excess of these alleles relative to their population prevalence has been reported.[205,212,213] However, a significant relationship between the size of CGG repeat within the GZ range and cognitive decline and hallucinations was reported in one of those studies.[213] Among 507 patients (253 male, 254 female) with MSA, 3.6% of males and 3.9% of females had a GZ of size 40-54, comparing with 3.2% and 4.2%, respectively, in controls;[204] in a smaller French series of 77 patients, five had a GZ of size 41-54, for a fraction of 6.5%, this being somewhat higher, although not significantly so, than the control figure of 3.4% (4/117).[208]

In the latest study,[214] 228 Australian males affected with parkinsonism were screened for GZ, as well as PM alleles. Importantly, this sample included broader diagnostic group of the primary parkinsonian syndromes according to the classification in Schapira et al.[277] The frequencies of either of these alleles were compared with the frequencies in a population-based sample of 578 Guthrie spots from consecutive Tasmanian male newborns (controls). Apart from a significant excess of PM, there was also a more than 2-fold increase in GZ carriers, with odds ratio (OR) = 2.36, and 95% confidence intervals (CI): 1.20-4.63, compared with controls.

These results have suggested that both PM and GZ alleles may contribute to the risk of parkinsonism, and that the 'toxic' effect of expanded mRNA linked to the PM alleles may also occur when the expansions are smaller. It is possible that the modest elevation of 'toxic' RNA, such as is the case in the GZ and lower-end PM carriers, may present an additional risk for neurodegeneration by synergizing with other susceptibility genes for parkinsonism. This notion is supported by common occurrence of parkinsonian manifestations amongst patients primarily diagnosed as FXTAS (as detailed in previous section).

Detailed clinical and genetic testing was performed in 14 carriers of the small expansion *FMR1* alleles encompassing the GZ and the lower-end PM range.[240] This was in order to determine whether these alleles present a significant modifying factor for the manifestations of parkinsonian disorders. The results were compared with the same data obtained from 24 noncarriers. Both GZ and PM carriers presented with more severe symptoms than a sample of noncarriers matched for age, diagnosis, and treatment. The Parkinson's (UPDRS) motor score and the measures of cognitive decline (MMSE and/or ACE-R scores) were significantly correlated with the size of the CGG repeat, and the (elevated) levels of antisense *FMR1* and Cytochrome C1 mRNAs in blood leucocytes. In addition, the carriers showed a significant depletion of NADH dehydrogenase (ND) subunit 1 (ND1) gene mitochondrial DNA in whole blood. These data have shown that both small CGG expansion *FMR1* alleles may play a significant role in the acquisition of the parkinsonian phenotype, possibly through the cytotoxic effect of elevated sense and/or antisense *FMR1* transcripts involving mitochondrial dysfunction and leading to progressive neurodegeneration. Notably, evidence for mitochondrial involvement was also presented in another study based on fibroblasts and brain samples from PM carriers affected with FXTAS.[239]

In conclusion, although the results on phenotypic involvements of GZ alleles are still sketchy, there is sufficient evidence to postulate their role in the aetiology of certain human conditions affecting the brain and reproductive system. Since the expansion size and thus the elevation of the *FMR1* transcripts is less than in the PM carriers, the effect may be more subtle and more difficult to detect, especially as the carriers cannot be, as is the case with PM carriers, ascertained through fragile X families. Moreover, if the effect of *FMR1* mRNA toxicity in GZ carriers is not sufficient to be the sole determinant of a disease such as FXTAS, it may increase their risk of other disorders that are likely to involve mitochondrial dysfunction, such as Parkinson's disease or multiple sclerosis. This hypothesis requires testing via further comprehensive studies based on large samples.

Apart from the possible effect of small CGG expansions/elevation of *FMR1* mRNA on the phenotype, the results of the latest studies[239,240,278] have thrown new light on the ambiguity currently associated with the definition of GZ alleles. We determined a threshold of ~39 CGGs for the onset of a progressive increase in *FMR1* transcription.[266] Although more data are required to narrow the error margins, it appears that the lower bound for the GZ currently defined as 45CGGs[90] should be revisited in the light of these findings. Moreover, these data also suggested a continuous scale of involvement, especially at the lower end of PM range, rather than the clear-cut separation between

GZ and PM categories recommended[90] solely on the basis of the potential for CGG expansion. It appears both from the elevation of the *FMR1* transcript in GZ alleles, and the observed effects of these alleles on the phenotype that the difference between these two categories is quantitative in nature rather than qualitative. The phenotypic effects reported so far in GZ carriers, including POI, late onset neurological involvement, autism, and possibly other behavioural problems and learning problems are similar to these seen in PM carriers. The elevation of the *FMR1* transcript, as well as the linear relationship between the levels of mRNA and the number of CGGs, are other common features of both the GZ and PM *FMR1* alleles and thus the mechanism of this elevation must also be similar.

PRESENT STATUS OF TREATMENT OF FRAGILE X-RELATED DISORDERS

Over the years a variety of pharmacological interventions has been developed for treatment of FXS related to the full mutation[68] and also for PM associated disorders including FXTAS.[10,279] Typically treatment of both FXS and FXTAS involves multiple professionals and multimodality intervention that have been described at length elsewhere.[10,280]

For the psychopharmacology interventions in those with FXS both stimulants and alpha agonists, such as guanfacine and clonidine, can be helpful for treatment of ADHD.[68] However, the stimulants are usually not started until after 5 years of age because of frequent side effects in those under 5 years. Selective serotonin reuptake inhibitors (SSRIs) are commonly used to treat anxiety and hyperarousal and early treatment may help with language and socialization benefits.[68] In addition, the atypical antipsychotic aripiprasole (Abilify) has been frequently used with generally good success for anxiety and aggression.[281] Aripiprazole will also stabilize mood in addition to treating ADHD symptoms, but often a lower dose is better tolerated without activation.[68]

The most exciting development in the treatment of FXS is the introduction of targeted treatments that can reverse some of the neurobiological abnormalities that have been identified in the animal models of FXS. The up-regulation of the metabotropic glutamate receptor 5 system (mGluR5) leading to long term depression (LTD) described by Huber et al[282] and Bear et al[283] has lead to the use of mGluR5 antagonists that have been shown to be helpful in the animal models of FXS[40,284,285] and are now demonstrating benefit in individuals with FXS.[286] At least three mGluR5 antagonists have been developed and 2 are undergoing assessments in adults with FXS in phase II trials with plans for broad phase III trials in the near future. Another way of decreasing the level of glutamate at the synapse is with the use of a $GABA_B$ agonist, specifically Arbaclofen, the right isomer of baclofen. Studies in mice and a phase II trial in children and adults aged 6 to 40 years suggest benefit in those with FXS plus autism and FXS with low sociability.[287] These promising results in FXS and preliminary results in those with autism without FXS have lead to phase III trials in both disorders.

The use of minocycline for 1 month after birth in the KO mouse led to benefits in behaviour and cognition in addition to improvements in the immature dendritic structure that is typical for FXS.[288] Minocycline appears to work by lowering matrix metalloproteinase 9 (MMP9) levels in the KO mouse which are elevated compared to wild type mice. The results of the Bilousova trial stimulated use of minocycline by families of children with FXS and a subsequent survey of parents regarding the benefits and side effects of minocycline suggested that about 70% of children improved on minocycline particularly in language and behaviour.[289] These preliminary results are being followed by a controlled trial of minocycline in children with FXS between the ages of 3.5 years and 16 years. Side effects of minocycline include graying of the tooth enamel when used under 8 years old although occasional occurrence of graying of skin and nails has been reported in older individuals. Minocycline can rarely cause pseudotumor cerebri, a lupus-like rash and seropositivity of the antinuclear antibody (ANA), so regular follow-up of these patients is necessary.

Lastly the use of nutritional supplements including antioxidants have demonstrated benefit in the treatment of the KO mouse with improvements in behaviour and testicular size and this is related to the findings of oxidative stress in the KO mouse.[290] Melatonin which can be helpful for the sleep disturbances in FXS[291] also seems to have an antioxidant effect in the mouse.[292]

The targeted treatments will hopefully strengthen synaptic connections in FXS, but these patients will also need to use educational interventions before the intellectual disabilities of FXS can be reversed.[293] However, the future looks bright for the treatment of FXS and further studies have been planned. The treatment of PM disorders is currently symptomatic, including the treatment of FXTAS[10,279] although neuroprotective agents such as memantine are currently being studied. Further research into the molecular mechanisms leading to RNA toxicity will certainly lead to more trials in both animal models and humans.

CONCLUSION

The fragile X field has changed remarkably since the discovery of the *FMR1* gene in 1991. The nonpenetrant status of the premutation carrier has expanded into an array of clinical involvement influencing psychiatric, endocrinologic, neurocognitive and neurologic health. The clinician must be alert to clinical involvement in all generations of a family identified with a proband. Screening studies of high risk populations will identify many more individuals affected by the mutations in *FMR1* and the premutation will be an important additive component to many common problems of aging including motor and cognitive decline. We are living in an age of targeted treatments and the portal disorder of FXS will open the doors to new treatment endeavors in autism and other disorders that have overlapping molecular involvement because of FMRP's ubiquitous presence and influence. This is an exciting time for clinicians focused on targeted treatments and FXS is likely to be the first neurodevelopmental disorder whose phenotype will be reversed.

ACKNOWLEDGMENTS

This study was supported by the National Institutes of Child Health and Human Development Grant HD 036071 to Dr RJ Hagerman and Dr DZ Loesch, NINDS RL1AG032115 to Dr Hagerman and NHMRC project grant No 330400, to Dr DZ Loesch. We thank Jacky Au for his excellent work regarding the preparation of this manuscript.

REFERENCES

1. Verkerk AJ, Pieretti M, Sutcliffe JS et al. Identification of a gene (FMR-1) containing a CGG repeat coincident with a breakpoint cluster region exhibiting length variation in fragile X syndrome. Cell 1991; 65(5):905-914.
2. Lubs HA. A marker X chromosome. American Journal of Human Genetics 1969; 21(3):231-244.
3. Cronister A, Schreiner R, Wittenberger M et al. Heterozygous fragile X female: historical, physical, cognitive, and cytogenetic features. Am J Med Genet 1991; 38(2-3):269-274.
4. Sullivan AK, Marcus M, Epstein MP et al. Association of FMR1 repeat size with ovarian dysfunction. Hum Reprod 2005; 20(2):402-412.
5. Wittenberger MD, Hagerman RJ, Sherman SL et al. The FMR1 premutation and reproduction. Fertil Steril 2007; 87(3):456-465.
6. Welt CK. Primary ovarian insufficiency: a more accurate term for premature ovarian failure. Clin Endocrinol (Oxf) 2008; 68(4):499-509.
7. Hagerman RJ, Leehey M, Heinrichs W et al. Intention tremor, parkinsonism, and generalized brain atrophy in male carriers of fragile X. Neurology 2001; 57:127-130.
8. Jacquemont S, Hagerman RJ, Leehey M et al. Fragile X premutation tremor/ataxia syndrome: molecular, clinical, and neuroimaging correlates. American Journal of Human Genetics 2003; 72(4):869-878.
9. Berry-Kravis E, Abrams L, Coffey SM et al. Fragile X-associated tremor/ataxia syndrome: clinical features, genetics and testing guidelines. Mov Disord 2007; 22(14):2018-2030.
10. Tassone F, Berry-Kravis E, eds. Fragile X-associated Tremor Ataxia Syndrome (FXTAS). New York, NY: Springer-Verlag New York Inc.; 2011.
11. Tassone F, Hagerman RJ, Taylor AK et al. Elevated levels of FMR1 mRNA in carrier males: a new mechanism of involvement in the fragile-X syndrome. American Journal of Human Genetics 2000; 66(1):6-15.
12. Hagerman PJ, Hagerman RJ. The fragile-X premutation: a maturing perspective. American Journal of Human Genetics 2004; 74(5):805-816.
13. Garcia-Arocena D, Hagerman PJ. Advances in understanding the molecular basis of FXTAS. Hum Mol Genet 2010; 19(R1):R83-R89.
14. Farzin F, Perry H, Hessl D et al. Autism spectrum disorders and attention-deficit/hyperactivity disorder in boys with the fragile X premutation. J Dev Behav Pediatr 2006; 27(2 Suppl):S137-S144.
15. Aziz M, Stathopulu E, Callias M et al. Clinical features of boys with fragile X premutations and intermediate alleles. Am J Med Genet. in press 2003; 121B(1):119-127.
16. Clifford S, Dissanayake C, Bui QM et al. Autism spectrum phenotype in males and females with fragile X full mutation and premutation. Journal of Autism and Developmental Disorders 2007; 37(4):738-747.
17. Goodlin-Jones B, Tassone F, Gane LW et al. Autistic spectrum disorder and the fragile X premutation. J Dev Behav Pediatr 2004; 25(6):392-398.
18. Bailey DB Jr., Raspa M, Olmsted M et al. Co-occurring conditions associated with FMR1 gene variations: findings from a national parent survey. American Journal of Medical Genetics 2008; 146A(16):2060-2069.
19. Roberts JE, Bailey DB Jr., Mankowski J et al. Mood and anxiety disorders in females with the FMR1 premutation. Am J Med Genet B Neuropsychiatr Genet 2009; 150B(1):130-139.
20. Hessl D, Tassone F, Loesch DZ et al. Abnormal elevation of FMR1 mRNA is associated with psychological symptoms in individuals with the fragile X premutation. Am J Med Genet B Neuropsychiatr Genet 2005; 139B(1):115-121.
21. Cornish K, Kogan C, Turk J et al. The emerging fragile X premutation phenotype: evidence from the domain of social cognition. Brain and cognition 2005; 57(1):53-60.
22. Rodriguez-Revenga L, Madrigal I, Pagonabarraga J et al. Penetrance of FMR1 premutation associated pathologies in fragile X syndrome families. Eur J Hum Genet 2009; 17(10):1359-1362.
23. Bourgeois JA, Coffey SM, Rivera SM et al. A review of fragile X premutation disorders: expanding the psychiatric perspective. J Clin Psychiatry 2009; 70(6):852-862.
24. Bourgeois J, Seritan A, Casillas E et al. Lifetime prevalence of mood and anxiety disorders in fragile X premutation carriers. J Clin Psychiatry 2010: doi: 10.4088/JCP.4009m05407blu, Published online by Physicians Postgraduate Press, Inc.
25. Grigsby J, Brega AG, Engle K et al. Cognitive profile of fragile X premutation carriers with and without fragile X-associated tremor/ataxia syndrome. Neuropsychology 2008; 22(1):48-60.
26. Hashimoto RI, Backer KC, Tassone F et al. An fMRI study of the prefrontal activity during the performance of a working memory task in premutation carriers of the fragile X mental retardation 1 gene with and without fragile X-associated tremor/ataxia syndrome (FXTAS). Journal of Psychiatric Research 2010: doi:10.1016/j.jpsychires.2010.1004.1030.
27. Coffey SM, Cook K, Tartaglia N et al. Expanded clinical phenotype of women with the FMR1 premutation. American Journal of Medical Genetics 2008; 146A(8):1009-1016.

28. Hagerman RJ. Physical and behavioral phenotype. In: Hagerman RJ, Hagerman PJ, eds. Fragile X syndrome: Diagnosis, Treatment and Research, 3rd edition. Baltimore: The Johns Hopkins University Press; 2002:3-109.
29. Pieretti M, Zhang FP, Fu YH et al. Absence of expression of the FMR-1 gene in fragile X syndrome. Cell 1991; 66(4):817-822.
30. Tassone F, Hagerman RJ, Ikle DN et al. FMRP expression as a potential prognostic indicator in fragile X syndrome. Am J Med Genet 1999; 84(3):250-261.
31. Loesch DZ, Huggins RM, Hagerman RJ. Phenotypic variation and FMRP levels in fragile X. Ment Retard Dev Disabil Res Rev 2004; 10(1):31-41.
32. Comery TA, Harris JB, Willems PJ et al. Abnormal dendritic spines in fragile X knockout mice: maturation and pruning deficits. Proc Natl Acad Sci U S A 1997; 94(10):5401-5404.
33. Devys D, Lutz Y, Rouyer N et al. The FMR-1 protein is cytoplasmic, most abundant in neurons and appears normal in carriers of a fragile X premutation. Nature genetics 1993; 4(4):335-340.
34. Hinds HL, Ashley CT, Sutcliffe JS et al. Tissue specific expression of FMR-1 provides evidence for a functional role in fragile X syndrome. Nat Genet 1993; 3(1):36-43.
35. Irwin SA, Galvez R, Greenough WT. Dendritic spine structural anomalies in fragile-X mental retardation syndrome. Cereb Cortex 2000; 10(10):1038-1044.
36. Witt R, Kaspar B, Brazelton AD et al. Developmental localization of fragile X mRNA in rat brain. Soc Neurosci 1995; 21(1):293-296.
37. Weiler IJ, Greenough WT. Synaptic synthesis of the Fragile X protein: possible involvement in synapse maturation and elimination. Am J Med Genet 1999; 83(4):248-252.
38. Weiler IJ, Irwin SA, Klintsova AY et al. Fragile X mental retardation protein is translated near synapses in response to neurotransmitter activation PNAS 1997; 94(10):5395-5400.
39. Bassell GJ, Warren ST. Fragile X syndrome: loss of local mRNA regulation alters synaptic development and function. Neuron 2008; 60(2):201-214.
40. Dolen G, Carpenter RL, Ocain TD et al. Mechanism-based approaches to treating fragile X. Pharmacol Ther 2010; 127(1):78-93.
41. Oostra BA, Willemsen R. FMR1: a gene with three faces. Biochim Biophys Acta 2009; 1790(6):467-477.
42. Qin M, Kang J, Burlin TV et al. Postadolescent changes in regional cerebral protein synthesis: an in vivo study in the FMR1 null mouse. J Neurosci 2005; 25(20):5087-5095.
43. Darnell JC, van Dreische S, Zhang C et al. HITS-CLIP Identifies Specific Neuronal mRNA Targets of Translational Repression by the Fragile X Mental Retardation Protein, FMRP [abstract]. Paper presented at: Keystone Symposia 2010; Snowbird, UT.
44. Hagerman R, Hoem G, Hagerman P. Fragile X and autism: intertwined at the molecular level leading to targeted treatments. Mol Autism 2010; 1(1):12.
45. Zalfa F, Eleuteri B, Dickson KS et al. A new function for the fragile X mental retardation protein in regulation of PSD-95 mRNA stability. Nat Neurosci 2007; 10(5):578-587.
46. Dahlhaus R, El-Husseini A. Altered neuroligin expression is involved in social deficits in a mouse model of the fragile X syndrome. Behavioural Brain Research 2010; 208(1):96-105.
47. Miyashiro KY, Beckel-Mitchener A, Purk TP et al. RNA cargoes associating with FMRP reveal deficits in cellular functioning in Fmr1 null mice. Neuron 2003; 37(3):417-431.
48. Darnell JC, Mostovetsky O, Darnell RB. FMRP RNA targets: identification and validation. Genes, Brain and Behavior 2005; 4(6):341-349.
49. Sharma A, Hoeffer CA, Takayasu Y et al. Dysregulation of mTOR signaling in fragile X syndrome. J Neurosci 2010; 30(2):694-702.
50. Tassone F. mTOR up-regulation in patients with FXS [abstract]. Paper presented at: FRAXA Investigators meeting 2010; Durham, NH.
51. de Vries PJ. Targeted treatments for cognitive and neurodevelopmental disorders in tuberous sclerosis complex. Neurotherapeutics 2010; 7(3):275-282.
52. Harris SW, Hessl D, Goodlin-Jones B et al. Autism profiles of males with fragile X syndrome. Am J Ment Retard 2008; 113(6):427-438.
53. Hatton DD, Sideris J, Skinner M et al. Autistic behavior in children with fragile X syndrome: prevalence, stability and the impact of FMRP. American Journal of Medical Genetics 2006; 140(17):1804-1813.
54. Kaufmann WE, Cortell R, Kau AS et al. Autism spectrum disorder in fragile X syndrome: communication, social interaction and specific behaviors. Am J Med Genet 2004; 129A(3):225-234.
55. Cordeiro L, Ballinger E, Hagerman R et al. Clinical assessment of DSM-IV anxiety disorders in fragile X syndrome: prevalence and characterization. Journal of Neurodevelopmental Disorders. in press.
56. Munir F, Cornish KM, Wilding J. A neuropsychological profile of attention deficits in young males with fragile X syndrome. Neuropsychologia 2000; 38(9):1261-1270.
57. Sullivan K, Hatton D, Hammer J et al. ADHD symptoms in children with FXS. American Journal of Medical Genetics 2006; 140(21):2275-2288.

58. Cornish KM, Turk J, Hagerman R. The fragile X continuum: new advances and perspectives. J Intellect Disabil Res 2008; 52(Pt 6):469-482.

59. Butler MG, Pratesi R, Watson MS et al. Anthropometric and craniofacial patterns in mentally retarded males with emphasis on the fragile X syndrome. Clinical genetics 1993; 44(3):129-138.

60. Hagerman RJ, Hull CE, Safanda JF et al. High functioning fragile X males: demonstration of an unmethylated fully expanded FMR-1 mutation associated with protein expression. Am J Med Genet 1994; 51(4):298-308.

61. Wright-Talamante C, Cheema A, Riddle JE et al. A controlled study of longitudinal IQ changes in females and males with fragile X syndrome. Am J Med Genet 1996; 64(2):350-355.

62. Tassone F, Hagerman RJ, Loesch DZ et al. Fragile X males with unmethylated, full mutation trinucleotide repeat expansions have elevated levels of FMR1 messenger RNA. Am J Med Genet 2000; 94(3):232-236.

63. Hall D, Pickler L, Riley K et al. Parkinsonism and cognitive decline in a fragile X mosaic male. Mov Disord 2010; 25(10):1523-1524.

64. Utari A, Adams E, Berry-Kravis E et al. Aging in fragile X syndrome. Journal of Neurodevelopmental Disorders 2010; 2(2):70-76.

65. de Vries BB, Wiegers AM, Smits AP et al. Mental status of females with an FMR1 gene full mutation. American Journal of Human Genetics 1996; 58(5):1025-1032.

66. Bennetto L, Pennington BF. Neuropsychology. In: Hagerman RJ, Hagerman PJ, eds. Fragile X Syndrome: Diagnosis, Treatment and Research. 3rd edition. Baltimore: Johns Hopkins University Press; 2002:206-248.

67. Angkustsiri K, Wirojanan J, Deprey LJ et al. Fragile X syndrome with anxiety disorder and exceptional verbal intelligence. American Journal of Medical Genetics 2008; 146(3):376-379.

68. Hagerman RJ, Berry-Kravis E, Kaufmann WE et al. Advances in the treatment of fragile X syndrome. Pediatrics 2009; 123(1):378-390.

69. Epstein J, Riley K, Sobesky W. The treatment of emotional and behavioral problems. In: Hagerman R, Hagerman P, eds. Fragile X Syndrome: Diagnosis, Treatment and Research. 3rd edition. Baltimore: The Johns Hopkins University Press; 2002:339-362.

70. Pembrey ME, Winter RM, Davies KE. A premutation that generates a defect at crossing over explains the inheritance of fragile X mental retardation. Am J Med Genet 1985; 21:709-717.

71. Fu YH, Kuhl DP, Pizzuti A et al. Variation of the CGG repeat at the fragile X site results in genetic instability: resolution of the Sherman paradox. Cell 1991; 67(6):1047-1058.

72. Yu S, Mulley J, Loesch D et al. Fragile-X syndrome: unique genetics of the heritable unstable element. American Journal of Human Genetics 1992; 50(5):968-980.

73. Heitz D, Devys D, Imbert G et al. Inheritance of the fragile X syndrome: size of the fragile X premutation is a major determinant of the transition to full mutation. Journal of Medical Genetics 1992; 29(11):794-801.

74. Snow K, Doud LK, Hagerman R et al. Analysis of a CGG sequence at the FMR-1 locus in fragile X families and in the general population. Am J Hum Genet 1993; 53(6):1217-1228.

75. Vaisanen ML, Kahkonen M, Leisti J. Diagnosis of fragile X syndrome by direct mutation analysis. Hum Genet 1994; 93(2):143-147.

76. Nolin SL, Lewis FA 3rd, Ye LL et al. Familial transmission of the FMR1 CGG repeat. American Journal of Human Genetics 1996; 59(6):1252-1261.

77. Ashley-Koch AE, Robinson H, Glicksman AE et al. Examination of factors associated with instability of the FMR1 CGG repeat. Am J Hum Genet 1998; 63(3):776-785.

78. Nolin SL, Brown WT, Glicksman A et al. Expansion of the fragile X CGG repeat in females with premutation or intermediate alleles. American Journal of Human Genetics 2003; 72:454-464.

79. Brown WT, Houck GE Jr., Jeziorowska A et al. Rapid Fragile X carrier screening and prenatal diagnosis using a nonradioactive PCR test. JAMA 1993; 270(13):1569-1575.

80. Snow K, Tester DJ, Kruckeberg KE et al. Sequence analysis of the fragile X trinucleotide repeat: implications for the origin of the fragile X mutation. Hum Mol Genet 1994; 3(9):1543-1551.

81. Brown WT, Nolin S, Houck G Jr. et al. Prenatal diagnosis and carrier screening for fragile X by PCR. Am J Med Genet 1996; 64(1):191-195.

82. Youings SA, Murray A, Dennis N et al. FRAXA and FRAXE: the results of a five year survey. J Med Genet 2000; 37(6):415-421.

83. Dombrowski C, Levesque S, Morel ML et al. Premutation and intermediate-size FMR1 alleles in 10572 males from the general population: loss of an AGG interruption is a late event in the generation of fragile X syndrome alleles. Hum Mol Genet 2002; 11(4):371-378.

84. Mitchell RJ, Holden JJ, Zhang C et al. FMR1 alleles in Tasmania: a screening study of the special educational needs population. Clinical genetics 2005; 67(1):38-46.

85. Zhou Y, Lum JM, Yeo GH et al. Simplified molecular diagnosis of fragile X syndrome by fluorescent methylation-specific PCR and GeneScan analysis. Clin Chem 2006; 52(8):1492-1500.

86. Khaniani MS, Kalitsis P, Burgess T et al. An improved diagnostic PCR assay for identification of cryptic heterozygosity for CGG triplet repeat alleles in the Fragile X Gene (FMR1). Mol Cytogenet 2008; 1(1):5.

87. Filipovic-Sadic S, Sah S, Chen L et al. A novel FMR1 PCR method for the routine detection of low-abundance expanded alleles and full mutations in fragile X syndrome. Clin Chem 2010; 56(3):399-408.
88. Fernandez-Carvajal I, Lopez Posadas B, Pan R et al. Expansion of an FMR1 grey-zone allele to a full mutation in two generations. J Mol Diagn 2009; 11(4):306-310.
89. Sullivan AK, Crawford DC, Scott EH et al. Paternally transmitted FMR1 alleles are less stable than maternally transmitted alleles in the common and intermediate size range. Am J Hum Genet 2002; 70(6):1532-1544.
90. Maddalena A, Richards CS, McGinniss MJ et al. Technical standards and guidelines for fragile X: the first of a series of disease-specific supplements to the Standards and Guidelines for Clinical Genetics Laboratories of the American College of Medical Genetics. Quality Assurance Subcommittee of the Laboratory Practice Committee. Genet Med 2001; 3(3):200-205.
91. Loesch DZ, Huggins R, Petrovic V et al. Expansion of the CGG repeat in fragile X in the FMR1 gene depends on the sex of the offspring. American Journal of Human Genetics 1995; 57(6):1408-1413.
92. Huggins RM, Loesch DZ, Qian GQ et al. Hierarchical Bayes model for random haplotype and family effects in the transmission of fragile-X. Genet Epidemiol 2004; 26(4):294-304.
93. Ashley AE, Sherman SL. Population dynamics of a meiotic/mitotic expansion model for the fragile X syndrome. Am J Hum Genet 1995; 57(6):1414-1425.
94. Zhong N, Ju W, Pietrofesa J et al. Fragile X "gray zone" alleles: AGG patterns, expansion risks and associated haplotypes. Am J Med Genet 1996; 64(2):261-265.
95. Eichler EE, Holden JJ, Popovich BW et al. Length of uninterrupted CGG repeats determines instability in the FMR1 gene. Nature genetics 1994; 8(1):88-94.
96. Eichler EE, Macpherson JN, Murray A et al. Haplotype and interspersion analysis of the FMR1 CGG repeat identifies two different mutational pathways for the origin of the fragile X syndrome. Hum Mol Genet 1996; 5(3):319-330.
97. Eichler EE, Nelson DL. Genetic variation and evolutionary stability of the FMR1 CGG repeat in six closed human populations. Am J Med Genet 1996; 64(1):220-225.
98. Zhong N, Yang W, Dobkin C et al. Fragile X gene instability: anchoring AGGs and linked microsatellites. Am J Hum Genet 1995; 57(2):351-361.
99. Yrigollen C, Long K, Tong T et al. The role of AGG interruptions in the stability of FMR1 gene. Paper presented at: 12th International Fragile X Conference; 2010; Detroit, MI.
100. Richards RI, Holman K, Kozman H et al. Fragile X syndrome: genetic localisation by linkage mapping of two microsatellite repeats FRAXAC1 and FRAXAC2 which immediately flank the fragile site. J Med Genet 1991; 28(12):818-823.
101. Zhong N, Kajanoja E, Smits B et al. Fragile X founder effects and new mutations in Finland. Am J Med Genet 1996a; 64(1):226-233.
102. Faradz SMH, Pattiiha MZ, Leigh DA et al. Genetic diversity at the FMR1 locus in the Indonesian population. Annals of Human Genetics 2000; 64(4):329-339.
103. Kunst CB, Warren ST. Cryptic and polar variation of the fragile X repeat could result in predisposing normal alleles. Cell 1994; 77(6):853-861.
104. Gunter C, Paradee W, Crawford DC et al. Re-examination of factors associated with expansion of CGG repeats using a single nucleotide polymorphism in FMR1. Hum Mol Genet 1998; 7(12):1935-1946.
105. Brightwell G, Wycherley R, Waghorn A. SNP genotyping using a simple and rapid single-tube modification of ARMS illustrated by analysis of 6 SNPs in a population of males with FRAXA repeat expansions. Molecular and Cellular Probes 2002; 16(4):297-305.
106. Curlis Y, Zhang C, Holden JJ et al. Haplotype study of intermediate-length alleles at the fragile X (FMR1) gene: ATL1, FMRb and microsatellite haplotypes differ from those found in common-size FMR1 alleles. Hum Biol 2005; 77(1):137-151.
107. Toledano-Alhadef H, Basel-Vanagaite L, Magal N et al. Fragile-X carrier screening and the prevalence of premutation and full-mutation carriers in Israel. Am J Hum Genet 2001; 69(2):351-360.
108. Dawson AJ, Chodirker BN, Chudley AE. Frequency of FMR1 premutations in a consecutive newborn population by PCR screening of Guthrie blood spots. Biochem Mol Med 1995; 56(1):63-69.
109. Rousseau F, Rouillard P, Morel ML et al. Prevalence of carriers of premutation-size alleles of the FMRI gene—and implications for the population genetics of the fragile X syndrome. American Journal of Human Genetics 1995; 57(5):1006-1018.
110. Hagerman PJ. The fragile X prevalence paradox. Journal of Medical Genetics 2008; 45(8):498-499.
111. Fernandez-Carvajal I, Walichiewicz P, Xiaosen X et al. Screening for expanded alleles of the FMR1 gene in blood spots from newborn males in a Spanish population. J Mol Diagn 2009; 11(4):324-329.
112. Song FJ, Barton P, Sleightholme V et al. Screening for fragile X syndrome: a literature review and modelling study. Health Technol Assess 2003; 7(16):1-106.
113. Otsuka S, Sakamoto Y, Siomi H et al. Fragile X carrier screening and FMR1 allele distribution in the Japanese population. Brain and Development 2010; 32(2):110-114.

114. Loesch DZ, Huggins R, Hay DA et al. Genotype-phenotype relationships in fragile X syndrome: a family study. American Journal of Human Genetics 1993; 53(5):1064-1073.
115. Hagerman RJ, Staley LW, O'Conner R et al. Learning-disabled males with a fragile X CGG expansion in the upper premutation size range. Pediatrics 1996; 97(1):122-126.
116. Loesch D, Hay D, Mulley J. Transmitting males and carrier females in fragile X-revisited. Am J Med Genet 1994; 51(4):392-399.
117. Hull C, Hagerman RJ. A study of the physical, behavioral and medical phenotype, including anthropometric measures, of females with fragile X syndrome. Am J Dis Child 1993; 147(11):1236-1241.
118. Riddle JE, Cheema A, Sobesky WE et al. Phenotypic involvement in females with the FMR1 gene mutation. Am J Ment Retard 1998; 102(6):590-601.
119. Loesch DZ, Huggins RM, Bui QM et al. Effect of fragile X status categories and FMRP deficits on cognitive profiles estimated by robust pedigree analysis. Am J Med Genet A 2003; 122A(1):13-23.
120. Loesch DZ, Bui QM, Grigsby J et al. Effect of the fragile X status categories and the fragile X mental retardation protein levels on executive functioning in males and females with fragile X. Neuropsychology 2003; 17(4):646-657.
121. Grigsby J, Kaye K. Behavioral Dyscontrol Scale: Manual. 2nd edition. 1996.
122. Meyers JE, Meyers KR. Rey Complex Figure Test and Recognition Trial Manual. Odessa FI.: Psychological Assessment Resources, Inc.; 1995.
123. Moore CJ, Daly EM, Schmitz N et al. A neuropsychological investigation of male premutation carriers of fragile X syndrome. Neuropsychologia 2004; 42(14):1934-1947.
124. Lachiewicz AM, Dawson DV, Spiridigliozzi GA et al. Arithmetic difficulties in females with the fragile X premutation. American Journal of Medical Genetics 2006; 140(7):665-672.
125. Keri S, Benedek G. Visual pathway deficit in female fragile X premutation carriers: a potential endophenotype. Brain Cogn 2009; 69(2):291-295.
126. Keri S, Benedek G. The perception of biological and mechanical motion in female fragile X premutation carriers. Brain Cogn 2010; 72(2):197-201.
127. Hunsaker MR, Goodrich-Hunsaker NJ, Willemsen R et al. Temporal ordering deficits in female CGG KI mice heterozygous for the fragile X premutation. Behavioural Brain Research 2010; 213(2):263-268.
128. Mínguez M, Ibáñez B, Ribate MP et al. Risk of cognitive impairment in female premutation carriers of fragile X premutation: analysis by means of robust segmented linear regression models. Am J Med Genet B Neuropsychiatr Genet 2009; 150B(2):262-270.
129. Goodlin-Jones BL, Tassone F, Gane LW et al. Autistic spectrum disorder and the fragile X premutation. J Dev Behav Pediatr 2004; 25(6):392-398.
130. Loesch DZ, Huggins RM, Bui QM et al. Relationship of deficits of FMR1 gene specific protein with physical phenotype of fragile X males and females in pedigrees: a new perspective. Am J Med Genet A 2003; 118A(2):127-134.
131. Jakala P, Hanninen T, Ryynanen M et al. Fragile-X: neuropsychological test performance, CGG triplet repeat lengths and hippocampal volumes. J Clin Invest 1997; 100(2):331-338.
132. Murphy DGM, Mentis MJ, Pietrini P et al. Premutation female carriers of fragile X syndrome: a pilot study on brain anatomy and metabolism. Journal of the American Academy of Child and Adolescent Psychiatry 1999; 38(10):1294-1301.
133. Moore CJ, Daly EM, Tassone F et al. The effect of premutation of X chromosome CGG trinucleotide repeats on brain anatomy. Brain 2004; 127(Pt 12):2672-2681.
134. Adams PE, Adams JS, Nguyen DV et al. Psychological symptoms correlate with reduced hippocampal volume in fragile X premutation carriers. Am J Med Genet B Neuropsychiatr Genet 2010; 153B(3); 775-78.
135. Hessl D, Rivera S, Koldewyn K et al. Amygdala dysfunction in men with the fragile X premutation. Brain 2007; 130(2):404-416.
136. Koldewyn K, Hessl D, Adams J et al. Reduced hippocampal activation during recall is associated with elevated FMR1 mRNA and psychiatric symptoms in men with the fragile X premutation. Brain Imaging Behav 2008; 2(2):105-116.
137. Hagerman RJ, Hagerman PJ. The fragile X premutation: into the phenotypic fold. Current Opinion in Genetics and Development 2002; 12:278-283.
138. Cornish KM, Li L, Kogan CS et al. Age-dependent cognitive changes in carriers of the fragile X syndrome. Cortex 2008; 44(6):628-636.
139. Kogan C, Turk J, Hagerman RJ et al. Impact of the fragile X mental retardation 1 (FMR1) gene premutation on neuropsychiatric functioning in adult males without fragile X-associated tremor/ataxia syndrome: a controlled study. Am J Med Genet B Neuropsychiatr Genet 2008; 147B(6):859-872.
140. Cornish KM, Kogan CS, Li L et al. Lifespan changes in working memory in fragile X premutation males. Brain and Cognition 2009; 69(3):551-558.
141. Sevin M, Kutalik Z, Bergman S et al. Penetrance of marked cognitive impairment in older male carriers of the FMR1 gene premutation. Journal of Medical Genetics 2009; 46(12):818-824.

142. Hunter J, Abramowitz A, Rusin M et al. Is there evidence for neuropsychological and neurobehavioral phenotypes among adults without FXTAS who carry the FMR1 premutation? A review of current literature. Genet Med 2009; 11(2):79-89.

143. Reiss AL, Freund L, Abrams MT et al. Neurobehavioral effects of the fragile X premutation in adult women: a controlled study. American journal of human genetics 1993; 52(5):884-894.

144. Franke P, Leboyer M, Gänsicke M et al. Genotype-phenotype relationship in female carriers of the premutation and full mutation of FMR-1. Psychiatry Research 1998; 80(2):113-127.

145. Al-Semaan Y, Malla AK, Lazosky A. Schizoaffective disorder in a fragile-X carrier. Australian and New Zealand Journal of Psychiatry 1999; 33(3):436-440.

146. Rodriguez-Revenga L, Madrigal I, Alegret M et al. Evidence of depressive symptoms in fragile-X syndrome premutated females. Psychiatr Genet 2008; 18(4):153-155.

147. Roberts J, Bailey D Jr., Mankowski J et al. Mood and anxiety disorders in females with the FMR1 premutation. Am J Med Genet B Neuropsychiatr Genet 2009; 150B(1):130-139.

148. Bailey DJ, Raspa M, Olmsted M et al. Co-occurring conditions associated with FMR1 gene variations: findings from a national parent survey. Am J Med Genet A 2008; 146A(16):2060-2069.

149. Hunter J, Allen E, Abramowitz A et al. Investigation of phenotypes associated with mood and anxiety among male and female fragile X premutation carriers. Behavior Genetics 2008; 38(5):493-502.

150. Allen EG, He W, Yadav-Shah M et al. A study of the distributional characteristics of FMR1 transcript levels in 238 individuals. Hum Genet 2004; 114(5):439-447.

151. Tassone F, Iwahashi C, Hagerman PJ. FMR1 RNA within the intranuclear inclusions of fragile X-associated tremor/ataxia syndrome (FXTAS). RNA Biol 2004; 1(2):103-105.

152. Tassone F, Hagerman RJ, Chamberlain WD et al. Transcription of the FMR1 gene in individuals with fragile X syndrome. Am J Med Genet Fall 2000; 97(3):195-203.

153. Tassone F, Hagerman R, Taylor A et al. Clinical involvement and protein expression in individuals with the FMR1 premutation. American Journal of Medical Genetics 2000; 91(2):144-152.

154. Tassone F, Hagerman RJ, Garcia-Arocena D et al. Intranuclear inclusions in neural cells with premutation alleles in fragile X associated tremor/ataxia syndrome. Journal of Medical Genetics 2004; 41(4):e43.

155. Kenneson A, Zhang F, Hagedorn CH et al. Reduced FMRP and increased FMR1 transcription is proportionally associated with CGG repeat number in intermediate-length and premutation carriers. Hum Mol Genet 2001; 10(14):1449-1454.

156. Chen LS, Tassone F, Sahota P et al. The (CGG)n repeat element within the 5' untranslated region of the FMR1 message provides both positive and negative cis effects on in vivo translation of a downstream reporter. Hum Mol Genet 2003; 12(23):3067-3074.

157. Tassone F, Beilina A, Carosi C et al. Elevated FMR1 mRNA in premutation carriers is due to increased transcription. RNA 2007; 13(4):555-562.

158. García-Alegría E, Berta Ibáñez, Mónica Mínguez et al. Analysis of FMR1 gene expression in female premutation carriers using robust segmented linear regression models. RNA 2007; 13:756-762.

159. Şentürk D, Nguyen DV, Tassone F et al. Covariate adjusted correlation analysis with application to FMR1 premutation female carrier data. Biometrics 2009; 65(3):781-792.

160. Hagerman R, Greco CM, Chudley AE et al. Neuropathology and neurodegenerative features in some older male premutation carriers of fragile X syndrome. Am J Hum Genet 2001; 69(177).

161. Hagerman RJ, Hagerman PJ. Fragile X Syndrome and fragile X-associated tremor/ataxia syndrome. In: Robert D. Wells, Ashizawa T, eds. Genetic Instabilities and Neurological Diseases. 2nd edition. Houston, USA: Academic Press; 2006:165-174.

162. Berry-Kravis E, Lewin F, Wuu J et al. Tremor and ataxia in fragile X premutation carriers: blinded videotape study. Ann Neurol 2003; 53(5):616-623.

163. Leehey MA, Berry-Kravis E, Min SJ et al. Progression of tremor and ataxia in male carriers of the FMR1 premutation. Mov Disord 2007; 22(2):203-206.

164. Leehey MA, Hagerman RJ, Landau WM et al. Tremor/ataxia syndrome in fragile X carrier males. Movement Disorders 2002; 17(4):744-745.

165. Jacquemont S, Hagerman RJ, Leehey MA et al. Penetrance of the fragile X-associated tremor/ataxia syndrome in a premutation carrier population. JAMA 2004; 291(4):460-469.

166. Loesch DZ, Churchyard A, Brotchie P et al. Evidence for and a spectrum of, neurological involvement in carriers of the fragile X premutation: FXTAS and beyond. Clinical Genetics 2005; 67(5):412-417.

167. Pugliese P, Annesi G, Cutuli N et al. The fragile X premutation presenting as postprandial hypotension. Neurology 2004; 63(11):2188-2189.

168. Soontarapornchai K, Maselli R, Fenton-Farrell G et al. Abnormal nerve conduction features in fragile X premutation carriers. Archives of Neurology 2008; 65(4):495-498.

169. Adams JS, Adams PE, Nguyen D et al. Volumetric brain changes in females with fragile X-associated tremor/ataxia syndrome (FXTAS). Neurology 2007; 69(9):851-859.

170. Brunberg JA, Jacquemont S, Hagerman RJ et al. Fragile X premutation carriers: characteristic MR imaging findings of adult male patients with progressive cerebellar and cognitive dysfunction. Am J Neuroradiol 2002; 23(10):1757-1766.

171. Morís G, Arias M, López MV, Álvarez V. Hyperintensity in the basis pontis: atypical neuroradiological findings in a woman with FXTAS. Movement Disorders 2010; 25(5):649-650.

172. Loesch DZ, Kotschet K, Trost, NC et al. White matter changes in basic pontis in small expansion FMR1 allele carriers with parkinsonism. Neuropsychiatric Genetics 2011; in press.

173. Greco CM, Berman RF, Martin RM et al. Neuropathology of fragile X-associated tremor/ataxia syndrome (FXTAS). Brain 2006; 129(Pt 1):243-255.

174. Cohen S, Masyn K, Adams J et al. Molecular and imaging correlates of the fragile X-associated tremor/ataxia syndrome. Neurology 2006; 67(8):1426-1431.

175. Loesch DZ, Litewka L, Brotchie P et al. Magnetic resonance imaging study in older fragile X premutation male carriers. Annals of Neurology 2005; 58(2):326-330.

176. Greco CM, Hagerman RJ, Tassone F et al. Neuronal intranuclear inclusions in a new cerebellar tremor/ataxia syndrome among fragile X carriers. Brain 2002; 125(8):1760-1771.

177. Willemsen R, Hoogeveen-Westerveld M, Reis S et al. The FMR1 CGG repeat mouse displays ubiquitin-positive intranuclear neuronal inclusions; implications for the cerebellar tremor/ataxia syndrome. Hum Mol Genet 2003; 12(9):949-959.

178. Wenzel HJ, Hunsaker MR, Greco CM et al. Ubiquitin-positive intranuclear inclusions in neuronal and glial cells in a mouse model of the fragile X premutation. Brain Res 2010; 1318:155-166.

179. Grigsby J, Brega AG, Jacquemont S et al. Impairment in the cognitive functioning of men with fragile X-associated tremor/ataxia syndrome (FXTAS). J Neurol Sci 2006; 248(1-2):227-233.

180. Grigsby J, Brega AG, Leehey MA et al. Impairment of executive cognitive functioning in males with fragile X-associated tremor/ataxia syndrome. Mov Disord 2007; 22(5):645-650.

181. Bacalman S, Farzin F, Bourgeois JA et al. Psychiatric phenotype of the fragile X-associated tremor/ataxia syndrome (FXTAS) in males: newly described fronto-subcortical dementia. J Clin Psychiatry 2006; 67(1):87-94.

182. Bourgeois JA, Cogswell JB, Hessl D et al. Cognitive, anxiety and mood disorders in the fragile X-associated tremor/ataxia syndrome. Gen Hosp Psychiatry 2007; 29(4):349-356.

183. Seritan AL, Nguyen DV, Farias ST et al. Dementia in fragile X-associated tremor/ataxia syndrome (FXTAS): comparison with Alzheimer's disease. Am J Med Genet B Neuropsychiatr Genet 2008; 147B(7):1138-1144.

184. Loesch DZ, Cook M, Litewka L et al. A low symptomatic form of neurodegeneration in younger carriers of the FMR1 premutation, manifesting typical radiological changes. Journal of Medical Genetics 2008; 45(3):179-181.

185. Loesch DZ, Litewka L, Churchyard A et al. Tremor/ataxia syndrome and fragile X premutation: diagnostic caveats. J Clin Neurosci 2007; 14(3):245-248.

186. Capelli LP, Goncalves MR, Kok F et al. Fragile X-associated tremor/ataxia syndrome: intrafamilial variability and the size of the FMR1 premutation CGG repeat. Mov Disord 2007; 22(6):866-870.

187. Greco CM, Tassone F, Garcia-Arocena D et al. Clinical and neuropathologic findings in a woman with the FMR1 premutation and multiple sclerosis. Archives of Neurology 2008; 65(8):1114-1116.

188. Zhang L, Coffey S, Lua LL et al. FMR1 premutation in females diagnosed with multiple sclerosis. J Neurol Neurosurg Psychiatry 2009; 80(7):812-814.

189. Mothersead PK, Conrad K, Hagerman RJ et al. GRAND ROUNDS: an atypical progressive dementia in a male carrier of the fragile X premutation: an example of fragile X-associated tremor/ataxia syndrome. Appl Neuropsychol 2005; 12(3):169-178.

190. Yachnis AT, Roth HL, Heilman KM. Fragile X Dementia Parkinsonism Syndrome (FXDPS). Cognitive and Behavioral Neurology 2010; 23(1):39-43. doi: 10.1097/WNN.1090b1013e3181b1096e1091b1099.

191. Chonchaiya W, Nguyen DV, Au J et al. Clinical involvement in daughters of men with fragile X-associated tremor ataxia syndrome. Clinical Genetics 2010; 78(1):39-46.

192. Brouwer JR, Huizer K, Severijnen LA et al. CGG-repeat length and neuropathological and molecular correlates in a mouse model for fragile X-associated tremor/ataxia syndrome. J Neurochem 2008; 107(6):1671-1682.

193. Tassone F, Adams J, Berry-Kravis EM et al. CGG repeat length correlates with age of onset of motor signs of the fragile X-associated tremor/ataxia syndrome (FXTAS). Am J Med Genet B Neuropsychiatr Genet 2007; 144(4):566-569.

194. Leehey MA, Berry-Kravis E, Goetz CG et al. FMR1 CGG repeat length predicts motor dysfunction in premutation carriers. Neurology 2008; 70(16 Pt 2):1397-1402.

195. Berry-Kravis E, Goetz CG, Leehey MA et al. Neuropathic features in fragile X premutation carriers. Am J Med Genet A 2007; 143(1):19-26.

196. Tassone F, Hagerman RJ, Gane LW et al. Strong similarities of the FMR1 mutation in multiple tissues: postmortem studies of a male with a full mutation and a male carrier of a premutation. American Journal of Medical Genetics 1999; 84(3):240-244.
197. Milunsky JM, Maher TA. Fragile X carrier screening and spinocerebellar ataxia in older males. American Journal of Medical Genetics 2004; 125(3):320.
198. Macpherson J, Waghorn A, Hammans S et al. Observation of an excess of fragile-X premutations in a population of males referred with spinocerebellar ataxia. Hum Genet 2003; 112(5-6):619-620.
199. Van Esch H, Dom R, Bex D et al. Screening for FMR-1 premutations in 122 older Flemish males presenting with ataxia. Eur J Hum Genet 2005; 13(1):121-123.
200. Tan EK, Zhao Y, Puong KY et al. Fragile X premutation alleles in SCA, ET and parkinsonism in an Asian cohort. Neurology 2004; 63(2):362-363.
201. Brussino A, Gellera C, Saluto A et al. FMR1 gene premutation is a frequent genetic cause of late-onset sporadic cerebellar ataxia. Neurology 2005; 64(1):145-147.
202. Garcia Arocena D, Louis ED, Tassone F et al. Screen for expanded FMR1 alleles in patients with essential tremor. Mov Disord 2004; 19(8):930-933.
203. Deng H, Le W, Jankovic J. Premutation alleles associated with Parkinson disease and essential tremor. Jama 2004; 292(14):1685-1686.
204. Kamm C, Healy DG, Quinn NP et al. The fragile X tremor ataxia syndrome in the differential diagnosis of multiple system atrophy: data from the EMSA Study Group. Brain 2005; 128(Pt 8):1855-1860.
205. Toft M, Aasly J, Bisceglio G et al. Parkinsonism, FXTAS and FMR1 premutations. Mov Disord 2005; 20(2):230-233.
206. Zuhlke C, Budnik A, Gehlken U et al. FMR1 premutation as a rare cause of late onset ataxia—evidence for FXTAS in female carriers. J Neurol 2004; 251(11):1418-1419.
207. Garland EM, Vnencak-Jones CL, Biaggioni I et al. Fragile X gene premutation in multiple system atrophy. J Neurol Sci 2004; 227(1):115-118.
208. Biancalana V, Toft M, Le Ber I et al. FMR1 premutations associated with fragile X-associated tremor/ataxia syndrome in multiple system atrophy. Archives of Neurology 2005; 62(6):962-966.
209. Seixas AI, Maurer MH, Lin M et al. FXTAS, SCA10 and SCA17 in American patients with movement disorders. American Journal of Medical Genetics 2005; 136(1):87-89.
210. Tan EK, Zhao Y, Puong KY et al. Expanded FMR1 alleles are rare in idiopathic Parkinson's disease. Neurogenetics 2005; 6(1):51-52.
211. Jacquemont S, Farzin F, Hall D et al. Aging in individuals with the FMR1 mutation. Am J Ment Retard 2004; 109(2):154-164.
212. Kurz MW, Schlitter AM, Klenk Y et al. FMR1 alleles in Parkinson's disease: relation to cognitive decline and hallucinations, a longitudinal study. J Geriatr Psychiatry Neurol 2007; 20(2):89-92.
213. Kraff J, Tang HT, Cilia R et al. Screen for excess FMR1 premutation alleles among males with parkinsonism. Arch Neurol 2007; 64(7):1002-1006.
214. Loesch DZ, Khaniani MS, Slater HR et al. Small CGG repeat expansion alleles of FMR1 gene are associated with parkinsonism. Clinical Genetics 2009; 76(5):471-476.
215. Karmon Y, Gadoth N. Fragile X associated tremor/ataxia syndrome (FXTAS) with dementia in a female harbouring FMR1 premutation. J Neurol Neurosurg Psychiatry 2008; 79(6):738-739.
216. Hagerman RJ, Leavitt BR, Farzin F et al. Fragile-X-associated tremor/ataxia syndrome (FXTAS) in females with the FMR1 premutation. American Journal of Human Genetics 2004; 74(5):1051-1056.
217. Berry-Kravis E, Potanos K, Weinberg D et al. Fragile X-associated tremor/ataxia syndrome in sisters related to X-inactivation. Annals of Neurology 2005; 57(1):144-147.
218. Jacquemont S, Orrico A, Galli L et al. Spastic paraparesis, cerebellar ataxia and intention tremor: a severe variant of FXTAS? Journal of Medical Genetics 2005; 42(2):e14.
219. O'Dwyer JP, Clabby C, Crown J et al. Fragile X-associated tremor/ataxia syndrome presenting in a woman after chemotherapy. Neurology 2005; 65(2):331-332.
220. Al-Hinti JT, Nagan N, Harik SI. Fragile X premutation in a woman with cognitive impairment, tremor and history of premature ovarian failure. Alzheimer Dis Assoc Disord 2007; 21(3):262-264.
221. Galvao R, Mendes-Soares L, Camara J et al. Triplet repeats, RNA secondary structure and toxic gain-of-function models for pathogenesis. Brain Res Bull 2001; 56(3-4):191-201.
222. Finsterer J. Myotonic dystrophy type 2. European Journal of Neurology 2002; 9(5):441-447.
223. Mankodi A, Thornton CA. Myotonic syndromes. Current Opinion in Neurology 2002; 15(5):545-552.
224. Khalili K, Del Valle L, Muralidharan V et al. Puralpha is essential for postnatal brain development and developmentally coupled cellular proliferation as revealed by genetic inactivation in the mouse. Mol Cell Biol 2003; 23(19):6857-6875.
225. Jin P, Duan R, Qurashi A et al. Pur alpha binds to rCGG repeats and modulates repeat-mediated neurodegeneration in a Drosophila model of fragile X tremor/ataxia syndrome. Neuron 2007; 55(4):556-564.

226. Garcia-Arocena D, Yang JE, Brouwer JR et al. Fibroblast phenotype in male carriers of FMR1 premutation alleles. Hum Mol Genet 2010; 19(2):299-312.
227. Paul R, Pessah IN, Gane L et al. Early onset of neurological symptoms in fragile X premutation carriers exposed to neurotoxins. Neurotoxicology 2010; 31(4):399-402.
228. Iwahashi CK, Yasui DH, An HJ et al. Protein composition of the intranuclear inclusions of FXTAS. Brain 2006; 129(Pt 1):256-271.
229. Arocena DG, Iwahashi CK, Won N et al. Induction of inclusion formation and disruption of lamin A/C structure by premutation CGG-repeat RNA in human cultured neural cells. Hum Mol Genet 2005; 14(23):3661-3671.
230. Sellier C, Rau F, Liu Y et al. Sam68 sequestration and partial loss of function are associated with splicing alterations in FXTAS patients. The EMBO Journal 2010; 29(7):1248-1261.
231. Sellier C, Hagerman P, Willemsen R et al. DROSHA/DGCR8 sequestration by expanded CGG repeats leads to global micro-RNA processing alteration in FXTAS patients [abstract]. Paper presented at: 12th International Fragile X Conference; 2010; Detroit, MI.
232. Jin P, Zarnescu DC, Zhang F et al. RNA-mediated neurodegeneration caused by the fragile X premutation rCGG repeats in Drosophila. Neuron 2003; 39(5):739-747.
233. Van Dam D, Errijgers V, Kooy RF et al. Cognitive decline, neuromotor and behavioural disturbances in a mouse model for fragile-X-associated tremor/ataxia syndrome (FXTAS). Behavioural Brain Research 2005; 162(2):233-239.
234. Hashem V, Galloway JN, Mori M et al. Ectopic expression of CGG containing mRNA is neurotoxic in mammals. Hum Mol Genet 2009; 18(13):2443-2451.
235. Greco CM, Soontrapornchai K, Wirojanan J et al. Testicular and pituitary inclusion formation in fragile X associated tremor/ataxia syndrome. The Journal of Urology 2007; 177(4):1434-1437.
236. Brouwer JR, Severijnen E, de Jong FH et al. Altered hypothalamus-pituitary-adrenal gland axis regulation in the expanded CGG-repeat mouse model for fragile X-associated tremor/ataxia syndrome. Psychoneuroendocrinology 2008; 33(6):863-873.
237. Hunsaker MR, Greco C, Tassone F et al. Rare intranuclear inclusions in the brains of three older adult males with fragile X syndrome: implications for the spectrum of fragile X-associated disorders. JNEN. under review.
238. Chen Y, Tassone F, Berman RF et al. Murine hippocampal neurons expressing Fmr1 gene premutations show early developmental deficits and late degeneration. Hum Mol Genet 2010; 19(1):196-208.
239. Ross-Inta C, Omanska-Klusek A, Wong S et al. Evidence of mitochondrial dysfunction in fragile X-associated tremor/ataxia syndrome. Biochem J 2010; 429(3):545-552.
240. Loesch DZ, Godler DE, Evans A et al. Evidence for the toxicity of bidirectional transcripts and mitochondrial dysfunction in blood associated with small CGG expansions in the FMR1 gene in patients with parkinsonism. Genet Med 2011; Jan 25. [Epub ahead of print].
241. Ladd PD, Smith LE, Rabaia NA et al. An antisense transcript spanning the CGG repeat region of FMR1 is upregulated in premutation carriers but silenced in full mutation individuals. Hum Mol Genet 2007; 16(24):3174-3187.
242. Khalil AM, Faghihi MA, Modarresi F et al. A novel RNA transcript with antiapoptotic function is silenced in fragile X syndrome. PLoS ONE 2008; 3(1):e1486.
243. Welt CK, Smith PC, Taylor AE. Evidence of early ovarian aging in fragile X premutation carriers. J Clin Endocrinol Metab 2004; 89(9):4569-4574.
244. De Caro JJ, Dominguez C, Sherman SL. Reproductive health of adolescent girls who carry the FMR1 premutation: expected phenotype based on current knowledge of fragile x-associated primary ovarian insufficiency. Ann N Y Acad Sci 2008; 1135:99-111.
245. Schwartz CE, Dean J, Howard-Peebles PN et al. Obstetrical and gynecological complications in fragile X carriers: a multicenter study. Am J Med Genet 1994; 51(4):400-402.
246. Allingham-Hawkins DJ, Babul-Hirji R, Chitayat D et al. Fragile X premutation is a significant risk factor for premature ovarian failure: the international collaborative POF in fragile X study- preliminary data. Am J Med Genet 1999; 83:322-325.
247. Murray A, Ennis S, MacSwiney F et al. Reproductive and menstrual history of females with fragile X expansions. Eur J Hum Genet 2000; 8(4):247-252.
248. Vianna-Morgante AM, Costa SS. Premature ovarian failure is associated with maternally and paternally inherited premutation in Brazilian families with fragile X [see comments] [letter]. American Journal of Human Genetics 2000; 67(1):254-255; discussion 256-258.
249. Sherman SL. Premature ovarian failure in the fragile X syndrome. Am J Med Genet (Semin Med Genet) 2000; 97(3):189-194.
250. Tejada MI, Garcia-Alegria E, Bilbao A et al. Analysis of the molecular parameters that could predict the risk of manifesting premature ovarian failure in female premutation carriers of fragile X syndrome. Menopause 2008; 15(5):945-949.

251. Ennis S, Ward D, Murray A. Nonlinear association between CGG repeat number and age of menopause in FMR1 premutation carriers. Eur J Hum Genet 2006; 14(2):253-255.
252. Allen EG, Sullivan AK, Marcus M et al. Examination of reproductive aging milestones among women who carry the FMR1 premutation. Hum Reprod 2007:dem148.
253. Hunter JE, Epstein MP, Tinker SW et al. Fragile X-associated primary ovarian insufficiency: evidence for additional genetic contributions to severity. Genetic Epidemiology 2008; 32(6):553-559.
254. Pankratz VS, de Andrade M, Therneau TM. Random-effects Cox proportional hazards model: general variance components methods for time-to-event data. Genetic Epidemiology 2005; 28(2):97-109.
255. Lin Y-S, Yang M-L. Familial premature ovarian failure in female premutated carriers of fragile X syndrome: a Case Report and Literature Review. Taiwanese Journal of Obstetrics and Gynecology 2006; 45(1):60-63.
256. Miano M, Laperuta C, Chiurazzi P et al. Ovarian dysfunction and FMR1 alleles in a large Italian family with POF and FRAXA disorders: case report. BMC Medical Genetics 2007; 8(1):18.
257. Hundscheid RDL, Braat DDM, Kiemeney LALM et al. Increased serum FSH in female fragile X premutation carriers with either regular menstrual cycles or on oral contraceptives. Hum Reprod 2001; 16(3):457-462.
258. Murray A, Webb J, MacSwiney F et al. Serum concentrations of follicle stimulating hormone may predict premature ovarian failure in FRAXA premutation women. Hum Reprod 1999; 14(5):1217-1218.
259. Hundscheid RD, Smits AP, Thomas CM et al. Female carriers of fragile X premutations have no increased risk for additional diseases other than premature ovarian failure. American Journal of Medical Genetics 2003; 117(1):6-9.
260. Allen EG, Sullivan AK, Marcus M et al. Examination of reproductive aging milestones among women who carry the FMR1 premutation. Hum Reprod 2007; 22(8):2142-2152.
261. Peprah E, He W, Allen E et al. Examination of FMR1 transcript and protein levels among 74 premutation carriers. J Hum Genet 2010; 55(1):66-68.
262. Burgess B, Partington M, Turner G et al. Normal age of menarche in fragile X syndrome. American Journal of Medical Genetics 1996; 64(2):376-376.
263. Vianna-Morgante AM. Twinning and premature ovarian failure in premutation fragile X carriers. Am J Med Genet 1999; 83(4):326.
264. Van Esch H, Buekenhout L, Race V et al. Very early premature ovarian failure in two sisters compound heterozygous for the FMR1 premutation. Eur J Med Genet 2009; 52(1):37-40.
265. Patsalis PC, Sismani C, Hettinger JA et al. Frequencies of "grey-zone" and premutation-size FMR1 CGG-repeat alleles in patients with developmental disability in Cyprus and Canada. Am J Med Genet 1999; 84(3):195-197.
266. Loesch DZ, Bui QM, Huggins RM et al. Transcript levels of the intermediate size or grey zone fragile X mental retardation 1 alleles are raised and correlate with the number of CGG repeats. J Med Genet 2007; 44(3):198-204.
267. Murray A, Youings S, Dennis N et al. Population screening at the FRAXA and FRAXE loci: molecular analyses of boys with learning difficulties and their mothers. Hum Mol Genet 1996; 5(6):727-735.
268. Mazzocco MM, Sonna NL, Teisl JT et al. The FMR1 and FMR2 mutations are not common etiologies of academic difficulty among school-age children. J Dev Behav Pediatr 1997; 18(6):392-398.
269. Mazzocco MMM, Myers GF, Hamner JL et al. The prevalence of the FMR1 and FMR2 mutations among preschool children with language delay. The Journal of Pediatrics 1998; 132(5):795-801.
270. Haddad LA, Aguiar MJ, Costa SS et al. Fully mutated and gray-zone FRAXA alleles in Brazilian mentally retarded boys. Am J Med Genet 1999; 84(3):198-201.
271. Loesch DZ, Godler DE, Khaniani M et al. Linking the FMR1 alleles with small CGG expansions with neurodevelopmental disorders: preliminary data suggest an involvement of epigenetic mechanisms. American Journal of Medical Genetics 2009; 149A(10):2306-2310.
272. Bretherick KL, Fluker MR, Robinson WP. FMR1 repeat sizes in the gray zone and high end of the normal range are associated with premature ovarian failure. Hum Genet 2005; 117(4):376-382.
273. Bodega B, Bione S, Dalpra L et al. Influence of intermediate and uninterrupted FMR1 CGG expansions in premature ovarian failure manifestation. Hum Reprod 2006; 21(4):952-957.
274. Streuli I, Fraisse T, Ibecheole V et al. Intermediate and premutation FMR1 alleles in women with occult primary ovarian insufficiency. Fertil Steril 2009; 92(2):464-470.
275. Gleicher N, Weghofer A, Barad DH. A pilot study of premature ovarian senescence: I. Correlation of triple CGG repeats on the FMR1 gene to ovarian reserve parameters FSH and anti-Mullerian hormone. Fertil Steril 2009; 91(5):1700-1706.
276. Chatterjee S, Maitra A, Kadam S et al. CGG repeat sizing in the FMR1 gene in Indian women with premature ovarian failure. Reprod Biomed Online 2009; 19(2):281-286.
277. Schapira A, Hartmann A, Agid Y. Parkinsonian Disorders in Clinical Practice. Oxford, UK: Wiley-Blackwell; 2009.

278. Loesch DZ, Bui QM, Dissanayake C et al. Molecular and cognitive predictors of the continuum of autistic behaviours in fragile X. Neurosci Biobehav Rev 2007; 31(3):315-326.
279. Hagerman RJ, Hall DA, Coffey S et al. Treatment of fragile X-associated tremor ataxia syndrome (FXTAS) and related neurological problems. Clin Interv Aging 2008; 3(2):251-262.
280. Hagerman RJ, Hagerman PJ. Fragile X Syndrome: Diagnosis, Treatment and Research. 3rd edition. Baltimore: The Johns Hopkins University Press; 2002.
281. Erickson CA, Stigler KA, Posey DJ et al. Aripiprazole in autism spectrum disorders and fragile X syndrome. Neurotherapeutics 2010; 7(3):258-263.
282. Huber KM, Gallagher SM, Warren ST et al. Altered synaptic plasticity in a mouse model of fragile X mental retardation. Proc Natl Acad Sci USA 2002; 99(11):7746-7750.
283. Bear MF, Huber KM, Warren ST. The mGluR theory of fragile X mental retardation. Trends in Neurosciences 2004; 27(7):370-377.
284. Yan QJ, Rammal M, Tranfaglia M et al. Suppression of two major fragile X syndrome mouse model phenotypes by the mGluR5 antagonist MPEP. Neuropharmacology 2005; 49(7):1053-1066.
285. de Vrij FM, Levenga J, van der Linde HC et al. Rescue of behavioral phenotype and neuronal protrusion morphology in Fmr1 KO mice. Neurobiology of Disease 2008; 31(1):127-132.
286. Berry-Kravis E, Hessl D, Coffey S et al. A pilot open label, single dose trial of fenobam in adults with fragile X syndrome. Journal of Medical Genetics 2009; 46(4):266-271.
287. Berry-Kravis E, Cherubini M, Zarevics P et al. Arbaclofen for the Treatment of Children and Adults with Fragile X Syndrome: Results of a Phase 2, Randomized, Double-Blind, Placebo-Controlled, Crossover Study [Abstract]. Paper presented at: International Meeting for Autism Research; 2010; Philadelphia, PA.
288. Bilousova TV, Dansie L, Ngo M et al. Minocycline promotes dendritic spine maturation and improves behavioural performance in the fragile X mouse model. Journal of Medical Genetics 2009; 46(2):94-102.
289. Utari A, Chonchaiya W, Rivera SM et al. Side effects of minocycline treatment in patients with fragile x syndrome and exploration of outcome measures. Am J Intellect Dev Disabil 2010; 115(5):433-443.
290. de Diego-Otero Y, Romero-Zerbo Y, el Bekay R et al. Alpha-tocopherol protects against oxidative stress in the fragile X knockout mouse: an experimental therapeutic approach for the Fmr1 deficiency. Neuropsychopharmacology 2009; 34(4):1011-1026.
291. Wirojanan J, Jacquemont S, Diaz R et al. The efficacy of melatonin for sleep problems in children with autism, fragile X syndrome, or autism and fragile X syndrome. J Clin Sleep Med 2009; 5(2):145-150.
292. Romero-Zerbo Y, Decara J, el Bekay R et al. Protective effects of melatonin against oxidative stress in Fmr1 knockout mice: a therapeutic research model for the fragile X syndrome. J Pineal Res 2009; 46(2):224-234.
293. Wang LW, Berry-Kravis E, Hagerman RJ. Fragile X: leading the way for targeted treatments in autism. Neurotherapeutics 2010; 7(3):264-274.

CHAPTER 7

MOLECULAR PATHWAYS TO POLYGLUTAMINE AGGREGATION

Amy L. Robertson and Stephen P. Bottomley*

Department of Biochemistry and Molecular Biology, Monash University, Clayton, Victoria, Australia
*Corresponding Author: Stephen P. Bottomley—Email: steve.bottomley@monash.edu

Abstract: Over 100 human cellular proteins contain a repetitive polyglutamine tract, however, only nine of these proteins are associated with disease. In these proteins, the expanded polyQ tract perturbs the native conformation, resulting in an ordered aggregation process that leads to the formation of amyloid-like fibrils. The misfolding pathway involves the formation of prefibrillar oligomeric structures, which are proposed to be involved in cellular toxicity. Non-polyQ host protein regions modulate the misfolding pathway, suggesting an importance of protein context in aggregation. This chapter describes the current research regarding polyQ misfolding, with emphasis on the species populated during aggregation, suggesting an important role of protein context in modulating the aggregation pathway.

INTRODUCTION

There are over 100 polyglutamine (polyQ) containing proteins in the human proteome. It is not clear what the function of this tract is but its presence within some proteins can have a devastating effect. For nine proteins, expansion of the polyQ tract beyond a threshold of approximately 35 glutamines, causes a cascade of biological events that culminates in neurodegeneration. The presence of a pathological length polyQ repeat initiates a kinetically ordered aggregation process, involving a series of conformational changes within the native monomeric state culminating in the deposition of stable, fibrillar, amyloid-like aggregates within cytoplasmic and nuclear cellular compartments of specific neuronal cells. These aggregates associate with ubiquitin, chaperones and other important cellular components. Current research suggests that cellular toxicity arises from as yet unidentified prefibrillar species and that the mature fibrillar aggregates are nontoxic and possibly even protective.

Tandem Repeat Polymorphisms: Genetic Plasticity, Neural Diversity and Disease,
edited by Anthony J. Hannan ©2012 Landes Bioscience and Springer Science+Business Media.

Recent evidence suggests that the sequence and/or structural domains that surround the polyQ repeat influences the type and morphology of species formed. This chapter will discuss our current understanding of the polyQ-mediated misfolding pathways, highlighting the influence of protein context and avenues for potential therapeutic intervention.

PolyQ PROTEIN MISFOLDING

The ability of a protein to attain and maintain its correct native fold is essential for its functional activity and cellular homeostasis.[1,2] However, proteins are dynamic molecules with the propensity to populate a number of species within their conformational landscape (Fig. 1). Under normal cellular circumstances the native

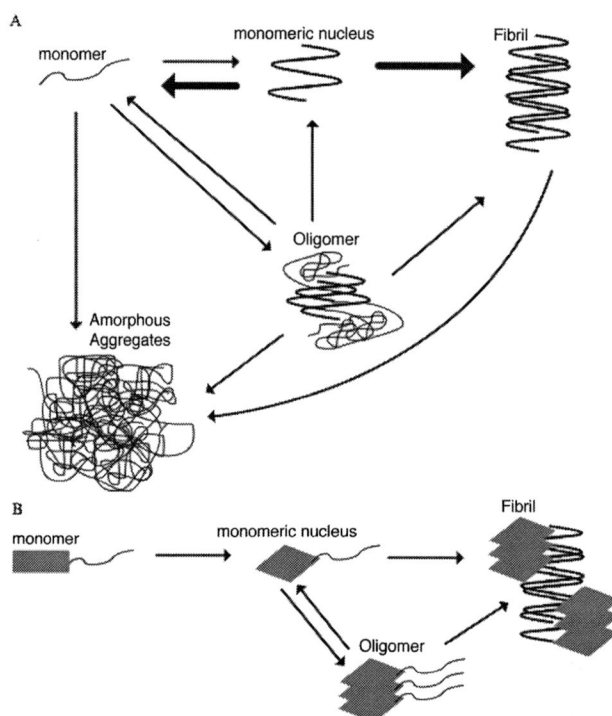

Figure 1. PolyQ aggregation pathways. A) Native polyQ exists in a random coil conformation, undergoing an unfavourable conformational change to a proposed β-sheet or β-hairpin monomer that acts as an aggregation nucleus. Rapid fibril elongation then occurs. Alternative pathways and/or stages in the aggregation pathway involve the formation of partially-structured oligomers. Oligomers may be on-pathway to fibril formation, or may be off-pathway species, requiring dissociation to the misfolded monomeric species prior to fibril formation or may be intermediates in the amorphous aggregation pathway. Under certain conditions, such as in the presence of excess molecular chaperones or EGCG, a pathway leading to the formation of amorphous aggregates is favoured over the fibrillar aggregation pathway. Addition of EGCG to amyloid fibrils can remodel the aggregates to an amorphous conformation. B) Multidomain Aggregation scheme of Ataxin-3. The Josephin domain (grey rectangle) undergoes a conformational change in the initial stage of aggregation to form an aggregation-prone monomeric nucleus prone to oligomerisation by intermolecular interactions between Josephin domains. This leads to subsequent polyQ aggregation and fibril growth by monomer addition.

state is maintained, and the formation of nonnative species is prevented by a large kinetic barrier.[3,4] This barrier can be overcome when a mutation, such as polyQ expansion, is present within the primary sequence of the protein. The presence of the expansion leads to an increased ability of the protein to populate nonnative species that lead to the formation of aberrant intermolecular interactions, ultimately resulting in fibril formation.

Several studies have shown that the addition of a pathological length polyQ tract to an otherwise innocuous protein leads to the formation of amyloid-like fibrils and cellular toxicity.[5-7] Amyloid fibrils typically display a nucleated growth polymerization mechanism (Fig. 2).[8,9] The rate of aggregation is determined by a nucleation phase, involving a slow transition from the native state to a nonnative conformer during the lag time of the reaction.[8] Nucleation involves a thermodynamically unfavourable transition to a high-energy conformer; either an altered monomeric state or a multimeric nucleus formed by condensation of several misfolded monomers.[8] Kinetic models describing the aggregation of a number of polyQ-containing proteins suggest that the nucleus is an unfavourable monomeric conformation.[10-12] The β-sheet rich fibril architecture has lead to the hypothesis that the misfolded monomeric conformer is β-sheet rich. The estimated equilibrium constant describing nucleation of a pathological length polyQ peptide is in the nanomolar range and therefore a misfolded conformer is not readily detectable using conventional ensemble approaches.[13] Detection of such low abundance species would require super-resolution single molecule approaches. Ensemble techniques have however, provided considerable insight into some of the species populated during polyQ misfolding, and suggest that monomeric, fibrillar and oligomeric conformations are populated and/or copopulated in various stages of aggregation (Fig. 2).

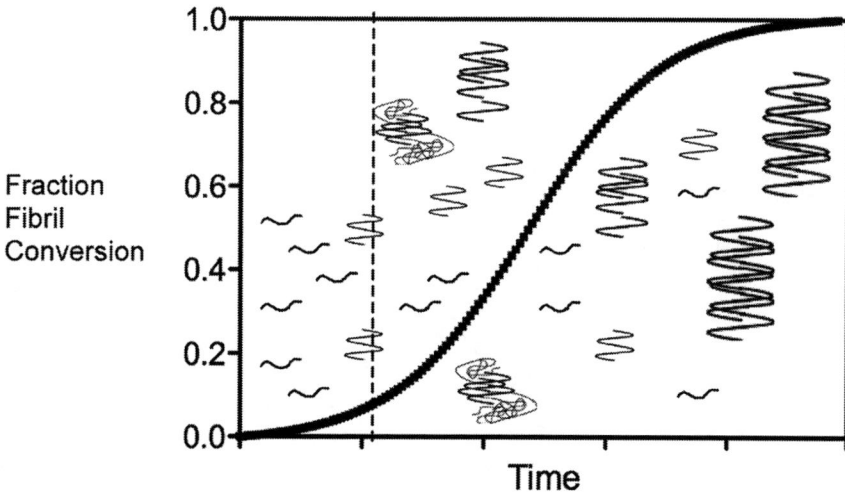

Figure 2. Nucleation-dependent aggregation kinetics. A representative trace of nucleation-dependent aggregation kinetics. The lag time in which competing nucleation-independent oligomerisation events may occur is divided from the aggregate/fibril elongation phase by the dashed line. The species proposed to be present during the stages of fibril formation are represented by the schematic drawings introduced in Figure 1.

MONOMERIC PolyQ

Experimental and theoretical models suggest that polyQ aggregation is entropically-driven.[14,15] The lowest energy conformation of a monomeric polyQ tract is random coil, which is stabilized by high configurational entropy.[14] A highly unfavourable conformational conversion from random coil to a β-sheet rich monomer is thought to be the nucleation step that initiates aggregation (Fig. 1A).[10,13,14] This reaction involves the formation of ordered structure, requiring a change in solvation state, in which the backbone solvates itself (rather than water as occurs in the random coil state).[14] Such a disorder to order transition requires the entropic barrier to be overcome. The ability to overcome this barrier is dependent on repeat length, as proteins containing a longer polyQ tract have a reduced aggregation lag time and hence faster rate of nucleation.[7] Other factors, including characteristics of the host protein and cellular factors can modulate the kinetic barrier to fibril formation and/or alter the molecular pathways leading to aggregation.

Experimental studies of the misfolding and aggregation behaviour of folded proteins have shown that aggregation occurs *via* a pathway including native state destabilization due to the presence of a mutation. This reduces the barrier between the native and partially-unfolded states, enhancing favourability for partially-unfolded conformations that have exposed aggregation-prone regions which interact and ultimately form aggregates.[1,4,16,17] PolyQ regions, however, are generally located in solvent-exposed, flexible protein regions, and as such native state destabilization is unnecessary for exposure of the aggregation-prone region.[18] Several studies have investigated the effects of a pathological length polyQ tract on protein stability finding varying effects depending upon the protein context.[6,7,12,19] Together, these studies indicate that native state destabilization is unnecessary for aggregation, however, perturbations to neighbouring protein regions can alter the misfolding pathway. Flanking domains appear to have an important role in polyQ aggregation, further highlighted by observations that a number of polyQ disease proteins undergo proteolytic cleavage to initiate aggregation, suggesting a suppressive effect of the host protein context on aggregation.[20-22] Recent studies in the polyQ field have highlighted an increasingly important role of protein context in aggregation and toxicity that will be discussed in a later section.

PolyQ FIBRILS

It is essential to determine the properties and structure of amyloid in order to understand disease-related misfolding. Although amyloid-like fibrils formed by different proteins have widely varying sidechain topologies, they appear to have a number of generic biochemical properties, including high β-sheet content and detergent insolubility.[23,24] The polyQ diseases are somewhat of an anomaly amongst the group of amyloid diseases, being caused by a homogenous repeating glutamine segment, which is intrinsically capable of aggregation.[10] Similar to other amyloid-like fibrils, polyQ fibrils are β-sheet rich.[25] Consistent with the β-sheet fibril architecture, Max Perutz suggested a polar zipper model for polyQ aggregates, whereby polyQ fibrils are stabilized by a hydrogen bonding network between the glutamine main-chain and side-chain amide groups.[25]

The polypepide backbone has an intrinsic propensity for aggregation in situations where native interactions are unfavoured. Proteins and peptides that are natively unstructured, including polyQ, α-synuclein, and other homopolymeric amino acids, have a high propensity for stabilization by fibril formation.[1,26] PolyQ peptides as short as seven glutamines in length form fibrillar aggregates, whilst the threshold to disease typically ranges from 37 to 40 glutamines, indicating that the protein context significantly modulates aggregation.[27-31]

PolyQ OLIGOMERS

Native and fibrillar polyQ species have been well characterized due to their high stability. It is, however, much more difficult to study species that are only transiently populated. Prefibrillar and oligomeric conformations are typical intermediates populated by amyloid-forming proteins, and contrasting the nucleated-growth kinetic model of polyQ aggregation shown by Wetzel and colleagues, which suggests a simplistic equilibrium involving only monomer and fibril, are a number of reports indicating oligomer formation by polyQ peptides and proteins. Using atomic force microscopy to detect oligomers in real-time, Muchowski and colleagues detected oligomers formed by polyQ peptides.[27] Nagai and colleagues also used fluorescence correlation spectroscopy to show that YFP- and CFP-polyQ fusion proteins form polyQ-dependent oligomers.[32] The oligomeric species formed from polyQ-only peptides, however, appeared to be less stable than those formed by peptides that also contained the Huntington (Htt) exon-1 flanking residues.[27] These data suggest that the polyQ flanking regions have a role in stabilizing oligomeric populations. There are several other reports suggesting that Htt exon-1 peptides can form oligomers and protofibrils in vivo and they have also been detected in HD patient brains and a mouse model, however, the relevance of such structures to HD pathology is yet to be determined. Htt exon-1 oligomers and protofibrils cross-react with the generic A11 and OC antibodies produced by the Glabe laboratory.[33] These antibodies were developed against oligomers of Aβ, and have been shown to bind to oligomers of many other proteins, indicating similar structural organization. Similarities in the morphology of protofibrillar and oligomeric species suggest that they may have similar mechanisms of toxicity, and it is possible that a number of different species are toxic. A recent in vitro study showed that Htt exon-1 can form a range of different oligomeric aggregates varying in size and stability, some of which are on- and off-pathway to endpoint fibril formation.[27] Considering the complexity of the in vitro aggregation pathways, it is conceivable that the situation in the cell is considerably more complex. The precise mechanisms by which oligomeric species induce toxicity are unknown, however, are thought to include membrane permeabilization, perturbations to redox status, impairment of calcium metabolism and disruption of proteasomal activity, ultimately leading to apoptosis.[4,34,35]

Oligomers are unfavourable species, hence are usually only transiently populated. As is the case for fibril formation, the initial monomeric conformation must undergo a considerable entropic penalty in order to form oligomers. Unlike the endpoint fibrils, however, that are highly stable due to their hydrogen-bonding network, subunits within oligomers are often loosely associated via few interactions.

Therefore, their lifetimes are relatively short-lived, due to the entropic driving force for dissociation and/or the growth and further stabilization leading to fibril formation.

Limited proteolysis, hydrogen-deuterium exchange and FTIR studies suggest that oligomers are generally less ordered than fibrils, have a reduced β-sheet content and enhanced exposure of hydrophobic regions compared to fibrils.[36-38] The transient nature of oligomeric populations and conformational heterogeneity has, however, limited their high-resolution structural characterization. Further spectroscopic techniques, including FRET,[38] NMR techniques[39-41] and single-molecule fluorescence[42] also show promise in detecting and characterizing small populations of oligomeric complexes.

INFLUENCE OF FLANKING DOMAINS ON PolyQ AGGREGATION

There are a number of lines of evidence suggesting that domains flanking the polyQ tract can modulate aggregation.[43,44] The composition of regions flanking repeats can alter the biophysical behavior of the polyQ region and adjust the kinetic barrier to amyloid formation.[45,46] Host protein regions have also been shown to self-associate and form oligomers, independent or dependent of the presence of a pathological length polyQ tract.[43,44,47,48] Together these data indicate an important role of protein context in aggregation and cellular toxicity.

A number of the disease proteins undergo proteolytic cleavage prior to aggregation.[20-22] Nuclear inclusions found in HD, SCA7, DRPLA and SBMA patients contain cleaved polyQ-containing fragments of disease proteins.[20-22,49,50] These data suggest that the protein context is "protective", acting to suppress polyQ-mediated aggregation prior to proteolysis. This could be due to the intrinsic solubility and/or stability of the host protein, suppressing the propensity of the polyQ tract to undergo the conformational changes necessary to initiate aggregation, or due to bulky flanking domains causing steric hindrance to block inter-polyglutamine interactions. Indeed the ability of large, soluble flanking domains to suppress polyQ aggregation is utilized in the production of recombinant variants of pathological length polyQ proteins.[51,52] The effect of folded flanking domains on the overall chain thermodynamics has not been described, however data obtained thus far clearly show that folded flanking domains alter the kinetic barrier to fibril formation. Larger proteins have slower translational diffusion and rotational relaxation rates compared to smaller proteins due to altered Brownian motion. In systems where proteins aggregate, altered molecular motion of larger proteins would affect the rate of conformational change and therefore alter nucleation, and would affect the frequency of collisions in solution, modulating the elongation rate. Proteolysis is a mechanism by which proteins can overcome the protective properties of the host protein.

There is an increasing body of literature that the native interactions of flanking domains can modulate aggregation. A recent study has shown that interdomain interactions between the N- and C-terminal domains of the androgen receptor can enhance the rate of polyglutamine-mediated aggregation.[52] This has lead to the hypothesis that stable interactions between domains N- and C-terminal to the polyQ tract can restrict the conformational freedom of the polyQ tract, decreasing the entropy and enhancing the propensity for the formation of species on-pathway to aggregation. This is not a widely applicable hypothesis, as the polyQ tracts of seven of the nine disease proteins are located close to the N- or C-termini of the proteins, and not in an interdomain location.[18] A number of these proteins, however, form intermolecular interactions in the native state. For example, the Josephin domain, which is present at the N-terminus of ataxin-3, has the ability to self-associate and form small, SDS-soluble aggregates.[53,54] In the context of ataxin-3 with a pathological length polyQ tract, this initial Josephin self-association precedes the subsequent deposition of polyQ-dependent SDS-insoluble fibrils (Fig. 1B).[44] There is a complex interplay between the Josephin and polyQ domains and importantly the Josephin self-association stage lowers the kinetic barrier to polyQ aggregation. The mechanistic rationale for this may be that the formation of small Josephin aggregates enhances the local polyQ concentration, and/or the increased molecular size entropically favours polyQ aggregation. Similar multidomain aggregation mechanisms have been described for Htt exon-1[43] and ataxin-1,[47,48] in addition to the polyalanine-containing protein PABPN1[55] and the yeast prion Sup35p.[56] Further investigations of the aggregation pathways of the remainder of the polyQ disease proteins will be required to determine whether multidomain aggregation mechanisms are generally applicable to the polyQ disease family.

INHIBITING AND DIVERTING PolyQ AGGREGATION: THERAPEUTIC PROSPECTS

Considering stabilization of the polyQ tract would adjust the entropic barrier in favour of aggregation, molecular tools directly targeting the polyQ tract would need to ensure that conformational restriction does not enhance the favourability of a pathological length polyQ tract for exploring regions of the conformational landscape on-pathway to fibril formation. The polyQ binding peptide, QBP1, interacts with pathological length polyQ repeats, suppressing aggregation and toxicity.[57] The precise mechanism by which QBP1 inhibits aggregation is unknown, however, it would be interesting to determine how this peptide affects the thermodynamics of the polyQ chain. The recent example showing enhanced aggregation of the androgen receptor when the N- and C-terminal domains are interacting suggests that a protein's native intramolecular contacts also have an important role in determining the susceptibility for polyQ-dependent aggregation.[58]

The expanding evidence that a number of polyQ proteins have multidomain aggregation mechanisms provides an alternative route by which the formation of aberrant polyQ species could be suppressed. A recent study from our laboratory showed that a small heat shock protein can suppress the aggregation of ataxin-3 by inhibiting the Josephin aggregation stage.[54] Similarly, intrabodies targeting the polyproline region and N-terminal domain flanking the polyQ tract in Htt exon-1 suppress aggregation and toxicity.[59-61] In these cases, suppressing the initial non-polyQ interactions involved in the aggregation pathways reduces the kinetic benefit of these steps to polyQ aggregation. Opposing aggregation by avoiding the native interactions of host protein domains could provide targets for rational therapeutic approaches.

The ultimate aim of therapeutic approaches is to avoid the formation of toxic species that are formed during fibrillar aggregation, including oligomeric and/or prefibrillar intermediates. Overexpression of the molecular chaperones Hsp70 and Hsp40 leads to avoidance of fibrillar aggregation and toxicity by a mechanism ultimately leading to the deposition of soluble amorphous aggregates.[62] Similarly the cellular chaperonin TRiC diverts pathological Htt exon-1 proteins along an alternative pathway involving the deposition of large nontoxic oligomers, which are clearly distinct from the toxic small oligomers that form under native conditions.[63] Geldanamycin and similar derivatives, which inhibit Hsp90, leading to upregulated expression of Hsp70 and Hsp40, are effective in suppressing toxicity in polyQ and other amyloid disease models.[64,65]

A number of screening studies have identified small molecules that modulate polyQ aggregation.[66-68] A number of polyphenolic compounds suppress fibril formation and toxicity, and in a similar manner to molecular chaperones, lead to the deposition of off-pathway amorphous aggregates. The green tea polyphenol, EGCG, reduces the formation of SDS-insoluble aggregates and suppresses toxicity in yeast and *Drosophila* models. In vitro analyses suggest that EGCG prevents the initial comformational changes involved in the misfolding of Htt exon-1 and leads to the formation of large off-pathway aggregates.[68] Considering the significant alterations in the misfolding pathway, EGCG could provide a good therapy for the polyQ and amyloid diseases in general.

CONCLUSION

The initial events leading to polyQ aggregation occur within the monomeric proteins, leading to a series of conformational changes with the ultimate deposition and accumulation of fibrillar amyloid-like aggregates. Accumulating evidence suggests that oligomeric conformations are also populated during the misfolding pathway and observations within the general amyloid field suggest that such prefibrillar intermediates may be involved in toxicity. A number of recent studies have highlighted a role of host protein regions in modulating the species populated during aggregation and the rates of their accumulation. This paradigm suggests that therapeutic approaches that target the flanking regions, in addition to the polyQ repeat may therefore prove to be effective in suppressing aggregation and toxicity.

REFERENCES

1. Chiti F, Dobson CM. Protein misfolding, functional amyloid, and human disease. Annu Rev Biochem 2006; 75:333-366.
2. Anfinsen CB. Principles that govern the folding of protein chains. Science 1973; 181(96):223-230.
3. Jahn TR, Radford SE. Folding versus aggregation: polypeptide conformations on competing pathways. Arch Biochem Biophys 2008; 469(1):100-117.
4. Stefani M, Dobson CM. Protein aggregation and aggregate toxicity: new insights into protein folding, misfolding diseases and biological evolution. J Mol Med 2003; 81(11):678-699.
5. Ordway JM, Tallaksen-Greene S, Gutekunst CA et al. Ectopically expressed CAG repeats cause intranuclear inclusions and a progressive late onset neurological phenotype in the mouse. Cell 1997; 91(6):753-763.
6. Tanaka M, Morishima I, Akagi T et al. Intra- and intermolecular beta-pleated sheet formation in glutamine-repeat inserted myoglobin as a model for polyglutamine diseases. J Biol Chem 2001; 276(48):45470-45475.
7. Robertson AL, Horne J, Ellisdon AM et al. The structural impact of a polyglutamine tract is location-dependent. Biophys J 2008; 95(12):5922-5930.
8. Ferrone F. Analysis of protein aggregation kinetics. Methods Enzymol 1999; 309:256-274.
9. Kheterpal I, Cook KD, Wetzel R. Hydrogen/deuterium exchange mass spectrometry analysis of protein aggregates. Methods Enzymol 2006; 413:140-166.
10. Chen S, Ferrone FA, Wetzel R. Huntington's disease age-of-onset linked to polyglutamine aggregation nucleation. Proc Natl Acad Sci U S A 2002; 99(18):11884-11889.
11. Ellisdon AM, Pearce MC, Bottomley SP. Mechanisms of ataxin-3 misfolding and fibril formation: kinetic analysis of a disease-associated polyglutamine protein. J Mol Biol 2007; 368(2):595-605.
12. Ignatova Z, Gierasch LM. Extended polyglutamine tracts cause aggregation and structural perturbation of an adjacent beta barrel protein. J Biol Chem 2006; 281(18):12959-12967.
13. Bhattacharyya AM, Thakur AK, Wetzel R. polyglutamine aggregation nucleation: thermodynamics of a highly unfavorable protein folding reaction. Proc Natl Acad Sci U S A 2005; 102(43):15400-15405.
14. Wang X, Vitalis A, Wyczalkowski MA et al. Characterizing the conformational ensemble of monomeric polyglutamine. Proteins 2006; 63(2):297-311.
15. Vitalis A, Lyle N, Pappu RV. Thermodynamics of beta-sheet formation in polyglutamine. Biophys J 2009; 97(1):303-311.
16. Dumoulin M, Canet D, Last AM et al. Reduced global co-operativity is a common feature underlying the amyloidogenicity of pathogenic lysozyme mutations. J Mol Biol 2005; 346(3):773-788.
17. Booth DR, Sunde M, Bellotti V et al. Instability, unfolding and aggregation of human lysozyme variants underlying amyloid fibrillogenesis. Nature 1997; 385(6619):787-793.

18. Faux NG, Bottomley SP, Lesk AM et al. Functional insights from the distribution and role of homopeptide repeat-containing proteins. Genome Res 2005; 15(4):537-551.
19. Chow MK, Ellisdon AM, Cabrita LD et al. Polyglutamine expansion in ataxin-3 does not affect protein stability: implications for misfolding and disease. J Biol Chem 2004; 279(46):47643-47651.
20. Wellington CL, Ellerby LM, Hackam AS et al. Caspase cleavage of gene products associated with triplet expansion disorders generates truncated fragments containing the polyglutamine tract. J Biol Chem 1998; 273(15):9158-9167.
21. Goti D, Katzen SM, Mez J et al. A mutant ataxin-3 putative-cleavage fragment in brains of Machado-Joseph disease patients and transgenic mice is cytotoxic above a critical concentration. J Neurosci 2004; 24(45):10266-10279.
22. Nucifora FC Jr, Ellerby LM, Wellington CL et al. Nuclear localization of a noncaspase truncation product of atrophin-1, with an expanded polyglutamine repeat, increases cellular toxicity. J Biol Chem 2003; 278(15):13047-13055.
23. Rambaran RN, Serpell LC. Amyloid fibrils: abnormal protein assembly. Prion 2008; 2(3):112-117.
24. Wanker EE, Scherzinger E, Heiser V et al. Membrane filter assay for detection of amyloid-like polyglutamine-containing protein aggregates. Methods Enzymol 1999; 309:375-386.
25. Perutz MF, Johnson T, Suzuki M et al. Glutamine repeats as polar zippers: their possible role in inherited neurodegenerative diseases. Proc Natl Acad Sci U S A 1994; 91(12):5355-5358.
26. Uversky VN. Amyloidogenesis of natively unfolded proteins. Curr Alzheimer Res 2008; 5(3):260-287.
27. Legleiter J, Mitchell E, Lotz GP et al. Mutant huntingtin fragments form oligomers in a polyglutamine length-dependent manner in vitro and in vivo. J Biol Chem 2010; 285(19):14777-14790.
28. La Spada AR, Wilson EM, Lubahn DB et al. Androgen receptor gene mutations in X-linked spinal and bulbar muscular atrophy. Nature 1991; 352(6330):77-79.
29. Banfi S, Servadio A, Chung MY et al. Identification and characterization of the gene causing type 1 spinocerebellar ataxia. Nat Genet 1994; 7(4):513-520.
30. Benomar A, Krols L, Stevanin G et al. The gene for autosomal dominant cerebellar ataxia with pigmentary macular dystrophy maps to chromosome 3p12-p21.1. Nat Genet 1995; 10(1):84-88.
31. David G, Abbas N, Stevanin G et al. Cloning of the SCA7 gene reveals a highly unstable CAG repeat expansion. Nat Genet 1997; 17(1):65-70.
32. Takahashi Y, Okamoto Y, Popiel HA et al. Detection of polyglutamine protein oligomers in cells by fluorescence correlation spectroscopy. J Biol Chem 2007; 282(33):24039-24048.
33. Kayed R, Head E, Thompson JL et al. Common structure of soluble amyloid oligomers implies common mechanism of pathogenesis. Science 2003; 300(5618):486-489.
34. Ding TT, Lee SJ, Rochet JC et al. Annular alpha-synuclein protofibrils are produced when spherical protofibrils are incubated in solution or bound to brain-derived membranes. Biochemistry 2002; 41(32):10209-10217.
35. Hyun DH, Lee M, Hattori N et al. Effect of wild-type or mutant Parkin on oxidative damage, nitric oxide, antioxidant defenses, and the proteasome. J Biol Chem 2002; 277(32):28572-28577.
36. Kheterpal I, Chen M, Cook KD et al. Structural differences in Abeta amyloid protofibrils and fibrils mapped by hydrogen exchange—mass spectrometry with on-line proteolytic fragmentation. J Mol Biol 2006; 361(4):785-795.
37. Myers SL, Thomson NH, Radford SE et al. Investigating the structural properties of amyloid-like fibrils formed in vitro from beta2-microglobulin using limited proteolysis and electrospray ionisation mass spectrometry. Rapid Commun Mass Spectrom 2006; 20(11):1628-1636.
38. Kaylor J, Bodner N, Edridge S et al. Characterization of oligomeric intermediates in alpha-synuclein fibrillation: FRET studies of Y125W/Y133F/Y136F alpha-synuclein. J Mol Biol 2005; 353(2):357-372.
39. O'Sullivan DB, Jones CE, Abdelraheim SR et al. NMR characterization of the pH 4 beta-intermediate of the prion protein: the N-terminal half of the protein remains unstructured and retains a high degree of flexibility. Biochem J 2007; 401(2):533-540.
40. Yu L, Edalji R, Harlan JE et al. Structural characterization of a soluble amyloid beta-peptide oligomer. Biochemistry 2009; 48(9):1870-1877.
41. Huang H, Milojevic J, Melacini G. Analysis and optimization of saturation transfer difference NMR experiments designed to map early self-association events in amyloidogenic peptides. J Phys Chem B 2008; 112(18):5795-5802.
42. Orte A, Birkett NR, Clarke RW et al. Direct characterization of amyloidogenic oligomers by single-molecule fluorescence. Proc Natl Acad Sci U S A 2008; 105(38):14424-14429.
43. Thakur AK, Jayaraman M, Mishra R et al. Polyglutamine disruption of the huntingtin exon 1 N terminus triggers a complex aggregation mechanism. Nat Struct Mol Biol 2009; 16(4):380-389.
44. Ellisdon AM, Thomas B, Bottomley SP. The two-stage pathway of ataxin-3 fibrillogenesis involves a polyglutamine-independent step. J Biol Chem 2006; 281(25):16888-16896.
45. Nozaki K, Onodera O, Takano H et al. Amino acid sequences flanking polyglutamine stretches influence their potential for aggregate formation. Neuroreport 2001; 12(15):3357-3364.

46. Bhattacharyya A, Thakur AK, Chellgren VM et al. Oligoproline effects on polyglutamine conformation and aggregation. J Mol Biol 2006; 355(3):524-535.

47. De Chiara C, Menon RP, Adinolfi S et al. The AXH domain adopts alternative folds the solution structure of HBP1 AXH. Structure 2005; 13(5):743-753.

48. Burright EN, Davidson JD, Duvick LA et al. Identification of a self-association region within the SCA1 gene product, ataxin-1. Hum Mol Genet 1997; 6(4):513-518.

49. DiFiglia M, Sapp E, Chase KO et al. Aggregation of huntingtin in neuronal intranuclear inclusions and dystrophic neurites in brain. Science 1997; 277(5334):1990-1993.

50. Ellerby LM, Andrusiak RL, Wellington CL et al. Cleavage of atrophin-1 at caspase site aspartic acid 109 modulates cytotoxicity. J Biol Chem 1999; 274(13):8730-8736.

51. Klein FA, Pastore A, Masino L et al. Pathogenic and nonpathogenic polyglutamine tracts have similar structural properties: towards a length-dependent toxicity gradient. J Mol Biol 2007; 371(1):235-244.

52. Busch A, Engemann S, Lurz R et al. Mutant huntingtin promotes the fibrillogenesis of wild-type huntingtin: a potential mechanism for loss of huntingtin function in Huntington's disease. J Biol Chem 2003; 278(42):41452-41461.

53. Masino L, Nicastro G, Menon RP et al. Characterization of the structure and the amyloidogenic properties of the Josephin domain of the polyglutamine-containing protein ataxin-3. J Mol Biol 2004; 344(4):1021-1035.

54. Robertson AL, Headey SJ, Saunders HM et al. Small heat-shock proteins interact with a flanking domain to suppress polyglutamine aggregation. Proc Natl Acad Sci U S A 2010; 107(23):10424-10429.

55. Klein AF, Ebihara M, Alexander C et al. PABPN1 polyalanine tract deletion and long expansions modify its aggregation pattern and expression. Exp Cell Res 2008; 314(8):1652-1666.

56. Fernandez-Bellot E, Guillemet E, Baudin-Baillieu A et al. Characterization of the interaction domains of Ure2p, a prion-like protein of yeast. Biochem J 1999; 338(Pt 2):403-407.

57. Nagai Y, Tucker T, Ren H et al. Inhibition of polyglutamine protein aggregation and cell death by novel peptides identified by phage display screening. J Biol Chem 2000; 275(14):10437-10442.

58. Orr CR, Montie HL, Liu Y et al. An interdomain interaction of the androgen receptor is required for its aggregation and toxicity in spinal and bulbar muscular atrophy. J Biol Chem 2010.

59. Snyder-Keller A, McLear JA, Hathorn T et al. Early or Late-Stage Anti-N-Terminal Huntingtin Intrabody Gene Therapy Reduces Pathological Features in B6.HDR6/1 Mice. J Neuropathol Exp Neurol 2010.

60. Southwell AL, Khoshnan A, Dunn DE et al. Intrabodies binding the proline-rich domains of mutant huntingtin increase its turnover and reduce neurotoxicity. J Neurosci 2008; 28(36):9013-9020.

61. Legleiter J, Lotz GP, Miller J et al. Monoclonal antibodies recognize distinct conformational epitopes formed by polyglutamine in a mutant huntingtin fragment. J Biol Chem 2009; 284(32):21647-21658.

62. Muchowski PJ, Schaffar G, Sittler A et al. Hsp70 and hsp40 chaperones can inhibit self-assembly of polyglutamine proteins into amyloid-like fibrils. Proc Natl Acad Sci U S A 2000; 97(14):7841-7846.

63. Behrends C, Langer CA, Boteva R et al. Chaperonin TRiC promotes the assembly of polyQ expansion proteins into nontoxic oligomers. Mol Cell 2006; 23(6):887-897.

64. Herbst M, Wanker EE. Small molecule inducers of heat-shock response reduce polyQ-mediated huntingtin aggregation. A possible therapeutic strategy. Neurodegener Dis 2007; 4(2-3):254-260.

65. Waza M, Adachi H, Katsuno M et al. Modulation of Hsp90 function in neurodegenerative disorders: a molecular-targeted therapy against disease-causing protein. J Mol Med 2006; 84(8):635-646.

66. Heiser V, Engemann S, Brocker W et al. Identification of benzothiazoles as potential polyglutamine aggregation inhibitors of Huntington's disease by using an automated filter retardation assay. Proc Natl Acad Sci U S A 2002; 99 Suppl 4:16400-16406.

67. Zhang X, Smith DL, Meriin AB et al. A potent small molecule inhibits polyglutamine aggregation in Huntington's disease neurons and suppresses neurodegeneration in vivo. Proc Natl Acad Sci U S A 2005; 102(3):892-897.

68. Ehrnhoefer DE, Duennwald M, Markovic P et al. Green tea (-)-epigallocatechin-gallate modulates early events in huntingtin misfolding and reduces toxicity in Huntington's disease models. Hum Mol Genet 2006; 15(18):2743-2751.

CHAPTER 8

POLYGLUTAMINE AGGREGATION
IN HUNTINGTON AND RELATED DISEASES

Saskia Polling, Andrew F. Hill and Danny M. Hatters*

Department of Biochemistry and Molecular Biology, University of Melbourne, Melbourne, Victoria, Australia
*Corresponding Author: Danny M. Hatters—Email: dhatters@unimelb.edu.au

Abstract: Polyglutamine (polyQ)-expansions in different proteins cause nine neurodegenerative diseases. While polyQ aggregation is a key pathological hallmark of these diseases, how aggregation relates to pathogenesis remains contentious. In this chapter, we review what is known about the aggregation process and how cells respond and interact with the polyQ-expanded proteins. We cover detailed biophysical and structural studies to uncover the intrinsic features of polyQ aggregates and concomitant effects in the cellular environment. We also examine the functional consequences of polyQ aggregation and how cells may attempt to intervene and guide the aggregation process.

INTRODUCTION

Nine neurodegenerative diseases, including Huntington disease (HD), result from an expansion in a CAG-codon sequence, encoding polyglutamine (polyQ), within a gene specific to each disease.[1-8] The expanded polyQ sequences cause the host proteins to form large intracellular aggregates known as inclusions.[9] In HD, the polyQ expansion occurs within exon 1 of the huntingtin (Htt) protein.[1] For the other diseases, the polyQ expansions occur in different ataxin-proteins (for spinocerebellar ataxias (SCA) 1, 2, 3, 6, 7, and 17);[2-8] the androgen receptor (for spinobulbar muscular atrophy (SBMA))[10] and the atrophin-1 protein (for dentatorubral pallidoluysian atrophy (DRPLA)).[11,12] For all these diseases the penetrance, age of onset and severity correlates with polyQ length beyond a critical disease threshold of approximately 40 residues depending on the disease protein.[13] For HD, pathological disease threshold is approximately 37 glutamines.[9] Here we discuss

Tandem Repeat Polymorphisms: Genetic Plasticity, Neural Diversity and Disease,
edited by Anthony J. Hannan ©2012 Landes Bioscience and Springer Science+Business Media.

how polyQ expansions affect the structure and aggregation properties of the host proteins and proposed mechanisms for how this process relates to pathogenesis.

PolyQ STRUCTURE, ASSEMBLY AND AGGREGATION

Detailed insight into the core intrinsic features of polyQ and their effects on the host protein has been gleaned from in vitro studies on purified proteins. Purified polyQ-containing peptides and proteins form SDS-insoluble, high molecular weight protein aggregates.[14,15] These aggregates are similar in morphology to the amyloid fibrils formed by proteins in other neurodegenerative diseases such as the prion protein in Creutzfeldt-Jakob disease and Aβ peptide in Alzheimer's disease.[14-17] The formation of aggregates occurs when polyQ-lengths reach the disease threshold, which suggests aggregation is fundamental to disease mechanisms.[9,18]

Aggregation of small polyQ-containing peptides and polyQ-expanded Htt has been proposed to occur via a nucleated growth mechanism, which is similar to models generally described for amyloid fibril formation.[15,19-22] In this model, aggregation proceeds rapidly upon the formation of an energetically unfavourable nucleus.[19] A key feature of nucleated aggregation is the so called "lag phase", a characteristic slow stage of aggregate accumulation early in the aggregation pathway, which can be eliminated by seeding preformed fibril fragments of Htt and polyQ-expanded peptides into the reaction.[15,22] In addition, the aggregation rates are time- and concentration-dependent, which is a feature of nucleated kinetics.[23,24]

How the fibrils extend once nucleated is still not clearly resolved. One model posits fibril elongation to occur via cycles of monomer binding ("docking") and subsequent intramolecular re-arrangements ("locking").[20,21] However other studies have shown monomers form small soluble oligomers that result in the formation of rod-like particles prior to assembling into the mature insoluble fibrils (Fig. 1).[25-27] These soluble aggregates/oligomers are spherical structures with a diameter spanning 5-65 nm that assemble into rod-like clusters and more rarely annular clusters.[27,28] How oligomers transition to fibrils is unclear, however, they dissipate as fibrils assemble suggesting they directly convert to fibrils or they form transiently in a competing non-amyloidogenic pathway.[27]

The transient nature of the aggregation nucleus and early oligomeric intermediates has made their direct measurement a technical challenge. However, modelling of aggregation data of synthetic polyQ-peptides has suggested that the aggregation nucleus is an energetically unfavourable conformation of monomer.[23,29] This suggests that nucleation requires the monomer to change conformation into an aggregation-permissive form, rather than by condensation of multiple monomers into an oligomer.

More recent studies have further indicated polyQ monomers to have unusual structural attributes that challenge the conventional nucleated growth model for aggregation kinetics and suggest novel mechanisms of cellular toxicity.[25,30-32] First, while polyQ monomers are classically disordered by circular dichroism and NMR spectroscopic assessment,[33-35] they are unusually compact compared to other "disordered" proteins.[25,31,32] The monomers also are heterogeneous in conformation-like other disordered proteins, yet unlike classic disordered proteins seem to be more mechanically rigid.[30,31] In addition, fluorescence correlation spectroscopy (FCS) and fluorescence resonance energy transfer (FRET) experiments suggest polyQ monomers become increasingly compact upon increasing polyQ length.[25,32] These results have been proposed to result from water acting as an

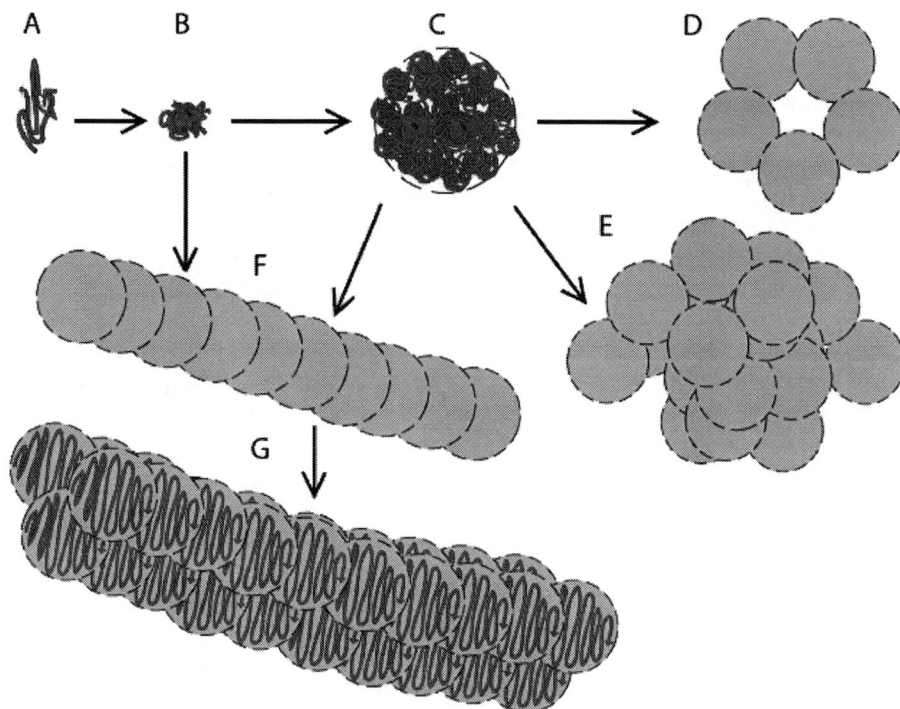

Figure 1. Model of polyQ aggregation. An expanded polyQ repeat in a protein monomer (A) leads to a compact conformation (B). Similar to compaction the monomer forms spherical oligomers (C) some of which assemble into rare annular structures (D), large amorphous structures (E) or premature fibrils (F). These fibrils (F) may arise from spherical oligomers or directly from monomers and continue to convert into mature insoluble fibril (G) with a high β-sheet content.

unfavourable solvent of the glutamine side chains relative to intra-chain interactions, which locks monomers into slowly-exchanging, compact globule conformations.[36,37] The consequence of rigid compact monomers in vivo is unknown, however, compact monomers may be difficult to degrade by the proteasome, which requires substrate proteins to be unfolded. Indeed, polyQ sequences have been reported to be resistant to proteasomal degradation.[38]

The formation of compact and rigid monomer conformations is likely to also influence the mode of aggregation because the forces driving compact monomers and aggregation are both likely to be governed by a preference to exclude interfaces with water.[32] Intramolecular glutamine side chain and peptide backbone interactions are proposed to be more favourable within a polyQ sequence than for chain-solvent interactions.[32] Hence longer polyQ lengths increasingly promote the favourability of distant parts of the polyQ chain to interact and stabilize a compact monomer structure.[25] However, glutamine residues can interact equivalently at many different points within a polyQ peptide sequence, which suggests a high level of heterogeneity in compact, largely rigid monomer conformations.[31]

While polyQ monomers have an overall disordered conformation, fibrils formed from these proteins are similar to those of other amyloid forming proteins and have a high β-sheet content.[14,39,40] β-Sheet formation seems to occur *after* the early steps of aggregation since soluble aggregates retain a classically disordered structure.[26] Molecular dynamic studies indicates β-sheet content to be energetically unfavourable in monomers of longer polyQ-lengths relative to nonspecific collapsed states.[36] However, β-content becomes more favourable than disordered monomers after monomers oligomerize into liquid-like clusters.[41] Initially the oligomers comprise of disordered monomers where polyQ-chains are solvated by other chains instead of solvent. However, conformational fluctuations within the liquid-like peptide-rich environment subsequently favour stabilization of β-sheet structures which subsequently leads to a shift into a more fibrillar architecture.[26,41]

It is still not clear as to how the β-structure of polyQ fibrils is arranged and it could be affected by parameters such as flanking residues and buffer conditions.[42,43] Various β-strand assembly models have been proposed including parallel and anti-parallel β-sheets, polar β-sheet zipper, β-helix nanotubes, superpleated or cross-β structures.[44-51] Regardless of the structure, the β-sheet content contributes to the robust stability of the aggregates, which are resistant to normally denaturing detergents such as SDS.[14,15,26,41,52]

PolyQ AGGREGATION IN CELLS

Htt aggregates assemble into micrometer-sized punctate structures in cells, known as inclusion bodies, in human brain,[53] transgenic mice[54] and transfected cell line models of HD expressing pathogenic Htt fragments.[15,55] Inclusions appear to consist of a mixture of aggregates with granular and fibrillar morphology similar in dimensions to oligomers and fibrils spontaneously formed by purified proteins.[27]

The mechanism of aggregation and assembly into inclusions in cells is not well understood, but seems to differ somewhat to that of purified proteins. For example, fluorescence detection sedimentation velocity experiments (FDS) were applied to study the population size and heterogeneity of Htt fragments of both purified protein and cell lysates.[56,57] It was found that cells formed an invariant pool of polyQ-expanded Htt oligomers with a sedimentation co-efficient in the order of 140S (i.e., around 30 nm in nominal diameter), which contrasts to the purified protein which formed an increasingly heterogeneous aggregate population over time in the range of 100-6,000S.[56,57] Another microscopy study suggested a two-tiered cascade of oligomer assembly in individual cells whereby a discrete population of oligomers formed above a certain threshold concentration of monomers, which then rapidly redistributed into inclusions.[58] This mode of aggregation is supported by another study using fluorescent sensors specific for monomers to show oligomers forming spontaneously and evenly through the cytosol of individual cells and seemed often coupled to the presence of small inclusions.[57] Collectively these papers indicate aggregation in cells to proceed via discrete subpopulations of oligomers, which raises the possibility that quality control mechanisms engage with the aggregates and shift them into inclusions.

Indeed, several lines of evidence indicate inclusion bodies are actively formed by cells as a protective mechanism to sequester and detoxify diffuse forms of the polyQ-expanded protein.[59-63] First is that microtubules and associated motor proteins dynein and dynactin are essential cofactors for inclusion formation and that disruption of the microtubular network substantially enhances the toxicity of aggregating-forms of polyQ.[59,64,65] Second is that cultured neurons expressing pathogenic Htt fragments are more likely to survive— and have lower levels of proteasome impairment—when inclusions have formed relative to neurons that retain a diffusely distributed pool of Htt, suggesting that the shift of diffuse Htt into inclusions alleviates proteotoxic stress to the cell.[61] Third, is the related phenomenon of the aggresome, which was described in 1998 as a new cellular structure comprising deposited aggregated proteins.[66]

The aggresome model was developed to describe the organized co-alescence of diffuse aggregates of the cystic fibrosis transmembrane conductance regulator into a single perinuclear location at the microtubule organizing center.[66] The aggresome forms via dynein-mediated transport, is enriched in proteasome components and ensheathed with the intermediate filament vimentin.[67] This model has since been applied more generally to describe the behavior of many aggregating proteins including Htt.[68-70] However, the aggresome model does not fully describe the diversity of inclusions since different proteins form structures with fundamental differences in biochemical properties, composition and cellular localization.[64,71] For example, polyQ-expanded ataxin-1 forms both small inclusions comprised of mobile forms of mutant ataxin-1 and large inclusions with mobile and immobile ataxin-1.[71] In addition, Htt forms a diversity of inclusion types and cellular localizations including intranuclearly, perinuclearly and in dendrites.[72] This raises questions as to whether all inclusions are the same, whether they are all protective and whether different cell types harness different strategies to form inclusions.

Recent studies have unearthed two distinct aggresome types that partially address the deficiencies of the aggresome model.[64] These structures were identified as separate quality control compartments, the insoluble protein deposit (IPOD) and the juxtanuclear quality control (JUNQ).[64] The IPOD was reported to be a perivacuolar inclusion of immobile protein aggregates (in yeast), of which Htt and yeast prion proteins preferably went to. The JUNQ by contrast contained primarily mobile, and diffusible proteins of which other misfolded proteins preferably were sorted to.[64] Both JUNQ and IPOD form with microtubule-dependent processes, yet neither, unlike the original aggresomes formed by the cystic fibrosis transmembrane conductance regulator, are localized to the microtubule organizing centre, indicating the possibility that there may be more distinct structures formed in cells.[64,66]

Further studies are needed to fully flesh out the behaviour and properties of aggresomes, IPOD and JUNQ structures. Questions remaining include how other elements of the cell, such as actin which can alter inclusion formation, are involved in mechanisms for sorting aggregates into inclusions.[73,74] A number of proteins with a role in actin-cytoskeleton organization have been shown to influence polyQ-aggregation such as α-Pix, which binds Htt and promotes soluble intermolecular Htt interactions.[73] Htt also interacts with profilin-2a, increases degradation of profilin-1 and 2a, reducing the F-actin to G-actin ratio.[75] Overexpression of profilin reduces aggregation of expanded Htt fragments, while overexpression of cofilin, an important modulator of the actin cytoskeleton, enhances aggregation.[74] Htt has been reported to bind actin directly in vitro through its N-terminus, although this has not been detected in cells.[76]

Htt aggregation in vivo involves a number of posttranslational modifications including proteolytic fragmentation, which may be part of normal cellular functions of Htt. One of the earliest in vivo processing events preceding inclusion formation is cleavage of Htt by caspases, calpains and/or other proteases.[77] N-terminal Htt fragments are readily detected in aggregates and are thought to comprise the majority of the protein in inclusion bodies.[78-80] Different Htt fragments display varying aggregation properties in cells, indicating that cleavage of full-length Htt influences its intracellular localisation, aggregation and toxicity.[81-84] In addition, cleavage is influenced by phosphorylation at residues such as ser-536, which is necessary for calpain cleavage and associated events leading to toxicity.[85]

Other posttranslational modifications may also be critical factors influencing Htt aggregation and inclusion formation. Htt has been shown to be phosphorylated, palmitoylated, ubiquitinated, SUMOylated and acetylated, and changes in modification status of polyQ-expanded Htt influences the ability of the protein to aggregate.[86-89] PolyQ-expanded Htt has been shown to be acetylated at lys-444 in human HD brain and mouse HD models which may reflect attempts to remove the mutant protein by macroautophagy, since enhanced lys-444 acetylation by CREB binding protein (CBP) facilitates trafficking to autophagosomes and increases macroautophagic clearance and neuroprotection.[90] Phosphorylation and palmitoylation may also play roles in mediating Htt shift to inclusions, removal or aggregation. For example, phosphorylation of serine 421 by Akt and serum- and glucocorticoid-induced kinase (SGK) correlates with a reduced number of nuclear inclusions.[91,92] Reducing palmitoylation of Htt via a decreased interaction of Htt with interacting protein 14 (HIP14), leads to an increased inclusion formation and neuronal toxicity.[89]

CELLULAR TOXICITY AND DISEASE MECHANISM

Some of the abnormal events known to occur in HD include trancriptional dysregulation, mitochondrial dysfunction, impaired vesicle transport, dysfunctional axonal transport and aberrant neuronal signaling, including excitotoxicity.[93,94] However, there is no consensus to a single or combination of multiple mechanisms by which polyQ expansion definitively leads to toxicity which suggests a multifactorial basis for toxicity (or an insufficient understanding of the molecular pathways involved).

Most mechanisms proposed involve a toxic gain-of-function from polyQ expansion although loss-of-function has also been suggested to contribute to disease by separate mechanisms to the toxic gain of function.[95,96] This is primarily based on Htt being essential for viability since mouse knockouts are embryonic lethal,[97] but also on conditional knockouts in adult mice that lead to neurodegeneration.[98] However, mice can be rescued by mutant Htt, indicating that the polyQ-expanded protein is essentially functional, although this rescue may be incomplete due to these mice having testicular abnormalities.[99] When mice containing mutant Htt are backcrossed into lines with wild-type Htt, the additional gene dosage of the wild-type Htt partially supresses toxicity, which suggests that aggregation of the mutant Htt causes a loss of function and contributes to toxicity.[100-102]

Regardless of a potential role for a loss of Htt normal function in disease, the ectopic overexpression of pathogenic polyQ-peptides (but not nonpathogenic lengths) leads to neurodegeneration in *Drosophila*,[96] indicating that pathogenic lengths of polyQ are intrinsically toxic. The mechanism of toxicity may thus include two major components. One is that mutant Htt aggregation co-aggregates the wild-type allelic form (or other proteins—discussed below) and hence acts in a dominant negative manner to inactivate normal Htt function[103] and, two, the aggregates themselves are directly neurotoxic.[71,104-111] While the nine polyQ-diseases have different characteristics which could arise from expression and protein context differences, the phenotype of these diseases become more similar in the extreme polyQ lengths (e.g., juvenile disease cases) which suggests that a direct gain of toxic function from expanded polyQ is the dominant mode to pathogenesis.[95]

The models for the toxic gain-of-function of polyQ include aberrant interactions of the mutant Htt to other essential cellular protein and lipid components,[71,104-109] toxic aggregate intermediates and an impairment of protein homeostasis.[110,111] Abnormal interactions can be mediated through co-aggregation with other proteins rich in glutamine and impair their normal cellular functions[71,104-109] or cause the proteins to shift into abnormal cellular locations once they have co-aggregated.[112-114] For example, monomeric or small oligomeric Htt can interact with a polyQ sequence in the TATA-binding protein (TBP), which changes its conformation and impairs its function.[115] Alternatively, large protein aggregates may physically obstruct subcellular processes such as axonal transport.[116] PolyQ expansion may also be toxic by disrupting binding of Htt to its normal ligands, since Htt interacts with a large number of proteins many of which it may act as a scaffold for in signalling complexes.[117] Indeed, Htt is thought to have roles in vesicle transport, endocytosis, postsynaptic signalling, transcription, anti-apoptotic processes and cytoskeletal re-organization, which provides much scope for abnormal interactions interfering very broadly with important cellular activities.[94,117,118]

Another postulated gain of toxic function comes from polyQ-expansions altering interactions with membrane lipids. PolyQ peptides can interact with phospholipids and penetrate lipid bilayers to form channels, which could play roles in toxicity[119,120] or assist in the spread of protein aggregates to surrounding cells in a prion-like manner.[121] Moreover, mutant Htt has higher affinities to lipid rafts and penetration into lipid bilayers in vitro than wild-type forms of Htt.[122,123] These mechanisms may be shared more generally with other neurodegenerative disorders, such as Parkinson's disease, where α-synuclein assembles into annular oligomers that can form pores in membranes.[124]

Another proposed mechanism of toxicity involves the disruption of cellular protein quality control homeostasis mechanisms that keep the proteome folded, called proteostasis. For reasons that are not known, proteostasis naturally declines upon ageing, which also correlates with aggregation of misfolded proteins.[125-129] Aggregation-prone proteins also stress the proteostasis capacity, suggesting potential for a feedback loop to exacerbate damage.[110,130,131] This phenomenon was shown elegantly in C. elegans whereby expression of polyQ-expanded proteins led to aggregation of temperature-sensitive mutants at what was otherwise permissive temperatures, which suggests that the presence of polyQ exhausts proteostasis resources that are necessary to prevent other proteins in the proteome from misfolding.[110] The organismal capacity to process misfolding-prone proteins may become overloaded over time, leading to an accumulation of misfolded and abnormal proteins that further burden proteostasis mechanisms.[110,111] Weak folding mutations and polymorphisms throughout the genome can function as modifiers of proteostasis capacity, which may contribute to the wide phenotypic variability in the age of onset for a given pathogenic polyQ-length.[110]

An intriguing feature of this proteostasis mode of toxicity is that many proteins that modulate aggregation of polyQ-expanded proteins are components of protein quality control mechanisms that relate to proteostasis such as the unfolded protein response (UPR), heat shock response, the ubiquitin-proteasome system (UPS) and clearance via autophagy. Clearance of mutant Htt is mediated through both the UPS and autophagy.[132] Inhibition of UPS leads to the accumulation of mutant N-terminal fragments.[133] In addition, polyQ inclusions colocalized with heat shock chaperones Hdj2, Hdj1, Hsc70, Hsp70, Hsp90, Hsp84, and UPS components ubiquitin,[134] NEDD8, which is a ubiquitin-like protein, and 19S and 26S proteasomes.[135-137] The association of proteasome components with inclusion bodies suggests aggregation may impair

UPS and hinder proteostasis.[131] This conclusion was supported by studies in *C. elegans* showing polyQ-expanded SCA3 expression hinders UPS using a GFP reporter for UPS function.[138] However, the same UPS reporter did not reveal dysfunction of UPS in a mouse model of polyQ-expanded SCA7 or HD.[139,140] Interestingly, UPS impairment was observed in neuronal synapses of transgenic R6/2 HD mice, which express exon 1 of Htt with 150Q, but not in neuronal cell bodies, suggesting UPS defects are not global but localized, which may account for differences in dysfunction of specific cell populations in HD.[141]

Molecular chaperones are also involved in the aggregation process. Increased levels of heat shock protein (Hsp) 70 or 40 can rescue the neurodegenerative phenotype in *Drosophila* and cell culture models.[142,143] These chaperones appear to target specific aggregate populations, since Hsp70 and Hsp40 associate with a distinct set of oligomers.[144] Together they act to inhibit the formation of these oligomer types, which supresses caspase-3 activity and apoptosis.[144] However, these chaperones may also play roles in increasing aggregation or assembly into inclusions under certain situations. For example Hsp70 and Hsp40 can decrease oligomer levels, while increasing fibril formation in vitro.[28] In cells the Hsp70 family member Hsc70 also decreases oligomer levels in cells while increasing the rate cells form inclusions.[56] Further illustrating the complexity of chaperone action on polyQ-expanded proteins is that chaperonin TRiC in co-operation with Hsp40 and Hsp70 reduces the polyglutamine-induced compaction of Htt monomers and promotes the assembly of polyQ-expanded Htt into nontoxic soluble oligomers.[115,140,145]

Other elements of the protein quality control network, such as macroautophagy are likely pivotal players in HD disease mechanisms.[146-148] Macroautophagy regulates cellular clearance of organelles and aggregated proteins by engulfment of material into autophagosomes. Activation of macroautophagy ameliorates HD pathology in animal models while its loss increases the accumulation of misfolded proteins and enhances neurodegeneration.[149-151] Critically, there are polymorphisms in the autophagy-related gene 7 (Atg7) that lead to earlier HD onset demonstrating the importance of autophagy.[152] Components of the macroautophagy system such as LC3, which is involved in autophagosome formation, and cargo-receptor p62, are upregulated early in a sustained manner in a mutant Htt knock-in mouse model of HD.[146,147] PolyQ-expanded protein is enriched in autophagosomes in postmortem SCA3, DRPLA and HD brains which suggests cells attempt to clear the aggregates that have formed by macroautophagy or alternatively, that the aggregates impair macroautophagy and lead to a backlog of autophagosomes.[153-156] This conclusion is supported by studies showing polyQ-expanded Htt expression leads to inefficient cargo loading of autophagosomes.[148]

Chaperone-mediated autophagy (CMA), which involves the chaperone Hsc70 and lysosomal-associated membrane protein 2A (LAMP2A), may also be deregulated in HD.[157] PolyQ-expansion of Htt decreases Htt clearance via CMA,[157] which in part may be due to a decreased IκB kinase (IKK) S13 phosphorylation of mutant Htt.[157] However, decreased CMA clearance may also arise in a similar fashion as observed for α-synuclein where the pathogenic Parkinson's disease mutant protein blocks lysosomal translocation thus inhibiting α-synuclein degradation and that of other substrates,[158] which is reminiscent of the block in cargo loading due to expanded Htt demonstrated for macroautophagy.[148]

While most research has focused on the Htt protein as the toxic entity, additional data has implicated CAG trinucleotide repeat mRNA, which encodes the polyQ sequences, as directly toxic in addition to the protein.[159] RNA-mediated toxicity of noncoding CTG trinucleotide repeats is well established to cause dysfunction in myotonic dystrophy as well as SCA8.[160-166] Recently a similar mode of toxicity was demonstrated in *Drosophila* for SCA3.[159] Although the underlying mechanism is unknown, expression of ataxin-3 encoded with a CAACAG mixed codon sequence instead of a direct CAGCAG single codon repeat reduced toxicity and expression of noncoding CAG repeats induced progressive neural dysfunction.[159] RNA-mediated toxicity adds an extra dimension of complexity to the disease.

DIFFERENCES IN TOXICITY IN THE NERVOUS SYSTEM AND PERIPHERY

Finally, one area that remains largely unexplored is why different cell types seem to respond differently to the toxic effects of polyQ. The extent and nature of cellular dysfunction is likely to substantially differ across cell types and developmentally since HD pathology most severely affects the striatal neurons even though Htt is expressed ubiquitously from early embryogenesis.[167] Furthermore, the pathology outside the brain is complex and less well understood. Mutant Htt inclusion formation does occur outside the central nervous system[168,169] and there is peripheral toxicity and inclusion formation in the muscle cells,[170] β cells of the islets of Langerhans in the pancreas[171] and cardiac myocytes,[172] which lead to muscular atrophy, as well as diabetes[173] and heart failure in HD, respectively.[174]

Why striatal neurons seem most sensitive to aggregation remains a puzzle. Possible explanations lie in neurons being postmitotic, which leaves them incapable of using asymmetric cell division as a possible mechanism to lessen toxic burden.[175,176] Other studies have pointed to a decreased organismal capacity to handle protein aggregation upon ageing, which may more selectively sensitise neurons to damage.[127,177] In addition, early developmental effects in the brain may compound later damage from protein aggregation since there are signs of structural brain abnormalities occurring well before clinical diagnosis of HD and onset of major neurological deficits in knock-in mouse models of HD.[178-180]

CONCLUSION

While aggregation caused by polyQ-expansions seems central to the pathogenesis of HD, much remains to be discovered as to the precise means by which this occurs. We believe that the instrinsic aggregation and the core biophysical properties of polyQ-expanded proteins fundamentally shape the mechanisms underpinning pathogenesis and new approaches are needed to better understand this aspect in situ. Knowledge of these mechanisms will likely be of paramount importance to informing us of invaluable targets for therapeutic intervention, especially for emerging approaches that stimulate quality control mechanisms against misfolded and aggregating proteins.[181] These principles also apply more broadly to other major human diseases, including Alzheimer's, Parkinson's, frontotemporal dementia and amyotrophic lateral sclerosis (Lou Gehrig's disease), which are also associated with problems in protein misfolding and aggregation.

REFERENCES

1. A novel gene containing a trinucleotide repeat that is expanded and unstable on Huntington's disease chromosomes. The Huntington's disease collaborative research group. Cell 1993; 72(6):971-983.
2. Orr HT, Chung MY, Banfi S et al. Expansion of an unstable trinucleotide CAG repeat in spinocerebellar ataxia type 1. Nat Genet 1993; 4(3):221-226.
3. Imbert G, Saudou F, Yvert G et al. Cloning of the gene for spinocerebellar ataxia 2 reveals a locus with high sensitivity to expanded CAG/glutamine repeats. Nat Genet 1996; 14(3):285-291.
4. Sanpei K, Takano H, Igarashi S et al. Identification of the spinocerebellar ataxia type 2 gene using a direct identification of repeat expansion and cloning technique, DIRECT. Nat Genet 1996; 14(3):277-284.
5. Kawaguchi Y, Okamoto T, Taniwaki M et al. CAG expansions in a novel gene for Machado-Joseph disease at chromosome 14q32.1. Nat Genet 1994; 8(3):221-228.
6. Zhuchenko O, Bailey J, Bonnen P et al. Autosomal dominant cerebellar ataxia (SCA6) associated with small polyglutamine expansions in the alpha 1A-voltage-dependent calcium channel. Nat Genet 1997; 15(1):62-69.
7. David G, Abbas N, Stevanin G et al. Cloning of the SCA7 gene reveals a highly unstable CAG repeat expansion. Nat Genet 1997; 17(1):65-70.
8. Nakamura K, Jeong SY, Uchihara T et al. SCA17, a novel autosomal dominant cerebellar ataxia caused by an expanded polyglutamine in TATA-binding protein. Hum Mol Genet 2001; 10(14):1441-1448.
9. Andrew SE, Goldberg YP, Kremer B et al. The relationship between trinucleotide (CAG) repeat length and clinical features of Huntington's disease. Nat Genet 1993; 4(4):398-403.
10. La Spada AR, Wilson EM, Lubahn DB et al. Androgen receptor gene mutations in X-linked spinal and bulbar muscular atrophy. Nature 1991; 352(6330):77-79.
11. Koide R, Ikeuchi T, Onodera O et al. Unstable expansion of CAG repeat in hereditary dentatorubral-pallidoluysian atrophy (DRPLA). Nat Genet 1994; 6(1):9-13.
12. Nagafuchi S, Yanagisawa H, Ohsaki E et al. Structure and expression of the gene responsible for the triplet repeat disorder, dentatorubral and pallidoluysian atrophy (DRPLA). Nat Genet 1994; 8(2):177-182.
13. Koshy BT, Zoghbi HY. The CAG/polyglutamine tract diseases: gene products and molecular pathogenesis. Brain Pathol 1997; 7(3):927-942.
14. Scherzinger E, Lurz R, Turmaine M et al. Huntingtin-encoded polyglutamine expansions form amyloid-like protein aggregates in vitro and in vivo. Cell 1997; 90(3):549-558.
15. Scherzinger E, Sittler A, Schweiger K et al. Self-assembly of polyglutamine-containing huntingtin fragments into amyloid-like fibrils: implications for Huntington's disease pathology. Proc Natl Acad Sci U S A 1999; 96(8):4604-4609.
16. Caputo CB, Fraser PE, Sobel IE et al. Amyloid-like properties of a synthetic peptide corresponding to the carboxy terminus of beta-amyloid protein precursor. Arch Biochem Biophys 1992; 292(1):199-205.
17. Prusiner SB, McKinley MP, Bowman KA et al. Scrapie prions aggregate to form amyloid-like birefringent rods. Cell 1983; 35(2 Pt 1):349-358.
18. Huang CC, Faber PW, Persichetti F et al. Amyloid formation by mutant huntingtin: threshold, progressivity and recruitment of normal polyglutamine proteins. Somat Cell Mol Genet 1998; 24(4):217-233.
19. Ferrone F. Analysis of protein aggregation kinetics. Methods Enzymol 1999; 309:256-274.
20. Berthelier V, Hamilton JB, Chen S et al. A microtiter plate assay for polyglutamine aggregate extension. Anal Biochem 2001; 295(2):227-236.
21. Esler WP, Stimson ER, Jennings JM et al. Alzheimer's disease amyloid propagation by a template-dependent dock-lock mechanism. Biochemistry 2000; 39(21):6288-6295.
22. Jarrett JT, Lansbury PT Jr. Seeding "one-dimensional crystallization" of amyloid: a pathogenic mechanism in Alzheimer's disease and scrapie? Cell 1993; 73(6):1055-1058.
23. Chen S, Ferrone FA, Wetzel R. Huntington's disease age-of-onset linked to polyglutamine aggregation nucleation. Proc Natl Acad Sci U S A 2002; 99(18):11884-11889.
24. Bernacki JP, Murphy RM. Model discrimination and mechanistic interpretation of kinetic data in protein aggregation studies. Biophys J 2009; 96(7):2871-2887.
25. Walters RH, Murphy RM. Examining polyglutamine peptide length: a connection between collapsed conformations and increased aggregation. J Mol Biol 2009; 393(4):978-992.
26. Lee CC, Walters RH, Murphy RM. Reconsidering the mechanism of polyglutamine peptide aggregation. Biochemistry 2007; 46(44):12810-12820.
27. Legleiter J, Mitchell E, Lotz GP et al. Mutant huntingtin fragments form oligomers in a polyglutamine length-dependent manner in vitro and in vivo. J Biol Chem 2010; 285(19):14777-14790.
28. Wacker JL, Zareie MH, Fong H et al. Hsp70 and Hsp40 attenuate formation of spherical and annular polyglutamine oligomers by partitioning monomer. Nat Struct Mol Biol 2004; 11(12):1215-1222.
29. Bhattacharyya AM, Thakur AK, Wetzel R. polyglutamine aggregation nucleation: thermodynamics of a highly unfavorable protein folding reaction. Proc Natl Acad Sci U S A 2005; 102(43):15400-15405.

30. Dougan L, Li J, Badilla CL et al. Single homopolypeptide chains collapse into mechanically rigid conformations. Proc Natl Acad Sci U S A 2009; 106(31):12605-12610.
31. Wang X, Vitalis A, Wyczalkowski MA et al. Characterizing the conformational ensemble of monomeric polyglutamine. Proteins 2006; 63(2):297-311.
32. Crick SL, Jayaraman M, Frieden C et al. Fluorescence correlation spectroscopy shows that monomeric polyglutamine molecules form collapsed structures in aqueous solutions. Proc Natl Acad Sci U S A 2006; 103(45):16764-16769.
33. Altschuler EL, Hud NV, Mazrimas JA et al. Random coil conformation for extended polyglutamine stretches in aqueous soluble monomeric peptides. J Pept Res 1997; 50(1):73-75.
34. Chen S, Berthelier V, Yang W et al. Polyglutamine aggregation behavior in vitro supports a recruitment mechanism of cytotoxicity. J Mol Biol 2001; 311(1):173-182.
35. Masino L, Kelly G, Leonard K et al. Solution structure of polyglutamine tracts in GST-polyglutamine fusion proteins. FEBS Lett 2002; 513(2-3):267-272.
36. Vitalis A, Wang X, Pappu RV. Atomistic simulations of the effects of polyglutamine chain length and solvent quality on conformational equilibria and spontaneous homodimerization. J Mol Biol 2008; 384(1):279-297.
37. Pappu RV, Wang X, Vitalis A et al. A polymer physics perspective on driving forces and mechanisms for protein aggregation. Arch Biochem Biophys 2008; 469(1):132-141.
38. Venkatraman P, Wetzel R, Tanaka M et al. Eukaryotic proteasomes cannot digest polyglutamine sequences and release them during degradation of polyglutamine-containing proteins. Mol Cell 2004; 14(1):95-104.
39. Poirier MA, Li H, Macosko J et al. Huntingtin spheroids and protofibrils as precursors in polyglutamine fibrilization. J Biol Chem 2002; 277(43):41032-41037.
40. Chen S, Berthelier V, Hamilton JB et al. Amyloid-like features of polyglutamine aggregates and their assembly kinetics. Biochemistry 2002; 41(23):7391-7399.
41. Vitalis A, Lyle N, Pappu RV. Thermodynamics of beta-sheet formation in polyglutamine. Biophys J 2009; 97(1):303-311.
42. Bhattacharyya A, Thakur AK, Chellgren VM et al. Oligoproline effects on polyglutamine conformation and aggregation. J Mol Biol 2006; 355(3):524-535.
43. Nekooki-Machida Y, Kurosawa M, Nukina N et al. Distinct conformations of in vitro and in vivo amyloids of huntingtin-exon1 show different cytotoxicity. Proc Natl Acad Sci U S A 2009; 106(24):9679-9684.
44. Bevivino AE, Loll PJ. An expanded glutamine repeat destabilizes native ataxin-3 structure and mediates formation of parallel beta -fibrils. Proc Natl Acad Sci U S A 2001; 98(21):11955-11960.
45. Perutz MF, Finch JT, Berriman J et al. Amyloid fibers are water-filled nanotubes. Proc Natl Acad Sci U S A 2002; 99(8):5591-5595.
46. Singer SJ, Dewji NN. Evidence that Perutz's double-beta-stranded subunit structure for beta-amyloids also applies to their channel-forming structures in membranes. Proc Natl Acad Sci U S A 2006; 103(5):1546-1550.
47. Perutz MF, Johnson T, Suzuki M et al. Glutamine repeats as polar zippers: their possible role in inherited neurodegenerative diseases. Proc Natl Acad Sci U S A 1994; 91(12):5355-5358.
48. Sharma D, Shinchuk LM, Inouye H et al. Polyglutamine homopolymers having 8-45 residues form slablike beta-crystallite assemblies. Proteins 2005; 61(2):398-411.
49. Tanaka M, Morishima I, Akagi T et al. Intra- and intermolecular beta-pleated sheet formation in glutamine-repeat inserted myoglobin as a model for polyglutamine diseases. J Biol Chem 2001; 276(48):45470-45475.
50. Sikorski P, Atkins E. New model for crystalline polyglutamine assemblies and their connection with amyloid fibrils. Biomacromolecules 2005; 6(1):425-432.
51. Thakur AK, Wetzel R. Mutational analysis of the structural organization of polyglutamine aggregates. Proc Natl Acad Sci U S A 2002; 99(26):17014-17019.
52. Makin OS, Atkins E, Sikorski P et al. Molecular basis for amyloid fibril formation and stability. Proc Natl Acad Sci U S A 2005; 102(2):315-320.
53. DiFiglia M, Sapp E, Chase KO et al. Aggregation of huntingtin in neuronal intranuclear inclusions and dystrophic neurites in brain. Science 1997; 277(5334):1990-1993.
54. Davies SW, Turmaine M, Cozens BA et al. Formation of neuronal intranuclear inclusions underlies the neurological dysfunction in mice transgenic for the HD mutation. Cell 1997; 90(3):537-548.
55. Lunkes A, Trottier Y, Mandel JL. Pathological mechanisms in Huntington's disease and other polyglutamine expansion diseases. Essays Biochem 1998; 33:149-163.
56. Olshina MA, Angley LM, Ramdzan YM et al. Tracking mutant huntingtin aggregation kinetics in cells reveals three major populations that include an invariant oligomer pool. J Biol Chem 2010; 285(28):21807-21816.
57. Ramdzan YM, Nisbet RM, Miller J et al. Conformation sensors that distinguish monomeric proteins from oligomers in live cells. Chem Biol 2010; 17(4):371-379.
58. Ossato G, Digman MA, Aiken C et al. A two-step path to inclusion formation of huntingtin peptides revealed by number and brightness analysis. Biophys J 2010; 98(12):3078-3085.

59. Muchowski PJ, Ning K, D'Souza-Schorey C et al. Requirement of an intact microtubule cytoskeleton for aggregation and inclusion body formation by a mutant huntingtin fragment. Proc Natl Acad Sci U S A 2002; 99(2):727-732.

60. Taylor JP, Tanaka F, Robitschek J et al. Aggresomes protect cells by enhancing the degradation of toxic polyglutamine-containing protein. Hum Mol Genet 2003; 12(7):749-757.

61. Mitra S, Tsvetkov AS, Finkbeiner S. Single neuron ubiquitin-proteasome dynamics accompanying inclusion body formation in Huntington disease. J Biol Chem 2009; 284(7):4398-4403.

62. Ortega Z, Diaz-Hernandez M, Maynard CJ et al. Acute polyglutamine expression in inducible mouse model unravels ubiquitin/proteasome system impairment and permanent recovery attributable to aggregate formation. J Neurosci 2010; 30(10):3675-3688.

63. Arrasate M, Mitra S, Schweitzer ES et al. Inclusion body formation reduces levels of mutant huntingtin and the risk of neuronal death. Nature 2004; 431(7010):805-810.

64. Kaganovich D, Kopito R, Frydman J. Misfolded proteins partition between two distinct quality control compartments. Nature 2008; 454(7208):1088-1095.

65. Johnston JA, Illing ME, Kopito RR. Cytoplasmic dynein/dynactin mediates the assembly of aggresomes. Cell Motil Cytoskeleton 2002; 53(1):26-38.

66. Johnston JA, Ward CL, Kopito RR. Aggresomes: a cellular response to misfolded proteins. J Cell Biol 1998; 143(7):1883-1898.

67. Kopito RR. Aggresomes, inclusion bodies and protein aggregation. Trends Cell Biol 2000; 10(12):524-530.

68. Garcia-Mata R, Bebok Z, Sorscher EJ et al. Characterization and dynamics of aggresome formation by a cytosolic GFP-chimera. J Cell Biol 1999; 146(6):1239-1254.

69. Waelter S, Boeddrich A, Lurz R et al. Accumulation of mutant huntingtin fragments in aggresome-like inclusion bodies as a result of insufficient protein degradation. Mol Biol Cell 2001; 12(5):1393-1407.

70. Wang H, Strandin T, Hepojoki J et al. Degradation and aggresome formation of the Gn tail of the apathogenic Tula hantavirus. J Gen Virol 2009; 90(Pt 12):2995-3001.

71. Stenoien DL, Mielke M, Mancini MA. Intranuclear ataxin1 inclusions contain both fast- and slow-exchanging components. Nat Cell Biol 2002; 4(10):806-810.

72. DiFiglia M, Sapp E, Chase KO et al. Aggregation of huntingtin in neuronal intranuclear inclusions and dystrophic neurites in brain. Science 1997; 277(5334):1990-1993.

73. Eriguchi M, Mizuta H, Luo S et al. Alpha Pix enhances mutant huntingtin aggregation. J Neurol Sci 2010; 290(1-2):80-85.

74. Shao J, Welch WJ, Diprospero NA et al. Phosphorylation of profilin by ROCK1 regulates polyglutamine aggregation. Mol Cell Biol 2008; 28(17):5196-5208.

75. Burnett BG, Andrews J, Ranganathan S et al. Expression of expanded polyglutamine targets profilin for degradation and alters actin dynamics. Neurobiol Dis 2008; 30(3):365-374.

76. Angeli S, Shao J, Diamond MI. F-actin binding regions on the androgen receptor and huntingtin increase aggregation and alter aggregate characteristics. PLoS One 2010; 5(2):e9053.

77. Kim YJ, Yi Y, Sapp E et al. Caspase 3-cleaved N-terminal fragments of wild-type and mutant huntingtin are present in normal and Huntington's disease brains, associate with membranes, and undergo calpain-dependent proteolysis. Proc Natl Acad Sci U S A 2001; 98(22):12784-12789.

78. Landles C, Sathasivam K, Weiss A et al. Proteolysis of mutant huntingtin produces an exon 1 fragment that accumulates as an aggregated protein in neuronal nuclei in Huntington disease. J Biol Chem 2010; 285(12):8808-8823.

79. Li X, Li H, Li XJ. Intracellular degradation of misfolded proteins in polyglutamine neurodegenerative diseases. Brain Res Rev 2008; 59(1):245-252.

80. Sieradzan KA, Mechan AO, Jones L et al. Huntington's disease intranuclear inclusions contain truncated, ubiquitinated huntingtin protein. Exp Neurol 1999; 156(1):92-99.

81. Martindale D, Hackam A, Wieczorek A et al. Length of huntingtin and its polyglutamine tract influences localization and frequency of intracellular aggregates. Nat Genet 1998; 18(2):150-154.

82. Hackam AS, Singaraja R, Wellington CL et al. The influence of huntingtin protein size on nuclear localization and cellular toxicity. J Cell Biol 1998; 141(5):1097-1105.

83. Warby SC, Doty CN, Graham RK et al. Activated caspase-6 and caspase-6-cleaved fragments of huntingtin specifically colocalize in the nucleus. Hum Mol Genet 2008; 17(15):2390-2404.

84. Graham RK, Deng Y, Slow EJ et al. Cleavage at the caspase-6 site is required for neuronal dysfunction and degeneration due to mutant huntingtin. Cell 2006; 125(6):1179-1191.

85. Schilling B, Gafni J, Torcassi C et al. Huntingtin phosphorylation sites mapped by mass spectrometry. Modulation of cleavage and toxicity. J Biol Chem 2006; 281(33):23686-23697.

86. Aiken CT, Steffan JS, Guerrero CM et al. Phosphorylation of threonine 3: implications for Huntingtin aggregation and neurotoxicity. J Biol Chem 2009; 284(43):29427-29436.

87. Rockabrand E, Slepko N, Pantalone A et al. The first 17 amino acids of Huntingtin modulate its sub-cellular localization, aggregation and effects on calcium homeostasis. Hum Mol Genet 2007; 16(1):61-77.

88. Steffan JS, Agrawal N, Pallos J et al. SUMO modification of Huntingtin and Huntington's disease pathology. Science 2004; 304(5667):100-104.
89. Yanai A, Huang K, Kang R et al. Palmitoylation of huntingtin by HIP14 is essential for its trafficking and function. Nat Neurosci 2006; 9(6):824-831.
90. Jeong H, Then F, Melia TJ Jr et al. Acetylation targets mutant huntingtin to autophagosomes for degradation. Cell 2009; 137(1):60-72.
91. Humbert S, Bryson EA, Cordelieres FP et al. The IGF-1/Akt pathway is neuroprotective in Huntington's disease and involves Huntingtin phosphorylation by Akt. Dev Cell 2002; 2(6):831-837.
92. Rangone H, Poizat G, Troncoso J et al. The serum- and glucocorticoid-induced kinase SGK inhibits mutant huntingtin-induced toxicity by phosphorylating serine 421 of huntingtin. Eur J Neurosci 2004; 19(2):273-279.
93. Zuccato C, Valenza M, Cattaneo E. Molecular mechanisms and potential therapeutical targets in Huntington's disease. Physiol Rev 2010; 90(3):905-981.
94. Li SH, Li XJ. Huntingtin-protein interactions and the pathogenesis of Huntington's disease. Trends Genet 2004; 20(3):146-154.
95. Zoghbi HY, Orr HT. Glutamine repeats and neurodegeneration. Annu Rev Neurosci 2000; 23:217-247.
96. Marsh JL, Walker H, Theisen H et al. Expanded polyglutamine peptides alone are intrinsically cytotoxic and cause neurodegeneration in Drosophila. Hum Mol Genet 2000; 9(1):13-25.
97. Zeitlin S, Liu JP, Chapman DL et al. Increased apoptosis and early embryonic lethality in mice nullizygous for the Huntington's disease gene homologue. Nat Genet 1995; 11(2):155-163.
98. Dragatsis I, Levine MS, Zeitlin S. Inactivation of Hdh in the brain and testis results in progressive neurodegeneration and sterility in mice. Nat Genet 2000; 26(3):300-306.
99. Leavitt BR, Guttman JA, Hodgson JG et al. Wild-type huntingtin reduces the cellular toxicity of mutant huntingtin in vivo. Am J Hum Genet 2001; 68(2):313-324.
100. Van Raamsdonk JM, Pearson J, Murphy Z et al. Wild-type huntingtin ameliorates striatal neuronal atrophy but does not prevent other abnormalities in the YAC128 mouse model of Huntington disease. BMC Neurosci 2006; 7:80.
101. Leavitt BR, van Raamsdonk JM, Shehadeh J et al. Wild-type huntingtin protects neurons from excitotoxicity. J Neurochem 2006; 96(4):1121-1129.
102. Ho LW, Brown R, Maxwell M et al. Wild type Huntingtin reduces the cellular toxicity of mutant Huntingtin in mammalian cell models of Huntington's disease. J Med Genet 2001; 38(7):450-452.
103. Busch A, Engemann S, Lurz R et al. Mutant huntingtin promotes the fibrillogenesis of wild-type huntingtin: a potential mechanism for loss of huntingtin function in Huntington's disease. J Biol Chem 2003; 278(42):41452-41461.
104. Perez MK, Paulson HL, Pendse SJ et al. Recruitment and the role of nuclear localization in polyglutamine-mediated aggregation. J Cell Biol 1998; 143(6):1457-1470.
105. Kazantsev A, Preisinger E, Dranovsky A et al. Insoluble detergent-resistant aggregates form between pathological and nonpathological lengths of polyglutamine in mammalian cells. Proc Natl Acad Sci U S A 1999; 96(20):11404-11409.
106. Steffan JS, Kazantsev A, Spasic-Boskovic O et al. The Huntington's disease protein interacts with p53 and CREB-binding protein and represses transcription. Proc Natl Acad Sci U S A 2000; 97(12):6763-6768.
107. Holbert S, Denghien I, Kiechle T et al. The Gln-Ala repeat transcriptional activator CA150 interacts with huntingtin: neuropathologic and genetic evidence for a role in Huntington's disease pathogenesis. Proc Natl Acad Sci U S A 2001; 98(4):1811-1816.
108. Chai Y, Wu L, Griffin JD et al. The role of protein composition in specifying nuclear inclusion formation in polyglutamine disease. J Biol Chem 2001; 276(48):44889-44897.
109. Kim S, Nollen EA, Kitagawa K et al. Polyglutamine protein aggregates are dynamic. Nat Cell Biol 2002; 4(10):826-831.
110. Gidalevitz T, Ben-Zvi A, Ho KH et al. Progressive disruption of cellular protein folding in models of polyglutamine diseases. Science 2006; 311(5766):1471-1474.
111. Prahlad V, Morimoto RI. Integrating the stress response: lessons for neurodegenerative diseases from C. elegans. Trends Cell Biol 2009; 19(2):52-61.
112. Cornett J, Cao F, Wang CE et al. Polyglutamine expansion of huntingtin impairs its nuclear export. Nat Genet 2005; 37(2):198-204.
113. Lee WC, Yoshihara M, Littleton JT. Cytoplasmic aggregates trap polyglutamine-containing proteins and block axonal transport in a Drosophila model of Huntington's disease. Proc Natl Acad Sci U S A 2004; 101(9):3224-3229.
114. Milnerwood AJ, Raymond LA. Corticostriatal synaptic function in mouse models of Huntington's disease: early effects of huntingtin repeat length and protein load. J Physiol 2007; 585(Pt 3):817-831.
115. Schaffar G, Breuer P, Boteva R et al. Cellular toxicity of polyglutamine expansion proteins: mechanism of transcription factor deactivation. Mol Cell 2004; 15(1):95-105.

116. Piccioni F, Pinton P, Simeoni S et al. Androgen receptor with elongated polyglutamine tract forms aggregates that alter axonal trafficking and mitochondrial distribution in motor neuronal processes. FASEB J 2002; 16(11):1418-1420.
117. Harjes P, Wanker EE. The hunt for huntingtin function: interaction partners tell many different stories. Trends Biochem Sci 2003; 28(8):425-433.
118. Strehlow AN, Li JZ, Myers RM. Wild-type huntingtin participates in protein trafficking between the Golgi and the extracellular space. Hum Mol Genet 2007; 16(4):391-409.
119. Monoi H, Futaki S, Kugimiya S et al. Poly-L-glutamine forms cation channels: relevance to the pathogenesis of the polyglutamine diseases. Biophys J 2000; 78(6):2892-2899.
120. Suopanki J, Gotz C, Lutsch G et al. Interaction of huntingtin fragments with brain membranes—clues to early dysfunction in Huntington's disease. J Neurochem 2006; 96(3):870-884.
121. Ren P-H, Lauckner JE, Kachirskaia I et al. Cytoplasmic penetration and persistent infection of mammalian cells by polyglutamine aggregates. Nat Cell Biol 2009; 11(2):219-225.
122. Kegel KB, Sapp E, Alexander J et al. Polyglutamine expansion in huntingtin alters its interaction with phospholipids. J Neurochem 2009; 110(5):1585-1597.
123. Kegel KB, Schewkunow V, Sapp E et al. Polyglutamine expansion in huntingtin increases its insertion into lipid bilayers. Biochem Biophys Res Commun 2009; 387(3):472-475.
124. Lashuel HA, Petre BM, Wall J et al. Alpha-synuclein, especially the Parkinson's disease-associated mutants, forms pore-like annular and tubular protofibrils. J Mol Biol 2002; 322(5):1089-1102.
125. Brignull HR, Morley JF, Garcia SM et al. Modeling polyglutamine pathogenesis in C. elegans. Methods Enzymol 2006; 412:256-282.
126. Diguet E, Petit F, Escartin C et al. Normal aging modulates the neurotoxicity of mutant huntingtin. PLoS One 2009; 4(2):e4637.
127. David DC, Ollikainen N, Trinidad JC et al. Widespread protein aggregation as an inherent part of aging in C. elegans. PLoS Biol 2010; 8(8):e1000450.
128. Tonkiss J, Calderwood SK. Regulation of heat shock gene transcription in neuronal cells. Int J Hyperthermia 2005; 21(5):433-444.
129. Cuervo AM, Dice JF. Regulation of lamp2a levels in the lysosomal membrane. Traffic 2000; 1(7):570-583.
130. Duennwald ML, Lindquist S. Impaired ERAD and ER stress are early and specific events in polyglutamine toxicity. Genes Dev 2008; 22(23):3308-3319.
131. Jana NR, Zemskov EA, Wang G et al. Altered proteasomal function due to the expression of polyglutamine-expanded truncated N-terminal huntingtin induces apoptosis by caspase activation through mitochondrial cytochrome c release. Hum Mol Genet 2001; 10(10):1049-1059.
132. Sarkar S, Rubinsztein DC. Huntington's disease: degradation of mutant huntingtin by autophagy. FEBS J 2008; 275(17):4263-4270.
133. Li X, Wang CE, Huang S et al. Inhibiting the ubiquitin-proteasome system leads to preferential accumulation of toxic N-terminal mutant huntingtin fragments. Hum Mol Genet 2010; 19(12):2445-2455.
134. Ciechanover A, Brundin P. The ubiquitin proteasome system in neurodegenerative diseases: sometimes the chicken, sometimes the egg. Neuron 2003; 40(2):427-446.
135. Abel A, Walcott J, Woods J et al. Expression of expanded repeat androgen receptor produces neurologic disease in transgenic mice. Hum Mol Genet 2001; 10(2):107-116.
136. Stenoien DL, Cummings CJ, Adams HP et al. Polyglutamine-expanded androgen receptors form aggregates that sequester heat shock proteins, proteasome components, and SRC-1, and are suppressed by the HDJ-2 chaperone. Hum Mol Genet 1999; 8(5):731-741.
137. Mitsui K, Nakayama H, Akagi T et al. Purification of polyglutamine aggregates and identification of elongation factor-1alpha and heat shock protein 84 as aggregate-interacting proteins. J Neurosci 2002; 22(21):9267-9277.
138. Khan LA, Bauer PO, Miyazaki H et al. Expanded polyglutamines impair synaptic transmission and ubiquitin-proteasome system in Caenorhabditis elegans. J Neurochem 2006; 98(2):576-587.
139. Bowman AB, Yoo SY, Dantuma NP et al. Neuronal dysfunction in a polyglutamine disease model occurs in the absence of ubiquitin-proteasome system impairment and inversely correlates with the degree of nuclear inclusion formation. Hum Mol Genet 2005; 14(5):679-691.
140. Bett JS, Cook C, Petrucelli L et al. The ubiquitin-proteasome reporter GFPu does not accumulate in neurons of the R6/2 transgenic mouse model of Huntington's disease. PLoS One 2009; 4(4):e5128.
141. Wang J, Wang CE, Orr A et al. Impaired ubiquitin-proteasome system activity in the synapses of Huntington's disease mice. J Cell Biol 2008; 180(6):1177-1189.
142. Warrick JM, Chan HY, Gray-Board GL et al. Suppression of polyglutamine-mediated neurodegeneration in Drosophila by the molecular chaperone HSP70. Nat Genet 1999; 23(4):425-428.
143. Chai Y, Koppenhafer SL, Bonini NM et al. Analysis of the role of heat shock protein (Hsp) molecular chaperones in polyglutamine disease. J Neurosci 1999; 19(23):10338-10347.

144. Lotz GP, Legleiter J, Aron R et al. Hsp70 and Hsp40 functionally interact with soluble mutant huntingtin oligomers in a classic ATP-dependent reaction cycle. J Biol Chem 2010.
145. Behrends C, Langer CA, Boteva R et al. Chaperonin TRiC promotes the assembly of polyQ expansion proteins into nontoxic oligomers. Mol Cell 2006; 23(6):887-897.
146. Heng MY, Detloff PJ, Paulson HL et al. Early alterations of autophagy in Huntington disease-like mice. Autophagy 2010; 6(8):1206-1208.
147. Tung YT, Hsu WM, Lee H et al. The evolutionarily conserved interaction between LC3 and p62 selectively mediates autophagy-dependent degradation of mutant huntingtin. Cell Mol Neurobiol 2010; 30(5):795-806.
148. Martinez-Vicente M, Talloczy Z, Wong E et al. Cargo recognition failure is responsible for inefficient autophagy in Huntington's disease. Nat Neurosci 2010; 13(5):567-576.
149. Ravikumar B, Vacher C, Berger Z et al. Inhibition of mTOR induces autophagy and reduces toxicity of polyglutamine expansions in fly and mouse models of Huntington disease. Nat Genet 2004; 36(6):585-595.
150. Hara T, Nakamura K, Matsui M et al. Suppression of basal autophagy in neural cells causes neurodegenerative disease in mice. Nature 2006; 441(7095):885-889.
151. Komatsu M, Waguri S, Chiba T et al. Loss of autophagy in the central nervous system causes neurodegeneration in mice. Nature 2006; 441(7095):880-884.
152. Metzger S, Saukko M, Van Che H et al. Age at onset in Huntington's disease is modified by the autophagy pathway: implication of the V471A polymorphism in Atg7. Hum Genet 2010; 128(4):453-459.
153. Kegel KB, Kim M, Sapp E et al. Huntingtin expression stimulates endosomal-lysosomal activity, endosome tubulation, and autophagy. J Neurosci 2000; 20(19):7268-7278.
154. Sapp E, Schwarz C, Chase K et al. Huntingtin localization in brains of normal and Huntington's disease patients. Ann Neurol 1997; 42(4):604-612.
155. Yamada M, Tsuji S, Takahashi H. Pathology of CAG repeat diseases. Neuropathology 2000; 20(4):319-325.
156. Yamada M, Tsuji S, Takahashi H. Involvement of lysosomes in the pathogenesis of CAG repeat diseases. Ann Neurol 2002; 52(4):498-503.
157. Thompson LM, Aiken CT, Kaltenbach LS et al. IKK phosphorylates Huntingtin and targets it for degradation by the proteasome and lysosome. J Cell Biol 2009; 187(7):1083-1099.
158. Cuervo AM, Stefanis L, Fredenburg R et al. Impaired degradation of mutant alpha-synuclein by chaperone-mediated autophagy. Science 2004; 305(5688):1292-1295.
159. Li LB, Yu Z, Teng X et al. RNA toxicity is a component of ataxin-3 degeneration in Drosophila. Nature 2008; 453(7198):1107-1111.
160. Mankodi A, Logigian E, Callahan L et al. Myotonic dystrophy in transgenic mice expressing an expanded CUG repeat. Science 2000; 289(5485):1769-1773.
161. Amack JD, Paguio AP, Mahadevan MS. Cis and trans effects of the myotonic dystrophy (DM) mutation in a cell culture model. Hum Mol Genet 1999; 8(11):1975-1984.
162. Savkur RS, Philips AV, Cooper TA. Aberrant regulation of insulin receptor alternative splicing is associated with insulin resistance in myotonic dystrophy. Nat Genet 2001; 29(1):40-47.
163. Mankodi A, Takahashi MP, Jiang H et al. Expanded CUG repeats trigger aberrant splicing of ClC-1 chloride channel pre-mRNA and hyperexcitability of skeletal muscle in myotonic dystrophy. Mol Cell 2002; 10(1):35-44.
164. Taneja KL, McCurrach M, Schalling M et al. Foci of trinucleotide repeat transcripts in nuclei of myotonic dystrophy cells and tissues. J Cell Biol 1995; 128(6):995-1002.
165. Liquori CL, Ricker K, Moseley ML et al. Myotonic dystrophy type 2 caused by a CCTG expansion in intron 1 of ZNF9. Science 2001; 293(5531):864-867.
166. Koob MD, Moseley ML, Schut LJ et al. An untranslated CTG expansion causes a novel form of spinocerebellar ataxia (SCA8). Nat Genet 1999; 21(4):379-384.
167. Buraczynska MJ, Van Keuren ML, Buraczynska KM et al. Construction of human embryonic cDNA libraries: HD, PKD1 and BRCA1 are transcribed widely during embryogenesis. Cytogenet Cell Genet 1995; 71(2):197-202.
168. Sathasivam K, Hobbs C, Turmaine M et al. Formation of polyglutamine inclusions in non-CNS tissue. Hum Mol Genet 1999; 8(5):813-822.
169. Bradford JW, Li S, Li XJ. Polyglutamine toxicity in non-neuronal cells. Cell Res 2010; 20(4):400-407.
170. Ribchester RR, Thomson D, Wood NI et al. Progressive abnormalities in skeletal muscle and neuromuscular junctions of transgenic mice expressing the Huntington's disease mutation. Eur J Neurosci 2004; 20(11):3092-3114.
171. Andreassen OA, Dedeoglu A, Stanojevic V et al. Huntington's disease of the endocrine pancreas: insulin deficiency and diabetes mellitus due to impaired insulin gene expression. Neurobiol Dis 2002; 11(3):410-424.
172. Mihm MJ, Amann DM, Schanbacher BL et al. Cardiac dysfunction in the R6/2 mouse model of Huntington's disease. Neurobiol Dis 2007; 25(2):297-308.

173. Hurlbert MS, Zhou W, Wasmeier C et al. Mice transgenic for an expanded CAG repeat in the Huntington's disease gene develop diabetes. Diabetes 1999; 48(3):649-651.

174. Moffitt H, McPhail GD, Woodman B et al. Formation of polyglutamine inclusions in a wide range of non-CNS tissues in the HdhQ150 knock-in mouse model of Huntington's disease. PLoS One 2009; 4(11):e8025.

175. Fuentealba LC, Eivers E, Geissert D et al. Asymmetric mitosis: unequal segregation of proteins destined for degradation. Proc Natl Acad Sci U S A 2008; 105(22):7732-7737.

176. Rujano MA, Bosveld F, Salomons FA et al. Polarised asymmetric inheritance of accumulated protein damage in higher eukaryotes. PLoS Biol 2006; 4(12):e417.

177. Morley JF, Brignull HR, Weyers JJ et al. The threshold for polyglutamine-expansion protein aggregation and cellular toxicity is dynamic and influenced by aging in Caenorhabditis elegans. Proc Natl Acad Sci U S A 2002; 99(16):10417-10422.

178. Hobbs NZ, Barnes J, Frost C et al. Onset and progression of pathologic atrophy in Huntington disease: a longitudinal MR imaging study. AJNR Am J Neuroradiol 2010; 31(6):1036-1041.

179. Nopoulos PC, Aylward EH, Ross CA et al. Smaller intracranial volume in prodromal Huntington's disease: evidence for abnormal neurodevelopment. Brain 2010.

180. Molero AE, Gokhan S, Gonzalez S et al. Impairment of developmental stem cell-mediated striatal neurogenesis and pluripotency genes in a knock-in model of Huntington's disease. Proc Natl Acad Sci U S A 2009; 106(51):21900-21905.

181. Powers ET, Morimoto RI, Dillin A et al. Biological and chemical approaches to diseases of proteostasis deficiency. Ann Rev Biochem 2009; 78(1):959-991.

CHAPTER 9

SELECTIVE NEURODEGENERATION, NEUROPATHOLOGY AND SYMPTOM PROFILES IN HUNTINGTON'S DISEASE

Henry J. Waldvogel,*,[1,2] Doris Thu,[3] Virginia Hogg,[2,4]
Lynette Tippett[2,4] and Richard L.M. Faull[1,2]

[1]Department of Anatomy with Radiology, Faculty of Medical and Health Sciences, University of Auckland, Auckland, New Zealand; [2]Centre for Brain Research, University of Auckland, Auckland, New Zealand; [3]Brain Mind Institute, Ecole Polytechnique Federale de Lausanne, Lausanne, Switzerland; [4]Department of Psychology, University of Auckland, Auckland, New Zealand
*Corresponding Author: Henry J. Waldvogel—Email: h.waldvogel@auckland.ac.nz

Abstract: Huntington's disease (HD) is an autosomal dominant inherited neurodegenerative disease caused by a CAG repeat expansion in exon 1 of the Huntington gene *(HD)* also known as *IT15*. Despite the disease being caused by dysfunction of a single gene, expressed as an expanded polyglutamine in the huntingtin protein, there is a major variability in the symptom profile of patients with Huntington's disease as well as great variability in the neuropathology. The symptoms vary throughout the course of the disease and vary greatly between cases. These symptoms present as varying degrees of involuntary movements, mood, personality changes, cognitive changes and dementia. To determine whether there is a morphological basis for this symptom variability, recent studies have investigated the cellular and neurochemical changes in the striatum and cerebral cortex in the human brain to determine whether there is a link between the pathology in these regions and the symptomatology shown by individual cases. These studies together revealed that cases showing mainly mood symptom profiles correlated with marked degeneration in the striosomal compartment of the striatum, or in the anterior cingulate gyrus of the cerebral cortex. In contrast, in cases with mainly motor symptoms neurodegeneration was especially marked in the primary motor cortex with variable degeneration in both the striosomes and matrix compartments of the striatum. These studies suggest that the variable degeneration of the striatum and cerebral cortex correlates with the variable profiles of Huntington's disease.

Tandem Repeat Polymorphisms: Genetic Plasticity, Neural Diversity and Disease,
edited by Anthony J. Hannan ©2012 Landes Bioscience and Springer Science+Business Media.

INTRODUCTION

Huntington's disease (HD) is an autosomal dominant inherited neurodegenerative disease caused by a mutation in exon 1 of the Huntington gene *(HD)* also known as the *IT15* gene on chromosome 4. In the *HD* gene a triplet repeat of CAGs which codes for a stretch of glutamines is present in the N-terminal region of the gene with the normal range being up to approximately 35 although this may vary.[1] In HD this repeat sequence is expanded to 36 repeats and above resulting in toxic protein causing the disease. The mutant gene causes neurodegeneration in the brain through a variety of proposed mechanisms such as transcriptional dysregulation, synaptic dysfunction, excitotoxic mechanisms, oxidative stress and energy depletion.[2] One of the most important features of the disease which has been the subject of recent studies is the highly variable nature of the symptoms and neuropathology in affected subjects. This chapter will focus on correlations between symptom profile and pattern of degeneration in the basal ganglia and cortex.

SYMPTOM VARIABILITY

Classically, HD expresses a triad of symptoms which include motor, mood and cognitive deficits. However, despite the single-gene etiology of HD there is remarkable variability in the types of behavioural, cognitive, and motor symptoms present in different HD patients both at clinical onset and thereafter during the course of the disease.[3,4] Some HD patients exhibit mainly motor dysfunction at clinical onset, and few if any changes in mood for extended periods of time while, at the other extreme, others show mainly mood and/or cognitive changes, with minimal involuntary movements until the late stages of the disease.[3,4] Still others experience marked motor, mood and cognitive symptoms simultaneously.[4-9] Interestingly, observations in monozygotic twins who inherited identical HD genes with the same repeat length exhibit marked differences in their behavioural symptoms.[10] The onset of clinical symptoms in individual HD patients is generally correlated with the number of CAG repeats,[11] as does the disease severity, but there is no consistent relationship between CAG repeat length and symptom subtype.[4,9,12,13] Thus the source of variability in symptom subtypes is not clear. The clinical diagnosis is usually based on the movement disorder termed Huntington's chorea. These characteristic motor symptoms are expressed as a severe "choreoathetotic" disorder which describes the rapid, irregular, involuntary movements of HD. In addition clumsiness and unsteadiness in walking are also early symptoms. Studies on HD populations have indicated that approximately 50-70% of patients at onset present with chorea[14,15] whereas another 30-50% present first with mood, cognitive and behavioural changes. The chorea may develop into rigidity and dystonia later in the disease. Approximately 30-50% of patients may present first with cognitive and emotional problems such as irritability, aggression, anxiety, and obsessive behaviour with the most common being reported as depression.[14,15] There is also considerable phenotypic variation in the pattern of symptomatology during the course of the disease. Therefore factors such as different genetic interactions of the HD gene with the individuals' genome, the environment effects and epigenetics are all possible influential factors on the course of the disease.

NEURODEGENERATION IN HUNTINGTON'S DISEASE

Previous studies have shown that neurodegeneration in HD occurs most prominently in the striatum while other regions of the brain are also affected including the "downstream" structures of the striatum, the globus pallidus, substantia nigra and thalamus. Also, various regions of the cerebral cortex, hippocampus and amygdala of the limbic system show variable degeneration.[16,17] In the striatum, it is principally the GABAergic medium spiny projection neurons containing the neuropeptides enkephalin and substance P which are affected in HD. The most vulnerable are the GABAergic enkephalin containing medium spiny neurons that project to the external segment of the globus pallidus which degenerate before the GABAergic substance P containing medium spiny neurons which project to the internal segment of the globus pallidus and substantia nigra.[18-23] Furthermore, at the regional level the neuronal degeneration appears to be especially marked in the early stages in the tail of the caudate nucleus, and then progresses from the dorsal to the ventral regions and from the rostral to the caudal regions of the striatum.[24,25] It is also evident that many neurochemical and neurotransmitter changes occur before any significant cell death can be identified. This has been established in studies on post mortem HD brains which show marked neurotransmitter receptor up and down regulation in the striatum and in the output nuclei of the striatum. In particular, in Grade 0 cases which have little or no cell death, the cannabinoid (CB1) receptors are markedly decreased in the striatum, globus pallidus and substantia nigra and are very low in higher grades. With increasing grades $GABA_A$ receptor subunits and $GABA_B$ receptor subunits are reduced in the striatum but upregulate dramatically in the globus pallidus.[20-22,26]

STRIOSOME-MATRIX COMPARTMENTS IN THE STRIATUM

The striatum is neurochemically organised into two different compartments, the smaller striosomes and the larger 'back-ground' matrix. Many different neurochemical markers including acetylcholinesterase, calcium binding proteins, neurotransmitter receptors and others delineate these two neurochemical compartments by differential intensity of labelling of each compartment. Neurochemicals such as acetylcholinesterase, calbindin and parvalbumin show different intensities of immunoreactivity in the striosomes and matrix compartments.[27-29] The striosomes are characterised by intense enkephalin immunoreactivity while the matrix compartment is characterised by dense staining of acetylcholinesterase, calbindin and parvalbumin. Studies of nonhuman primates and humans have revealed differential patterns of connectivity of these compartments with the cerebral cortex, thalamus and substantia nigra. For instance, the frontal and sensory-motor cortex is connected with the matrix whereas the orbitofrontal and limbic cortex is affiliated with the striosomes;[30] the striosomes in turn project to the substantia nigra pars compacta.[31] These findings regarding connectivity, in combination with preliminary studies investigating functions of the striosomal system in nonhuman primates and rodents suggest a functional role for these compartments (e.g., refs. 32,33). In particular the striosome compartment in the striatum may play a major role in limbic functions such as mood, whereas, the matrix compartment may be more involved with sensory-motor functions.

In HD several studies have reported selective degeneration of neurons belonging to either the striosome or matrix compartments.[24,27,34-36] Some studies report preferential loss of neurons or neurochemical markers in the matrix compartment in low grade HD[20,24,36,37], and others report that neurons or neurochemical markers are lost in striosomes, with clear sparing of the matrix, at least early on.[34,35,38] To address the question of the significance of this variable degeneration of neurons in the two major neurochemical compartments of the striatum, the striosomes and the extra striosomal matrix, we recently undertook a study correlating the variable pattern of degeneration of the striatum with the variable pattern of symptomatology in 35 HD cases. In this study we utilised GABA$_A$ receptor, calbindin and enkephalin immunohistochemistry, to label specifically the neurodegeneration in the striosomes and matrix compartments in the striatum of the 35 HD cases and 13 normal control cases.[39] This study showed a large variability in the loss of GABA$_A$ receptors, chemical markers and neurons with respect to their location in the striatal striosome or matrix compartment (Fig. 1). The results of this study on the post-mortem human brain showed that there was a continuum of loss across the different cases, from those which showed predominantly striosomal loss, (Fig. 1C) to those with predominantly matrix loss (Fig. 1B), with a middle group that had a mixed loss in the matrix and striosome compartments (Fig. 1D). To determine whether this variability in compartmental loss was related to the clinical symptomatology, the chemical neuropathology data was compared to the pattern of symptomatology experienced during the life-times of the HD cases, with data collected by researchers blind to the neuroanatomical analyses of the brains. A significant association was found between HD cases with pronounced mood dysfunction and loss of the GABA$_A$ receptors and cells in striosomes of the HD striatum. This association held for both clinical onset and end stage assessments of symptoms, suggesting that changes in striosome-related circuits in Huntington's disease brains may lead to mood dysfunction. The cases with accentuated striosome abnormality further exhibited later onset age, lower disease grade, and lower CAG repeat length in the HD gene. However no association was found between CAG repeat length or age of onset and mood dysfunction. No clear association of motor symptoms and matrix pathology was found, although there was a tendency for cases with matrix loss to have higher voluntary motor impairment at end-stage of the disease. Overall we suggest that variation in clinical symptomatology in HD is associated with variation in the relative abnormality of GABA$_A$ receptor loss in the striosome or matrix compartments of the striatum.

To determine whether degeneration in the other major region of the forebrain affected in HD, the cerebral cortex, also varied with symptomatology profile, in a separate study the pattern of degeneration in the primary motor and anterior cingulate cortex was investigated and compared with variable patterns of symptoms in 12 HD cases and 15 control brains. The primary motor cortex is known to be involved in the control of motor functions, whilst there is evidence to show that the anterior cingulate cortex is involved in emotional regulation and mood disturbances.[40-44] Detailed stereological studies of post-mortem HD brains were carried out and these revealed a major overall cell loss in these two functionally diverse cortical regions. Interestingly, however, they showed marked variation in the extent of cell loss in the motor and cingulate cortices between individual HD cases. When the pattern of cortical cell loss in the HD cases was compared with the pattern of motor and mood symptoms present during the disease for each case, it was found that motor and mood symptomatology corresponded with the heterogeneity of cell loss in the corresponding functional regions of the cerebral cortex.[45] That is, the cell loss in the cingulate gyrus corresponded with the cases with predominant mood symptoms (Figs. 2,4), and the cases with major cell loss in the motor cortex corresponded with those with predominant motor symptoms (Figs. 3,4).

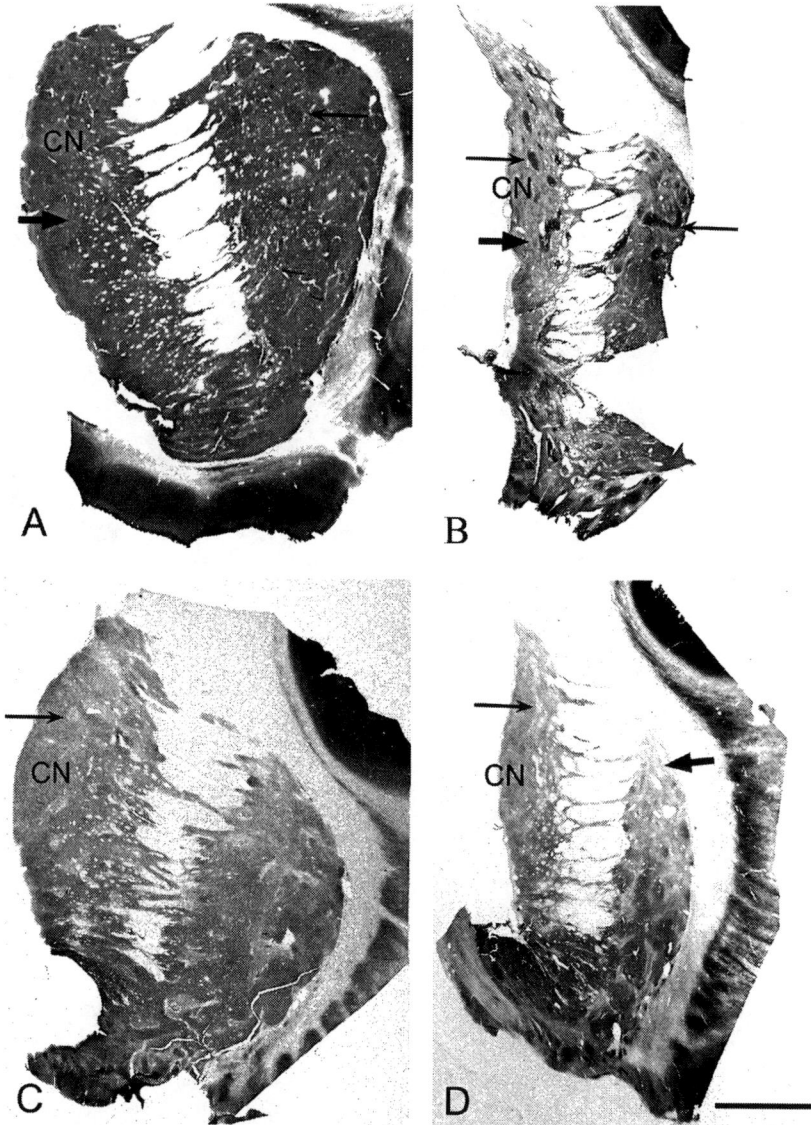

Figure 1. Comparison of the patterns of GABA$_A$ receptor $\beta_{2,3}$ subunit immunostaining in the striatum of a normal case (A) and three HD cases with predominant receptor loss in the matrix compartment and preservation of striosomes (B), in the striosomes compartment (C), or in both striatal compartments (D). Each photomicrograph illustrates a cross-section through the caudate nucleus and putamen from different cases (the matrix is indicated with a thickened arrow and the striosomes with a thin arrow). A) GABA$_A$ receptor immunostaining in the normal brain is distributed throughout the caudate nucleus and putamen with slightly higher densities of immunostaining, or encapsulated zones of staining, in the striosomes and moderately high densities of staining in the matrix. B) A Grade 3 matrix-loss case with extensive depletion of GABA$_A$ receptor immunostaining in the matrix compartment with preservation of striosomes Huntington's disease Grade 1, (C) a striosome-loss case with loss of GABA$_A$ receptor immunostaining mainly in the striosomal compartment, (D) Grade 3, exhibiting reduced GABA$_A$ receptor immunostaining in both striosome and matrix compartments. CN = caudate nucleus; IC = internal capsule; P = putamen. Scale bars = 5 mm, Modified from: Tippett et al. Brain 2007; 130(Pt 1):206-221;[39] ©2007 with permission of Oxford University Press.

Anterior cingulate cortex (BA24)

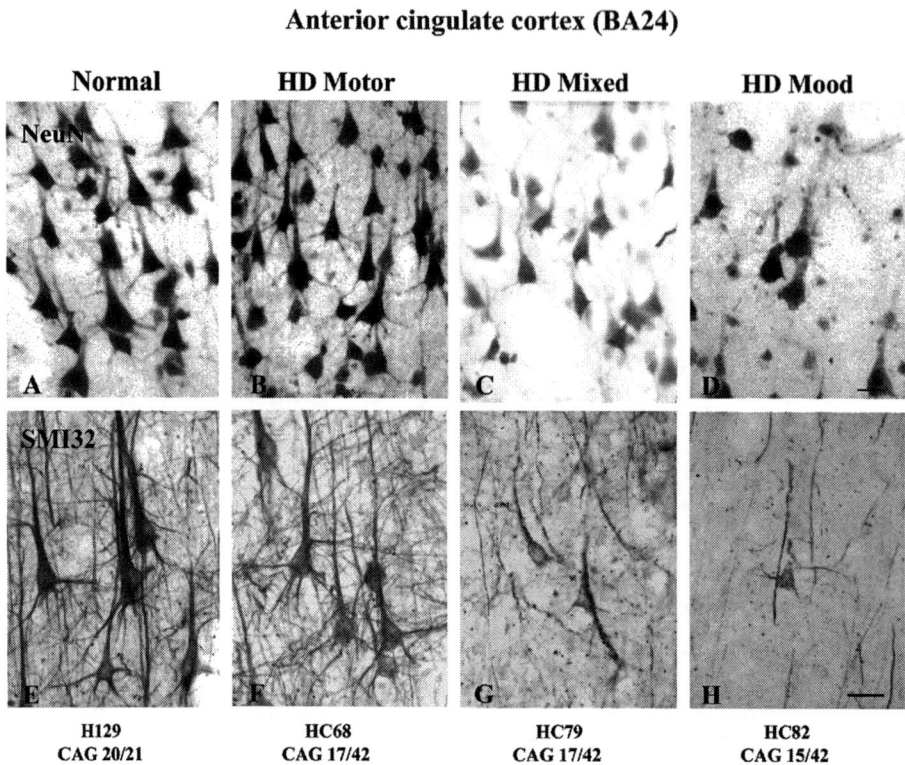

Normal	HD Motor	HD Mixed	HD Mood
H129	HC68	HC79	HC82
CAG 20/21	CAG 17/42	CAG 17/42	CAG 15/42

Figure 2. High power photomicrographs illustrating the pyramidal neurons in layer III in BA24 of the anterior cingulate cortex of normal (A,E) and Huntington's disease (B-D, F-H) cases with "mainly motor" (B,F), "mixed" (C,G) and "mainly mood" (D,H) symptoms. Note that the HD cases all have the same number of CAG repeats on the HD allele of the *IT15* gene. A-D) Pyramidal neurons in layer III of BA24 stained with NeuN. E-H) Pyramidal neurons in layer III of BA24 stained with SMI32. Scale bars, A-D and E-H = 30 μm. Reprinted from Thu et al. Brain 2010; 133(Pt 4):1094-1110;[45] with the permission of Oxford Univeristy Press.

These findings now help to explain the variable results shown in previous quantitative cell studies that have reported variable cell loss across widespread regions of the cerebral cortex.[46-51] For example, major losses of pyramidal projection neurons have been documented in HD cases in various regions of the cerebral cortex including the motor cortex,[49] superior frontal cortex, cingulate gyrus[46] and the angular gyrus of the parietal lobe.[50] These findings of variable cortical degeneration add to findings from recent studies using high resolution surface-based analysis of in vivo Magnetic Resonance Imaging (MRI) data to measure cortical thickness.[52-55] These in vivo studies in over 30 individuals with HD have shown a heterogeneous pattern of region-specific thinning of the cerebral cortex in Huntington's disease with some of the most marked changes occurring in the sensorimotor cortex and areas of the visual cortex.[55,56] In the motor cortex the more dorsal regions associated with the lower limbs showed the most thinning which is reminiscent of the neurodegeneration of the striatum which generally occurs mainly in the dorsal region. Whether the cortical and striatal degeneration is

Primary motor cortex (BA4)

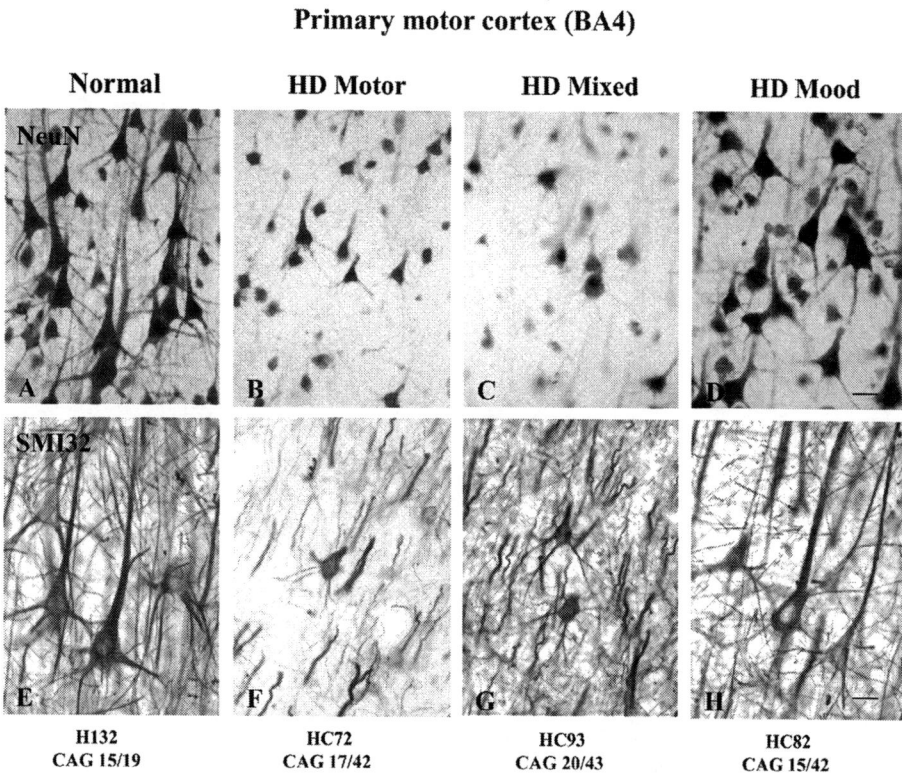

Normal	HD Motor	HD Mixed	HD Mood

| H132 | HC72 | HC93 | HC82 |
| CAG 15/19 | CAG 17/42 | CAG 20/43 | CAG 15/42 |

Figure 3. Photomicrographs illustrating the pyramidal neurons in layer III in Brodmann area 4 of the primary motor cortex of normal (A,E) and Huntington's disease (B-D, F-H) cases with "mainly motor" (B,F), "mixed" (motor/mood) (C,G) and "mainly mood" (D,H) symptom profiles. Note that the HD cases all have a similar number of CAG repeats on the HD allele of the *IT15* gene. A-D) Pyramidal neurons in layer III of Brodmann area 4 stained with NeuN. E-H) Pyramidal neurons in layer III of Brodmann area 4 stained with SMI32. Scale bars, A-D and E-H = 30 μm. Reprinted from Thu et al. Brain 2010; 133(Pt 4):1094-1110;[45] with the permission of Oxford Univeristy Press.

arranged topographically and whether the cortical thinning evident in the imaging data reflect cortical neuronal dysfunction or dysfunction from striatal alterations needs to be investigated further. Interestingly, however, the general pattern of cortical thinning in the MRI studies has been linked with distinct motor phenotypes determined using established clinical tests.[56] When the imaging studies are considered along side our findings demonstrating a significant association between patterns of neuronal degeneration in cerebral cortex and the symptom profile, it can be concluded that cortical changes begin early in HD, are regionally heterogeneous and that topologically selective changes in the cerebral cortex might explain much of the clinical heterogeneity found in this disease. The relationship demonstrated between symptom profiles and cortical degeneration provides a novel perspective on understanding the neural basis of clinical heterogeneity found in HD.

A   *Total neuronal population (NeuN)*

B   *Pyramidal neurons (SMI32)*

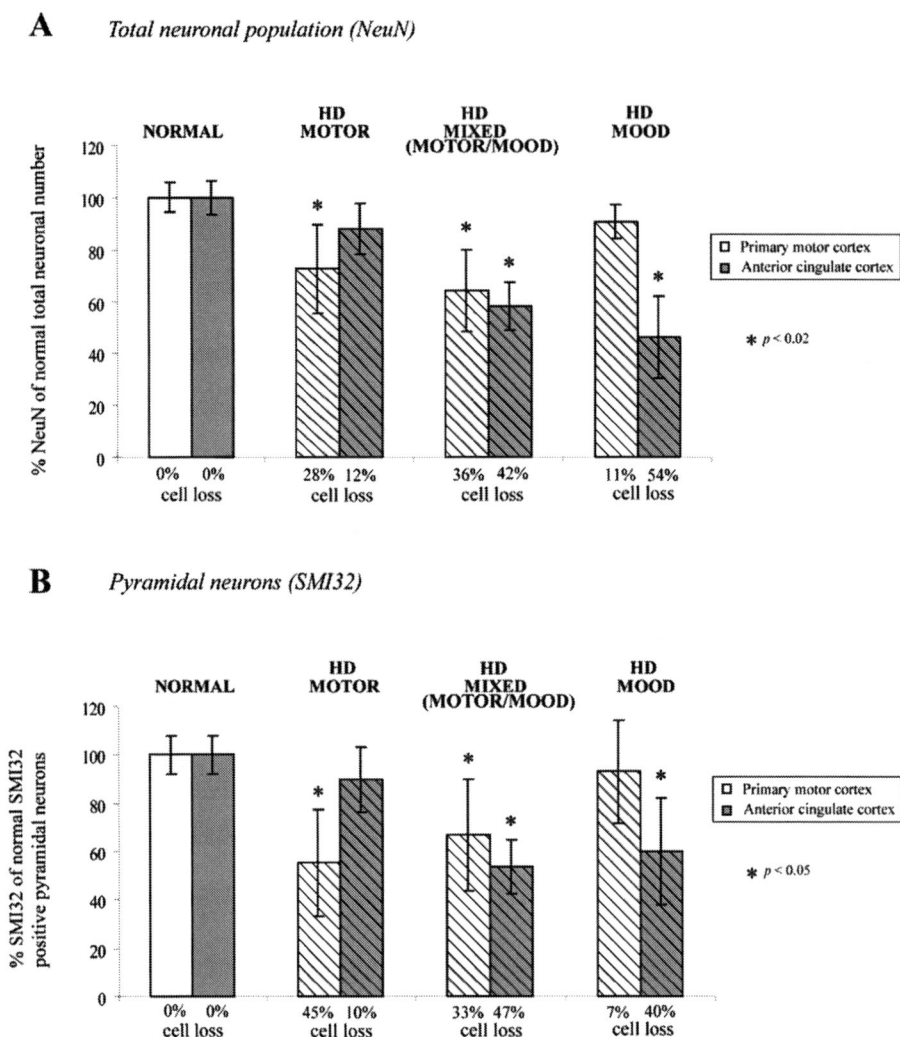

Figure 4. Graph showing the loss of neurons in different cortical regions correlated with the symptomatology. Graph showing (A) the total number of NeuN positive neurons and (B) SMI32 positive pyramidal neurons (expressed as a percentage of the normal) in the primary motor cortex and the anterior cingulate cortex of three different symptom profile groups of HD cases (motor, mixed and mood). Asterisks indicate statistically significant differences from the normal using the Bonferroni post-hoc test. Note that the cell loss in the primary motor cortex shows an opposite trend to the anterior cingulate cortex across the 3 HD symptom groups. Reprinted from Thu et al. Brain 2010; 133(Pt 4):1094-1110;[45] with permission from Oxford Univeristy Press.

MECHANISMS OF DISEASE

The exact mechanisms of neuronal cell death in HD are currently unclear. The expanded CAG repeat of the HD gene is expected to interact with large numbers of other genes as evidenced by the results of gene microarray studies showing large numbers of affected genes in studies on both post mortem HD tissue[57] and mouse models of HD.[58] These interactions lead to a complex set of parameters that may involve transcriptional dysregulation, excitotoxicity, oxidative stress, changes in neurotransmitters, disruption of cortical BDNF production, and breakdown of cellular and vesicular transport mechanisms in neurons of the striatum, cerebral cortex and other regions throughout the brain.[59-64] In the striatum it is the medium spiny neurons which are the most vulnerable, particularly the subset of enkephalin containing striatopallidal neurons. The regional death of these medium spiny neurons however can be quite variable in relation to the striosome-matrix compartments. Therefore other factors such as the connectivity of striatal neurons with other regions of the brain, for instance the topography of the excitatory projection from the cerebral cortex and thalamus may also play a role in this regional cell death. The role of BDNF in the cortico-striatal pathway has been implicated in either causing the death of glutamatergic pyramidal neurons and/or dysfunction of their firing which could be a primary factor in the death of striatal neurons.[64-66] It has long been known that the cerebral cortex is not a homogeneous structure as evidenced by the different morphological composition of the Brodmann areas. Furthermore genetic studies show that neurons in the different regions of the cerebral cortex have a variable genetic expression profile which defines their particular subtype.[67] Therefore neurons in different regions of the cortex may interact differently with the mutant *Htt* gene and cause degeneration in variable populations of pyramidal neurons and cortical interneurons.

Recent transgenic animal studies have implicated dysfunction of the cortex as one of the major indicators of phenotype; this may occur through cortical synaptic dysfunction even before cell death.[65,68] Dysfunction of the cortico-striatal neurons could lead to neurodegeneration of striatal neurons. Also, abnormal glutamate receptor functions in the cerebral cortex have been implicated in behavioural and motor impairments in transgenic mice with physiological and morphological cortical changes predicting the onset and severity of behavioural deficits.[69-71] Furthermore, studies in the conditional mouse model where cortical and/or striatal cells selectively express mutant *Htt*, dysfunction of the cortical neurons was essential to the development of significant behavioural and motor deficits.[72] Other transgenic mouse studies have implicated dysfunction of both the cortical projection and interneurons of the cerebral cortex in the development of HD pathology.[73,74] All of these animal studies provide accumulating mechanistic evidence that the cortex plays a major role in the initiation and development of the HD phenotype, and that dysfunction in the corticostriate neurons plays a major role in HD forebrain pathology.

In addition to the factors leading to cell death discussed above, the changes in receptor expression in the globus pallidus would also play a role in determining the symptom profile throughout the course of the disease. For example, the upregulation of $GABA_A$ and $GABA_B$ receptor subunits is thought to be mainly a compensatory mechanism for loss of GABA input to the output nuclei of the basal ganglia and this mechanism may contribute to ameliorating symptoms in HD by maintaining neurochemical balance despite major cell loss.[26]

CONCLUSION

Our recent studies on the post-mortem HD brain, using stereological counting methods as well as sophisticated MRI imaging techniques on Huntington's disease cases have collectively shown that there is a remarkable variability in the pattern of striatal and cortical degeneration in HD. This variability in neurodegeneration is also correlated with symptomatology. Thus the symptom profile expressed by particular HD cases can be correlated with a pattern of striatal compartmental loss and regional cortical cell loss. The striatum has long been regarded as the primary pathological region of HD causing the movement disorder of HD but the cerebral cortex can now also be regarded as a major contributor to HD symptomatology. Further investigations need to be carried out on the pattern of cell death in the basal ganglia-thalamo-cortical pathways to determine whether the precise pathology that occurs in the cortico-basal ganglia thalamo-cortical loop will correlate with symptomatology in each individual case. This will enable a more complete picture to be produced explaining how the cellular and morphological neurodegeneration of the brain correlates with specific symptom subtypes and these findings may well influence the therapeutic strategies for the treatment of HD in the future.

REFERENCES

1. Illarioshkin SN, Igarashi S, Onodera O et al. Trinucleotide repeat length and rate of progression of Huntington's disease. Ann Neurol 1994; 36(4):630-635.
2. Sharp AH, Ross CA. Neurobiology of Huntington's disease. Neurobiol Dis 1996; 3(1):3-15.
3. Andrew SE, Goldberg YP, Kremer B et al. The relationship between trinucleotide (CAG) repeat length and clinical features of Huntington's disease. Nat Genet 1993; 4(4):398-403.
4. Claes S, Van Zand K, Legius E et al. Correlations between triplet repeat expansion and clinical features in Huntington's disease. Arch Neurol 1995; 52(8):749-753.
5. Brandt J, Butters N. The neuropsychology of Huntington's disease. TINS 1986:118-120.
6. Folstein SE. Huntington's Disease: A Disorder of Families. Baltimore: John's Hopkins University Press; 1989.
7. Myers RH, Sax DS, Koroshetz WJ et al. Factors associated with slow progression in Huntington's disease. Arch Neurol 1991; 48(8):800-804.
8. Thompson JC, Snowden JS, Craufurd D et al. Behavior in Huntington's disease: dissociating cognition-based and mood-based changes. J Neuropsychiatry Clin Neurosci 2002; 14(1):37-43.
9. Zappacosta B, Monza D, Meoni C et al. Psychiatric symptoms do not correlate with cognitive decline, motor symptoms, or CAG repeat length in Huntington's disease. Arch Neurol 1996; 53(6):493-497.
10. Georgiou N, Bradshaw JL, Chiu E et al. Differential clinical and motor control function in a pair of monozygotic twins with Huntington's disease. Mov Disord 1999; 14(2):320-325.
11. Wexler NS, Lorimer J, Porter J et al. Venezuelan kindreds reveal that genetic and environmental factors modulate Huntington's disease age of onset. PNAS 2004; 101(10):3498-3503.
12. MacMillan JC, Snell RG, Tyler A et al. Molecular analysis and clinical correlations of the Huntington's disease mutation. Lancet 1993; 342(8877):954-958.
13. Telenius H, Kremer B, Goldberg YP et al. Somatic and gonadal mosaicism of the Huntington disease gene CAG repeat in brain and sperm. Nat Genet 1994; 6(4):409-414.
14. Witjes-Ane MN, Zwinderman AH, Tibben A et al. Behavioural complaints in participants who underwent predictive testing for Huntington's disease. J Med Genet 2002; 39(11):857-862.
15. Di Maio L, Squitieri F, Napolitano G et al. Onset symptoms in 510 patients with Huntington's disease. J Med Genet 1993; 30(4):289-292.
16. Vonsattel JP, Myers RH, Stevens TJ et al. Neuropathological classification of Huntington's disease. J Neuropath Exp Neurol 1985; 44:559-577.
17. Vonsattel JPG, Difiglia M. Huntington-disease. J Neuropath and Exp Neurol 1998; 57(5):369-384.
18. Albin RL, Makowiec RL, Hollingsworth ZR et al. Excitatory amino acid binding sites in the basal ganglia of the rat: a quantitative autoradiographic study. Neuroscience 1992; 46:35-48.
19. Deng YP, Albin RL, Penney JB et al. Differential loss of striatal projection systems in Huntington's disease: a quantitative immunohistochemical study. J Chem Neuroanat 2004; 27(3):143-164.
20. Faull RL, Waldvogel HJ, Nicholson LF et al. The distribution of GABAA-benzodiazepine receptors in the basal ganglia in Huntington's disease and in the quinolinic acid-lesioned rat. Prog Brain Res 1993; 99:105-123.
21. Glass M, Dragunow M, Faull RLM. The pattern of neurodegeneration in Huntington's disease: a comparative study of cannabinoid, dopamine, adenosine and GABA(A) receptor alterations in the human basal ganglia in Huntington's disease. Neuroscience 2000; 97(3):505-519.

22. Glass M, Faull RL, Dragunow M. Loss of cannabinoid receptors in the substantia nigra in Huntington's disease. Neuroscience 1993; 56(3):523-527.
23. Reiner A, Albin RL, Anderson KD et al. Differential loss of striatal projection neurons in Huntington disease. Proc Natl Acad Sci USA 1988; 85(15):5733-5737.
24. Ferrante RJ, Kowall NW, Beal MF et al. Morphologic and histochemical characteristics of a spared subset of striatal neurons in Huntington's disease. J Neuropathol Exp Neurol 1987; 46(1):12-27.
25. Vonsattel JP, Ge P, Kelly L. Huntington's disease. In: Esiri M, Morris JH, editors. The Neuropathology of Dementia. Cambridge: Cambridge University Press UK; 1997:219-240.
26. Allen KL, Waldvogel HJ, Glass M et al. Cannabinoid (CB(1)), GABA(A) and GABA(B) receptor subunit changes in the globus pallidus in Huntington's disease. J Chem Neuroanat 2009; 37(4):266-281.
27. Graybiel AM, Ragsdale CW Jr. Histochemically distinct compartments in the striatum of human, monkeys and cat demonstrated by acetylthiocholinesterase staining. Proc Nat Acad Sci USA 1978; 75(11):5723-5726.
28. Holt DJ, Graybiel AM, Saper CB. Neurochemical architecture of the human striatum. J Comp Neurol 1997; 384:1-25.
29. Waldvogel HJ, Faull RLM. Compartmentalization of parvalbumin immunoreactivity in the human striatum. Brain Res 1993; 610:311-316.
30. Eblen F, Graybiel AM. Highly restricted origin of prefrontal cortical inputs to striosomes in the macaque monkey. J Neurosci 1995; 15(9):5999-6013.
31. Gerfen CR. The neostriatal mosaic: multiple levels of compartmental organization. TINS 1992; 15:133-138.
32. Saka E, Goodrich C, Harlan P et al. Repetitive behaviors in monkeys are linked to specific striatal activation patterns. J Neurosci 2004; 24(34):7557-7565.
33. White NM, Hiroi N. Preferential localization of self-stimulation sites in striosomes/patches in the rat striatum. Proc Natl Acad Sci USA 1998; 95(11):6486-6491.
34. Hedreen JC, Folstein SE. Early loss of neostriatal striosome neurons in Huntington's disease. J Neuropathol Exp Neurol 1995; 54(1):105-120.
35. Morton AJ, Nicholson LF, Faull RL. Compartmental loss of NADPH diaphorase in the neuropil of the human striatum in Huntington's disease. Neuroscience 1993; 53(1):159-168.
36. Seto-Ohshima A, Emson PC, Lawson E et al. Loss of matrix calcium-binding protein-containing neurons in Huntington's disease. Lancet 1988; 1234:1252-1254.
37. Olsen JM, Penney JB, Shoulson I et al. Inhomogeneities of striatal receptor binding in Huntington's disease. Neurology 1986; 36:342.
38. Augood SJ, Faull RL, Love DR et al. Reduction in enkephalin and substance P messenger RNA in the striatum of early grade Huntington's disease: a detailed cellular in situ hybridization study. Neuroscience 1996; 72(4):1023-1036.
39. Tippett LJ, Waldvogel HJ, Thomas SJ et al. Striosomes and mood dysfunction in Huntington's disease. Brain 2007; 130(Pt 1):206-221.
40. Alexopoulos GS, Gunning-Dixon FM, Latoussakis V et al. Anterior cingulate dysfunction in geriatric depression. Int J Geriatr Psychiatry 2008; 23(4):347-355.
41. Davidson RJ, Pizzagalli D, Nitschke JB et al. Depression: perspectives from affective neuroscience. Annu Rev Psychol 2002; 53:545-574.
42. Ebert D, Ebmeier KP. The role of the cingulate gyrus in depression: from functional anatomy to neurochemistry. Biol Psychiatry 1996; 39(12):1044-1050.
43. Harrison PJ. The neuropathology of primary mood disorder. Brain 2002; 125(Pt 7):1428-1449.
44. Konarski JZ, McIntyre RS, Kennedy SH et al. Volumetric neuroimaging investigations in mood disorders: bipolar disorder versus major depressive disorder. Bipolar Disord 2008; 10(1):1-37.
45. Thu DC, Oorschot DE, Tippett LJ et al. Cell loss in the motor and cingulate cortex correlates with symptomatology in Huntington's disease. Brain 2010; 133(Pt 4):1094-1110.
46. Cudkowicz M, Kowall NW. Degeneration of pyramidal projection neurons in Huntington's disease cortex. Ann Neurol 1990; 27:200-204.
47. Hedreen JC, Peyser CE, Folstein SE et al. Neuronal loss in layers V and VI of cerebral cortex in Huntington's disease. Neurosci Lett 1991; 133(2):257-261.
48. Heinsen H, Strik M, Bauer M et al. Cortical and striatal neurone number in Huntington's disease. Acta Neuropathol 1994; 88(4):320-333.
49. Macdonald V, Halliday G. Pyramidal cell loss in motor cortices in Huntington's disease. Neurobiol Dis 2002; 10(3):378-386.
50. Macdonald V, Halliday GM, Trent RJ et al. Significant loss of pyramidal neurons in the angular gyrus of patients with Huntington's disease. Neuropathol Appl Neurobiol 1997; 23(6):492-495.
51. Selemon LD, Rajkowska G, Goldman-Rakic PS. Evidence for progression in frontal cortical pathology in late-stage Huntington's disease. J Comp Neurol 2004; 468(2):190-204.
52. Rosas HD, Feigin AS, Hersch SM. Using advances in neuroimaging to detect, understand and monitor disease progression in Huntington's disease. NeuroRx 2004; 1(2):263-272.

53. Rosas HD, Hevelone ND, Zaleta AK et al. Regional cortical thinning in preclinical Huntington disease and its relationship to cognition. Neurology 2005; 65(5):745-747.
54. Rosas HD, Koroshetz WJ, Chen YI et al. Evidence for more widespread cerebral pathology in early HD: an MRI-based morphometric analysis. Neurology 2003; 60(10):1615-1620.
55. Rosas HD, Liu AK, Hersch S et al. Regional and progressive thinning of the cortical ribbon in Huntington's disease. Neurology 2002; 58(5):695-701.
56. Rosas HD, Salat DH, Lee SY et al. Cerebral cortex and the clinical expression of Huntington's disease: complexity and heterogeneity. Brain 2008; 131(Pt 4):1057-1068.
57. Hodges A, Strand AD, Aragaki AK et al. Regional and cellular gene expression changes in human Huntington's disease brain. Hum Mol Genet 2006; 15(6):965-977.
58. Luthi-Carter R, Strand A, Peters NL et al. Decreased expression of striatal signaling genes in a mouse model of Huntington's disease. Hum Mol Genet 2000; 9(9):1259-1271.
59. Cattaneo E, Rigamonti D, Goffredo D et al. Loss of normal huntingtin function: new developments in Huntington's disease research. Trends Neurosci 2001; 24(3):182-188.
60. Cha JH. Transcriptional dysregulation in Huntington's disease. Trends Neurosci 2000; 23(9):387-392.
61. Morton AJ, Faull RL, Edwardson JM. Abnormalities in the synaptic vesicle fusion machinery in Huntington's disease. Brain Res Bull 2001; 56(2):111-117.
62. Petersen A, Mani K, Brundin P. Recent advances on the pathogenesis of Huntington's disease. Exp Neurol 1999; 157(1):1-18.
63. Rosas HD, Salat DH, Lee SY et al. Complexity and heterogeneity: what drives the ever-changing brain in Huntington's disease? Ann N Y Acad Sci 2008; 1147:196-205.
64. Zuccato C, Cattaneo E. Role of brain-derived neurotrophic factor in Huntington's disease. Prog Neurobiol 2007; 81(5-6):294-330.
65. Cepeda C, Wu N, Andre VM et al. The corticostriatal pathway in Huntington's disease. Prog Neurobiol 2007; 81(5-6):253-271.
66. Strand AD, Baquet ZC, Aragaki AK et al. Expression profiling of Huntington's disease models suggests that brain-derived neurotrophic factor depletion plays a major role in striatal degeneration. J Neurosci 2007; 27(43):11758-11768.
67. Molyneaux BJ, Arlotta P, Menezes JR et al. Neuronal subtype specification in the cerebral cortex. Nat Rev Neurosci 2007; 8(6):427-437.
68. Cummings DM, Andre VM, Uzgil BO et al. Alterations in cortical excitation and inhibition in genetic mouse models of Huntington's disease. J Neurosci 2009; 29(33):10371-10386.
69. Andre VM, Cepeda C, Venegas A et al. Altered cortical glutamate receptor function in the R6/2 model of Huntington's disease. J Neurophysiol 2006; 95(4):2108-2119.
70. Laforet GA, Sapp E, Chase K et al. Changes in cortical and striatal neurons predict behavioral and electrophysiological abnormalities in a transgenic murine model of Huntington's disease. J Neurosci 2001; 21(23):9112-9123.
71. Sapp E, Schwarz C, Chase K et al. Huntingtin localization in brains of normal and Huntington's disease patients. Ann Neurol 1997; 42(4):604-612.
72. Gu X, Andre VM, Cepeda C et al. Pathological cell-cell interactions are necessary for striatal pathogenesis in a conditional mouse model of Huntington's disease. Mol Neurodegener 2007; 2:8.
73. Gu X, Li C, Wei W et al. Pathological cell-cell interactions elicited by a neuropathogenic form of mutant Huntingtin contribute to cortical pathogenesis in HD mice. Neuron 2005; 46(3):433-444.
74. Spampanato J, Gu X, Yang XW et al. Progressive synaptic pathology of motor cortical neurons in a BAC transgenic mouse model of Huntington's disease. Neuroscience 2008; 157(3):606-620.
75. Van Roon-Mom WM, Hogg VM, Tippett LJ et al. Aggregate distribution in frontal and motor cortex in Huntington's disease brain. Neuroreport 2006; 17(6):667-670.

CHAPTER 10

KENNEDY'S DISEASE

Clinical Significance of Tandem Repeats in the Androgen Receptor

Jeffrey D. Zajac*[,1,2] and Mark Ng Tang Fui[2]

[1]Department of Medicine, University of Melbourne at Austin Health, Heidelberg, Victoria, Australia;
[2]Department of Endocrinology, Austin Health, Heidelberg, Victoria, Australia
*Corresponding Author: Jeffrey D. Zajac—Email: j.zajac@unimelb.edu.au

Abstract: Kennedy's disease (KD) or spinobulbar muscular atrophy is a hereditary X-linked, progressive neurodegenerative condition caused by an expansion of the CAG triplet repeat in the first exon of the androgen receptor gene. The phenotype in its full form is only expressed in males and presents as weakness and wasting of the upper and lower limbs and bulbar muscles associated with absent reflexes. Sensory disturbances are present. Various endocrine abnormalities including decreased fertility and gynecomastia are common and amongst the first features of KD. Animal models of KD have demonstrated improvement on withdrawal of testosterone, indicating that this agonist of the androgen receptor is required for the toxic effect. Potential therapies based on testosterone withdrawal in humans have shown some promise, but efficacy remains to be proven. Potential clinical factors, pathogenesis and future approaches to therapy are reviewed in this chapter.

INTRODUCTION

Kennedy's disease (KD), also known as spinobulbar muscular atrophy, is a hereditary X-linked, adult-onset progressive neurodegenerative condition with lower motor neuron weakness. It is one of a family of CAG repeat diseases. Expansion of the CAG repeat sequence in the first exon of the androgen receptor (AR) past 37 copies causes the neuronal degeneration characteristic of this disease (Fig. 1). The AR is a ligand-dependent transcription factor. The ligands, testosterone and dihydrotestosterone (DHT) are the major androgens in humans and regulate reproductive function, sexual function and a

Tandem Repeat Polymorphisms: Genetic Plasticity, Neural Diversity and Disease,
edited by Anthony J. Hannan ©2012 Landes Bioscience and Springer Science+Business Media.

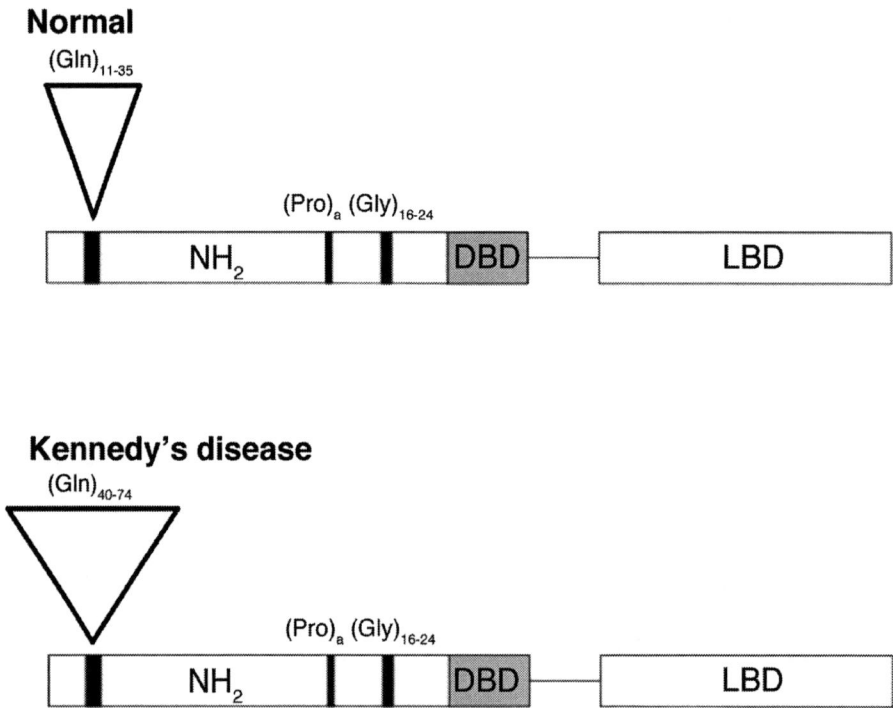

Figure 1. Structure of the androgen receptor gene. The androgen receptor is a steroid-dependent transcription factor consisting of four predominant domains, an N-terminal transactivation domain, a DNA-binding domain (DBD), a hinge region and a ligand-binding domain (LBD). The poly Q (Gln) tract resides in the N-terminal transactivation domain of the receptor. This domain contains two other repeat regions, a polyproline (Pro) repeat and a glycine (Gly) repeat region. The poly Q region is heterogenous in number within the normal population. Expansion of this region to above 40 glutamines causes Kennedy's disease.

host of secondary sexual characteristics including beard growth, scalp hair loss, muscle and bone structure and function. Androgens and the androgen receptor play a regulating role in the central nervous system, but this is much less well understood.

Understanding the clinical presentation, pathological changes, molecular pathogenesis and therapeutic targets of KD is facilitated by knowledge of the function and mechanism of action of the AR.

CLINICAL FEATURES

KD is characterised by gradually progressive upper limb, lower limb and bulbar weakness, as well as muscle atrophy with absent reflexes, fasciculations and cramping.[1] The prevalence of KD is estimated at 2.5 per 100,000 internationally[2] but is likely to be higher as this condition may be mistaken for motor neuron disease or other neurodegenerative disorders. KD is mostly diagnosed in the mid-adult years, yet despite being a progressive neurodegenerative disorder, deterioration is slow such that life expectancy is not drastically shortened.[3] Quality of life can be preserved through a long duration of illness. The most common cause of death is aspiration pneumonia.

KD has a variable age of onset, often in the 20's or 30's with muscle cramps. These initial symptoms may precede significant muscle weakness by as long as 30 years. Creatine kinase (CK) released from muscle is often elevated at diagnosis—around four times the upper limit of normal. Patients develop weakness and atrophy of the pelvic and shoulder muscles with weakness usually occurring first in the lower limbs. Spinobulbar involvement results in weakness and wasting of the facial muscles, tongue and jaw. Fasciculation is particularly common and usually obvious in the tongue and chin. There is no evidence of upper motor neuron signs. The tendon reflexes are absent and tremor is common. Higher CAG repeat number is strongly correlated with earlier onset of disease but does not appear to be correlated with a more rapid progression of disease.[4]

In a retrospective study of 223 Japanese subjects with KD followed-up for between 1-20 or more years, the earliest disease manifestation was hand tremor (median onset age 33y) followed by muscle weakness (44y), dysarthria and dysphagia (50y and 54y), use of a cane and wheelchair (59y and 61y), first episode of pneumonia (62y) then death (65y)[5] (Fig. 2). The initial presentation is often tremor and this is commonly postural in nature.[6] Weakness initially develops in the lower limbs in the majority of patients manifested by difficulty walking and climbing stairs, followed by weakness in the upper limbs then bulbar musculature resulting in progressive dyspnoea, dysphagia and dysarthria. Recurrent laryngospasm occurs in approximately 50% of subjects. Whilst this is usually transient and rarely fatal, it can be highly distressing due to the sensation of suffocation. No clear association has been found for this troublesome manifestation with age of onset of disease, bulbar muscle involvement or CAG repeat number.[7]

Signs of impaired androgen action develop because the CAG mutation causes a partial loss of AR function.[4] The features of androgen insensitivity include gynecomastia, testicular atrophy and reduced fertility. These features whilst subtle are usually apparent upon close examination.[4] Endocrine manifestations of androgen insensitivity largely occur before neurological symptoms. One French study reported gynecomastia in 16/22 patients with KD. Eleven of these 16 developed gynecomastia before the onset of any neurological symptoms.[8] Earlier onset of gynecomastia was associated with higher CAG repeats. 50% of these men reported mild symptoms of androgen insensitivity such as reduced libido, erectile dysfunction and decreased facial hair. One-third of subjects were sterile (either primary or secondary). Two-thirds of subjects had a testosterone level above the upper limit of normal, consistent with a defect of testosterone action at the receptor level and altered feedback regulation. Importantly, elevated levels of testosterone did not suppress luteinising hormone (LH) suggesting that androgen insensitivity occurs at the level of the pituitary (a central regular of testosterone) as well as in peripheral tissues. Overall, 19/22 patients had signs or biochemical findings consistent with partial androgen insensitivity. Testosterone levels have also been reported to be normal or low.[9,8] Metabolic derangements such as Type 2 diabetes mellitus and dyslipidemia may also be present but a causal relationship has not yet been determined given the high prevalence of these disorders in the general population.

In addition to motor dysfunction, KD subjects may develop a sensory neuropathy. Increasingly, sensory abnormalities and higher-centre deficits are being described, which were not noted as part of the phenotype in Kennedy's original paper.[10] In one case series 44% of subjects had sensory impairment to distal pinprick and vibration sensation.[11]

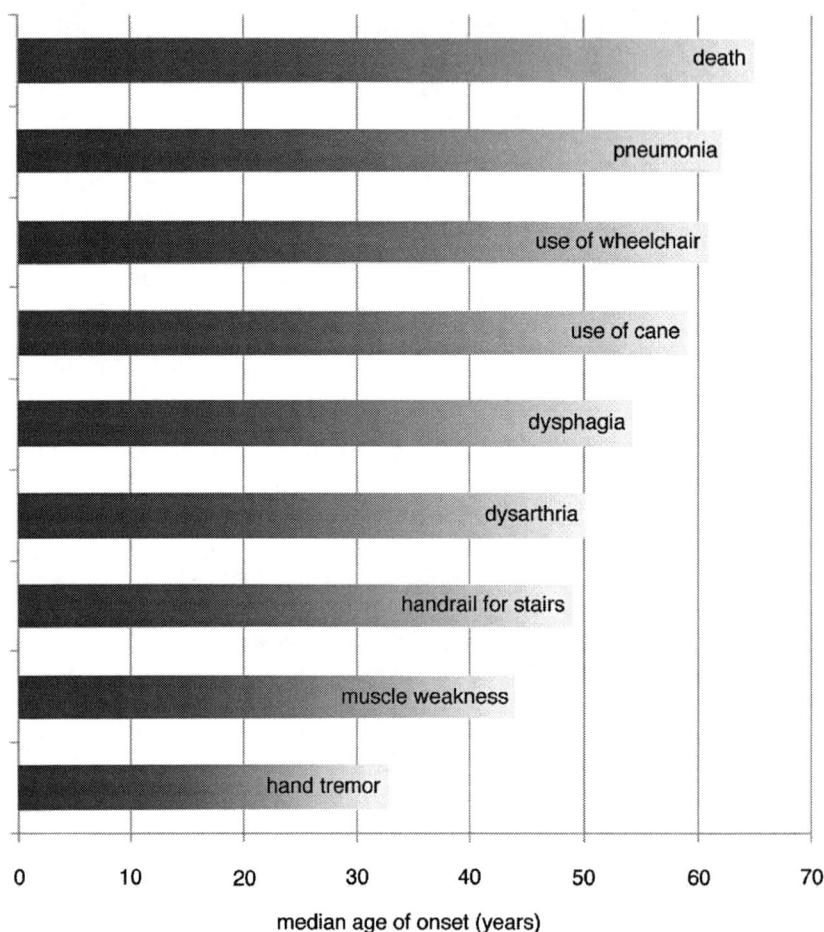

Figure 2. Timing of the development of the KD clinical phenotype, adapted from Atsuta et al, 2006.[5]

Patients with KD may also display cognitive dysfunction including impairment of executive function, memory and attention.[12]

Some case studies have reported clinical manifestations apart from those of the typical phenotype. Soragna et al reported a case associated with extrapyramidal features which were not thought to be idiopathic Parkinson's disease nor multisystem atrophy.[13] Shaw et al described a 58y.o man with rapidly progressive presenile dementia with relatively mild motor signs.[14] Autopsy revealed diffuse subcortical gliosis in the prefrontal region and neuronal loss in the hippocampus. Increasingly, KD is no longer thought of as a purely motor neuron disorder given both sensory and higher-centre involvement.

Heterozygous females may have very mild neurological symptoms and CAG repeat number does not correlate with either severity or age of onset of symptoms.[15] Since binding of androgens to the mutant AR is central to the pathogenesis, females have a mild phenotype due to their markedly lower levels of androgens compared to males. Homozygous females have been described. They also have mild symptoms and no detectable abnormalities on neuro-physiological testing.[16] In mouse models of KD, females exposed to androgens develop a typical KD phenotype which is attenuated when androgens are withdrawn.[17]

The electrophysiological features of KD include a high percentage (95%) of abnormal sensory nerve action potentials (SNAPs) with less frequent (37%) abnormal compound muscle action potentials (CMAPs) and ubiquitous abnormal needle examination.[18] SNAPs reveal low amplitude but not reduced conduction velocity whereas CMAPs commonly reveal low amplitudes and less frequently prolonged distal latencies or reduced conduction velocity. Needle electrode examination reveals acute and chronic motor axonal loss—fibrillation and fasciculation potentials with prominent neurogenic motor unit potential changes. As this is a slowly progressive disorder, chronic changes dominate acute changes. Longer CAG repeats (≥47) have been associated with motor-dominant neuropathy and decreased CMAPs whereas shorter CAG repeats (<47) associated with sensory-dominant neuropathy and decreased SNAPs.[19]

PATHOLOGY

There is a significant reduction in the number of anterior horn motor neurons in all spinal segments. Lower motor neurons through all spinal cord segments and brainstem motor nuclei show marked atrophy although cranial nerves III, IV and VI appear to be spared[20] (Fig. 3). Reduction of anterior horn cells in all examined spinal cord segments was the most striking finding in Kennedy's original paper. Some areas may be devoid of neurons. Neurons in the intermediate zone of the ventral horn are depleted and inclusion bodies are frequently seen. There is disaggregation of the polyribosomes in the endoplasmic reticulum of the affected anterior horn cells on electron microscopic evaluation. Pyramidal tracts, Clarke's column and long tracts of the spinal cord are preserved. Dorsal spinal roots can show segmental demyelination and remyelination. There is a marked loss of large myelinated fibres in the ventral spinal roots and large motor nerves (sciatic, tibial, sural) show demyelination. Regeneration clusters and axonal sprouting are seen uncommonly.

In contrast to motor impairment which is more marked proximally than distally, sensory impairment occurs distally.[21] Large sensory neurons may be extensively affected with neuronal and axonal atrophy. Regeneration of axons may occur peripherally with an increase in small myelinated fibres occasionally seen. Small myelinated and unmyelinated fibres appear to be protected from the disease process.[22]

Muscle atrophy may be widespread and muscle fibres may be replaced by adipose tissue. Muscle histology reveals both myopathic and neurogenic changes (Fig. 3). The histopathological features of KD include target and targetoid fibres, atrophic angulated fibres and clumping of subsarcolemmal nuclei. Muscle fibre size may be highly variable and fibres may undergo splitting and necrosis. Nuclei may be seen centrally located. A small degree of connective tissue proliferation may be seen.[23] It is controversial as to whether myopathic changes are due solely to denervation—in Soraru's study the severity of myogenic muscle involvement was not shown to be correlated with subject age, duration of disease nor CK value—suggesting myopathic changes may occur independently of denervation. The myopathic changes seen in heterozygous females without significant clinical or neuro-physiological abnormalities also point against myopathy solely due to denervation.

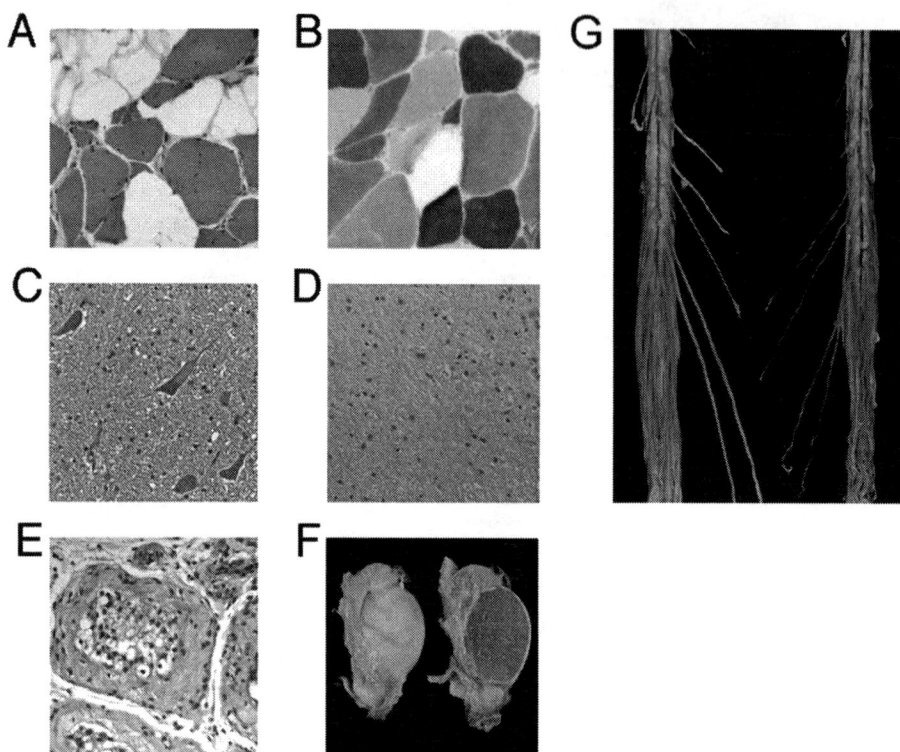

Figure 3. Post mortem findings in Kennedy's disease. A) Quadriceps femoris muscle, showing replacement of muscle fibres by fat, small angulated fibres and increased number of internal nuclei. B) Fibre typing in quadriceps femoris muscle, showing angulated, denervated fibres of Type I, Type 2A and Type 2B fibre types. C) Anterior horn region of cervical spinal cord from control subject (age-matched), showing large motor neuron cell bodies. D) Anterior horn region of cervical spinal cord from KD subject, showing marked loss of motor neurons. E) Testis, showing atrophy of seminiferous tubules, thickened basal lamina and no spermatogenesis. Original magnification 200× for (A) and (B), 100× for (C), (D), (E). Sections stained with hematoxylin and eosin (A), (C), (D), (E), or for ATPase activity (B). F) Testis showing atrophy. G) Ventral view of distal spinal cord showing atrophy of the anterior nerve roots compared to an age and sex matched control patient. (Right: Kennedy's. Left: Age matched control). Reprinted from *Neurobiology of Disease* ©2007, with permission of Elsevier.

Frontal white and grey matter changes have been described using 3D MRI,[12] however nuclear protein inclusions have not been found in the cortex and the pathologic mechanisms for the cognitive deficits have not yet been defined.[24]

DIFFERENTIAL DIAGNOSIS COMPARED WITH MND

As a progressive neurodegenerative disorder, KD may mimic motor neuron disease in its early stages, however with careful clinical assessment followed by targeted investigations, these two conditions are readily distinguishable. Clinicians are more likely to encounter MND as it has a higher prevalence (around 5 per 100,000) and seems better known than KD. MND presents later in life than KD, at around the age of 60.[25]

Patients with KD and MND may present to clinicians with the problem of muscle weakness, even though tremor, cramps and fasciculations may be the initial symptoms. Muscle weakness in MND commonly presents in an asymmetrical pattern with unilateral hand weakness and foot drop being common presentations. KD usually presents with bilateral limb girdle weakness. Hence, patients with MND commonly present with difficulties with fine motor tasks whereas KD presents with difficulty walking or climbing stairs. Bulbar weakness is a feature of both disorders. The progression of weakness is much more rapid with MND and median survival is 36 months. The rate of progression of KD is much slower than motor neuron disease with many patients surviving 20 or 30 years after the initial symptoms. As MND displays upper and lower motor neuron signs, spasticity and hyper-reflexia help differentiate it from KD where reflexes are reduced/absent.[26,27]

MND patients may display emotional lability of the pseudobulbar affect—inappropriate crying, laughing and yawning. Frontotemporal and executive symptoms may be seen in MND and although mutant AR is widespread in the cerebrum of KD subjects, the cortex appears to be generally spared. Sensory neuropathy is included in the KD phenotype and sensory abnormalities on neurophysiologic testing can be widespread whereas patients with MND have fewer or no sensory complaints. Sensory neuropathy is not currently recognised in the revised El Escorial diagnostic criteria of MND, however recent case reports have challenged this.[27-30] The endocrine findings of KD and family history of X-linked pedigree further assist in differentiation. Less than 10% of MND patients show an autosomal dominant inheritance. Features related to androgen insensitivity, such as gynecomastia, also differentiate KD from other forms of motor neuron disease.

Neuro-physiological needle electrode examination can help differentiate MND and KD—KD displays prominent chronic motor unit potential (MUP) changes with relatively fewer fibrillation potentials which is opposite to the findings in MND where fibrillation potentials are more prominent than chronic neurogenic MUP changes. As MND progresses at a faster rate, acute motor axonal loss (MAL) is more pronounced than chronic MAL compared with KD.[18] Motor unit number estimation (MUNE) is correlated with disease duration and muscle weakness in KD and has potential to act as a marker of disease progression. MUNE loss is slower in KD than in MND.[31]

MOLECULAR PATHOGENESIS

The genetic basis of KD has been elegantly attributed by La Spada to excess CAG repeats (encoding glutamine) in exon 1 of the AR gene resulting in mutant AR expression.[32] In KD the number of repeats is increased to 40-62, compared to 10-36 in healthy controls.[33] The human AR gene has eight exons and encodes a 919 amino acid protein of 110 kD. It is the receptor for testosterone and dihydrotestosterone—the major male sex hormones. It is a ligand dependent transcription factor which plays a role during foetal development as well as during puberty and adult life. Receptor binding is decreased in KD[4] but this loss of function is not the pathogenic mechanism of the motor neuron phenotype.

Post-mortem studies[34] have revealed very extensive distribution of mutant AR through neuronal and other tissues using a 1C2 antibody (1C2) against expanded poly Q. In neural tissues, mutant AR could be found in: the cerebrum (caudate, striatum, thalamus and hypothalamus, but not cerebral cortex), the pons, the medulla, the cerebellum and spinal cord (anterior and posterior horn nuclei), dorsal root ganglia and sympathetic ganglia. The strongest staining for mutant AR accumulation occurred in the anterior and posterior horn nuclei of the spinal cord, the dorsal root ganglia and the facial and trigeminal motor nuclei—in these areas neuron depletion was most marked. There are very few remaining spinal motor neurons that did not stain for mutant AR. Frequency of nuclear staining showed a strong positive correlation with CAG repeats but is not strongly associated with clinical phenotype. Peripheral nerves do not show staining for mutant AR.[35]

Mutant AR is also found in nonneuronal tissue including liver, kidney, testis, prostate and scrotum. Pancreatic islet cells demonstrate cytoplasmic staining. This raises the possibility that diabetes mellitus in KD may be due to insulin deficiency. However there is a strong association between diabetes mellitus, ageing and androgen deficiency. Hyperglycaemia in this cohort is thought to be mediated by insulin resistance, rather than insulin deficiency.[36] Mutant AR has also been found in the skeletal muscle suggesting the possibility of a direct effect on muscle.[17] In both neuronal and nonneuronal tissues, the distribution of mutant AR is very widespread compared to the selective clinical deficits. It is not known whether certain tissues are protected from mutant AR-mediated damage and destruction.

LOSS OF FUNCTION OF THE AR VERSUS TOXIC GAIN OF FUNCTION?

There is clear evidence that the poly Q expansion of the AR in KD causes a toxic gain of function dependent on the presence of androgens which leads to the neuronal degeneration rather than the effect being ablation of androgen receptor function. Clinical assessment of two homozygous females demonstrated minimal disease[16] and women carrying one copy of the poly Q expansion showed mild or no illness. XY individuals with complete ablation of the AR causing complete androgen insensitivity syndrome display a female phenotype with no neurological signs.[37] Furthermore, animal models with ablation of the androgen receptor (ARKO mice) show no significant neurological dysfunction.[38,39]

As opposed to the other poly Q diseases, the ligand dependent nature of the AR is thought to be the primary reason for the difference in phenotype between males and females. An initial suggestion attributed the paucity of features in females to random inactivation of the X chromosome,[40] however a much more straightforward and likely explanation is that testosterone or other androgens are necessary to activate the receptor and drive it into the nucleus to cause the subsequent cascade of events initiating neuronal degeneration. This is consistent with the above evidence and also consistent with androgen ablation experiments of genetically modified animal mouse models (see below).[41]

Nonetheless evidence presented by Cary[42] makes a strong case for some contribution to the phenotype related to a decrease in function of the androgen receptor. There is clear evidence of androgens having a trophic effect on motor neurons and Cary's paper reviews a potential mechanism for the decrease in AR function contributing to the KD phenotype. The role of the androgen receptor in neural tissues is reviewed by Poletti.[43] Androgens can promote growth of specific motor nuclei and regrowth of peripheral nerves after section can induce neurite outgrowth in immortalised motor neurons.

There are multiple potential mechanisms for the toxic effect of the abnormal protein, including:

1. Disruption of proteolytic processing of degradation of the AR.
2. Nuclear accumulation of poly Q expanded protein and proteolytic cleavage and aggregation of the mutant protein.[44]
3. Inhibitory effects on axonal transport.
4. Effects on transcriptional regulation by sequestration of transcription factors such as CBP.[45]
5. Disruption of histone acetylase as demonstrated by the fact that histone acetylase inhibitors can slow impairment in animal models.[46]

The mechanism by which mutant AR accumulation in neurons induces cell death is not yet resolved. AR-containing neuronal intranuclear inclusions (NI) are present in spinal motor neurons, and are increased by testosterone[47] and may be toxic to cells although the exact mechanism is to be determined. Mutant AR inclusions are found in many parts of the central nervous system but not all are thought to be clinically significant.

The ubiquitin/proteasome pathway may be involved as these proteins are commonly found in mutant AR NIs and inhibition of the proteasome pathway may enhance protein aggregation and neural toxicity.[48] Furthermore, cJun N-terminal kinase (JNK) has been implicated in modulating axonal transport and may play a role in neuronopathy and apoptosis, as it appears to be activated by the poly Q chain of AR and inhibitors of JNK attenuated these effects.[49] Suppression of TGF-β signalling has been found in a KD mouse model and in patients with KD. Mutant AR was shown to inhibit expression of TGF-β receptor Type II (TβRII) transcription and this was associated with neuronal dysfunction whereas TβRII overexpression attenuated the cytotoxic effects.[50] Caspase 3 is involved in apoptosis and cell death and also cleavage of the truncated AR. This process is important in the toxicity.[51]

ROLE OF AGGREGATES

Intracellular aggregates are a pathological hallmark of KD. They contain sequestered mutant AR as well as other transcription factors and proteosomal subunits such as ubiquitin involved in protein degradation.[52,53] These aggregates accumulate in nuclear inclusions (NIs). These NIs are present in the motor neurons of the spinal cord and brain stem but also in nonneural tissues, such as scrotum, testes, kidneys and heart.[54] As well as aggregates there is diffuse accumulation of mutant AR protein.[34]

The role of aggregates and NIs in causing neuronal degeneration still needs to be clarified. Are they causal, an epiphenomenon or even a protective mechanism sequestering toxic proteins? A number of animal studies reviewed below demonstrated dissociation between the presence of inclusions and toxicity.[55] Finally it is not clear why aggregates appear when the poly Q length passes a critical threshold of 38 to 40.

NIs ARE DYNAMIC

In a mouse model of KD removal of the testis and therefore testosterone reduced NI number in spinal cord and brain neurons in AR112 males.[56]

Animal models have demonstrated a number of other significant findings including mitochondrial abnormalities.[57] Overexpression of IGF1 in muscle attenuated the disease process[17] and overexpression of mutant AR in muscle caused a similar testosterone-dependent phenotype as AR overexpression in neural tissue.

A significant target of testosterone is the spinal nucleus of the bulbar cavernosus system known as ONUFs nucleus in humans.[58-60] This nucleus is normal in MND[61] but degenerates in KD.

ANIMAL MODELS

Initial experiments in mouse-models of KD down-regulated testosterone production through castration and this resulted in reduction of mutant AR accumulation and improvements on exercise testing.[62] Encouragingly, the improvement was observed in mice who displayed significant functional impairment and whilst functional state was not restored to that of castrated wide-type mice, significant gains were made.[56]

There is an interesting Drosophila model[63] in which overexpression of a poly Q expanded AR in the photoneurons of the eye resulted in DHT dependent degeneration of these neurons.

A number of full length AR and complete AR gene mouse models are reviewed[55] and in all of these there is a more severe expression of neural degeneration in males than females. In the McManamny model[64] there was a phenotype but no NIs or aggregates. In another study[56] there were NIs in neurons but no neuronal loss.

In contrast,[65] using a truncated AR model showed no gender differences but all the reported full length models demonstrated clear male predominance of the neurodegeneration.[55] Truncated forms of the mutated poly Q protein formed by a proteolytic cleavage may be more toxic than a full-length protein.

THERAPEUTIC TARGETS

In the 40 years since Kennedy's original description of the clinical phenotype, treatment for KD has largely been palliative, however potential therapies to halt disease progression have blossomed in the last decade. An accurate initial diagnosis of KD refuting other more morbid neurodegenerative disorders and subsequent genetic counselling are central to the management of these patients and their families. Physical and occupational therapies to assist with ambulation and daily activities have hitherto been the mainstay of patient management. Later in the disease, respiratory and nutritional support may infrequently be required through ventilator masks and enteral feeding.

In a human case study, testosterone administration resulted in worsening of gait[66] and another study revealed no improvement with 2 years of testosterone therapy.[67] These findings are consistent with cell and animal models demonstrating increased nuclear mutant AR and nuclear inclusions with testosterone therapy.

Following the description of the central pathogenic mechanism—androgen binding to mutant AR and translocation to the nucleus—initial animal studies focussed on down-regulating testosterone production.[68,69] Cell and animal-based studies are now exploring various mechanisms to induce destruction of excessive mutant AR and its toxic nuclear inclusions. Combination therapy with compounds reduces androgen formation (leuprorelin) and increase AR degradation. ASC-J9, 17-DMAG, CHIP, IGF-1[70-73] hold promise and may be more efficacious than either compound alone. However, given that such therapies are likely to have significant anti-androgen side effects whereas the disease in question displays a slowly progressing mild phenotype, careful consideration will be required for the absolute benefit of any therapy to patients with KD.

Leuprorelin, a LHRH agonist which down-regulates serum testosterone levels has been reported in a phase 2 study.[74] In a 48 week prospective double-blind randomised placebo-controlled trial, leuprorelin resulted in a marked reduction of CK levels and mutant AR as assessed by 1C2 staining of scrotal cells but no reduction on overall MND functional rating scale (ALSFRS) during this short follow-up. However at the conclusion of a subsequent 96-week open-label extension, leuprorelin was shown to arrest functional decline and improve swallowing as assessed by ALSFRS and videofluoroscopic studies. Anti-androgen side-effects included erectile dysfunction, fatigue, hot flushes and metabolic derangements through 48 weeks. One subject died from acute cardiac failure after 118 weeks of leuprorelin therapy. However although a subsequent larger double-blind randomised placebo-controlled trial confirmed the beneficial effects of leuprorelin on CK levels and mutant AR expression there was no functional benefit. In a short follow-up of 48 weeks this study did not show an improvement in functional outcomes including ALSFRS and pharyngeal swallowing as assessed by videofluorography.[57] Whilst leuprorelin may prove beneficial after larger and longer studies or in certain subgroups of patients, there are unresolved issues regarding the almost total deprivation of testosterone in a group of men already somewhat androgen insensitive. Extensive experience with androgen-deprivation therapy for prostate cancer has raised concerns regarding increased osteoporosis and fracture, muscle weakness, dyslipidemia, insulin resistance and hyperglycaemia, abdominal obesity and cardiovascular death.[75-77] Severe sexual dysfunction is almost universal.

In vitro and in vivo animal models are currently being used to examine pathways to increase degradation of mutant AR. This approach to therapy appears more targeted than depleting testosterone although human trials have not yet been conducted to assess efficacy and side-effects of such therapies in KD. In KD models, compounds that disrupt the coregulators of AR have been investigated, such as 5-hydroxy-1, 7-bis(3,4-dimethoxyphenyl)-1,4,6-heptatrien-3-one (ASC-J9). The hypothesis is to interrupt the interaction of excess mutant AR from its coregulators such that it will be vulnerable to degradation. In a cell model, ASC-J9 reduced the testosterone-induced nuclear translocation of mutant AR and formation of aggregates and reduced mutant AR-dependent cell death. Furthermore, in a mouse model, ASC-J9 reduced muscle atrophy, improved neurological and sexual function as well as prolonged the lifespan of mice.[70]

Another promising compound is 17-(dimethylaminoethylamino)-17-demethoxy-geldanamycin (17-DMAG). Heat shock protein 90 (Hsp90) is a cytosolic chaperone for many proteins including steroid receptors such as the AR. Inhibitors of Hsp90 allow the degradation of its target proteins by the ubiquitin-proteasome system. 17-allylamino-17-demethoxygeldanamycin (17-AAG) is an inhibitor of Hsp90 which has a favourable effect on AR accumulation[78] but in phase 1 human oncology trials was found to have significant side effects which may preclude its overall therapeutic benefit in less morbid conditions such as KD.[79] 17-DMAG is a derivative of 17-AAG but is not subjected to degradation to CYP3A4-mediated toxic metabolites that 17-AAG is. In a cell line over-expressing mutant AR, 17-DMAG enabled Hsp90 and its target protein (the mutant AR) to be degraded by the ubiquitin-proteasome system. In a mouse model of KD, 17-DMAG reduced mutant AR in spinal cord and skeletal muscle, ameliorated muscle atrophy and improved the motor function.[71] Basic side-effect profile appeared favourable.

The C terminus of Hsc70 (heat shock cognate protein 70)-interacting protein (CHIP) is a U-box Type E3 ubiquitin ligase which degrades proteins (such as mutant AR) trapped by the molecular chaperone Hsp90. Overexpression of CHIP in a cell model reduced the burden of mutant AR through ubiquination and degradation and in a mouse model, CHIP reduced nuclear accumulation of mutant AR in the spinal cord and ameliorated motor function. A very large range of proteins are CHIP substrates hence CHIP therapy will need to be targeted otherwise significant side effects may occur.[72]

Other potential targets include histone deacetalayse inhibitors;[46] caspase inhibitors;[80] inhibition of androgen receptor expression using RNAi;[81] and inhibition of nuclear localisation[63] and finally inhibition of nuclear aggregate formation.[52]

IGF-1 has trophic effects on muscle but also phosphorylates AR via Akt-dependent mechanisms therefore blocking androgen-AR binding and hence reducing nuclear translocation. In a mouse model of KD, IGF-1 overexpressed in muscle resulted in Akt phosphorylation of AR resulting in clearance of mutant AR via the ubiquitin-proteasome system.[73] Decreased AR aggregation and nuclear inclusions were reported in muscle resulting in less muscle atrophy and degeneration. Mice displayed improvements in functional motor tests. Interestingly, in these mice with IGF over-expressed in muscle, disease improvement was also seen in the spinal cord with less mutant AR expression, fewer nuclear inclusions as well as preservation of motor neurons. The mechanism is unknown—whether it is direct action of muscle-derived IGF on neurons or whether it is restored muscle releasing neurophic factors. In children and adults, growth hormone is currently prescribed for short stature and/or growth hormone deficiency. In adults, growth hormone therapy is titrated to achieve levels in the mid-normal range. Sustained excessive levels of growth hormone and IGF-1 lead to the condition of acromegaly which has a multitude of metabolic, soft-tissue, bone, cardiac, visceral and neoplastic complications. Hence future animal and human trials of IGF-1 must carefully assess for these serious side-effects.

Given the very gradual decline in functional ability in KD, monitoring disease progress and treatment effect in clinical trials through functional outcomes takes many years and even decades. In 35 subjects with clinically-significant KD for an average of 9.7 years, 6-minute walking distance was almost half that of age-matched healthy controls and deteriorated 11% after 1 year follow-up.[82] Within their limitations, intermediate biomarkers could provide more timely assessment of treatment response. Mutant AR accumulation can be assessed by 1C2 antibody staining against expanded poly Q of scrotal skin biopsies. This correlates well with anterior horn staining, CAG repeat length and functional state. Frequency and intensity of 1C2 staining of scrotal epithelial cells was diminished with testosterone depletion.[83] Scrotal skin is more accessible and practical than spinal motor neurons for assessment of this biomarker and although tissue is obtained by punch biopsy under local anaesthetic, its acceptance by large cohorts of men remains to be determined given that this biomarker has no data for longitudinal functional outcomes in human studies.

CONCLUSION

Pathogenic mechanisms for Kennedy's disease are being increasingly understood. It is possible that these developments may lead to novel approaches to therapy. Androgen withdrawal, although successful in animals, has not yet been proven to be effective in humans. Evidence from a number of human trials is promising but does not yet show a significant functional benefit. More research into androgen deprivation, androgen blockade and other emerging therapies is required.

ACKNOWLEDGMENTS

We would like to acknowledge Dr Kesha Rana, Department of Medicine, University of Melbourne, Austin Health, Australia, for the production of Figure 1 in this chapter. Supported by a grant from Eva and Les Erdi.

REFERENCES

1. MacLean HE, Warne GL, Zajac JD. Spinal and bulbar muscular atrophy: androgen receptor dysfunction caused by a trinucleotide repeat expansion. J Neurol Sci 1996; 135(2):149-157.
2. Lund A, Udd B, Juvonen V et al. Multiple founder effects in spinal and bulbar muscular atrophy (SBMA, Kennedy disease) around the world. Eur J Hum Genet 2001; 9(6):431-436.
3. Chahin N, Klein C, Mandrekar J et al. Natural history of spinal-bulbar muscular atrophy. Neurology 2008; 70(21):1967-1971.
4. MacLean HE, Choi WT, Rekaris G et al. Abnormal androgen receptor binding affinity in subjects with Kennedy's disease (spinal and bulbar muscular atrophy). J Clin Endocrinol Metab 1995; 80(2):508-516.
5. Atsuta N, Watanabe H, Ito M et al. Natural history of spinal and bulbar muscular atrophy (SBMA): a study of 223 Japanese patients. Brain 2006; 129(Pt 6):1446-1455.
6. Hanajima R, Terao Y, Nakatani-Enomoto S et al. Postural tremor in X-linked spinal and bulbar muscular atrophy. Mov Disord 2009; 24(14):2063-2069.
7. Sperfeld AD, Hanemann CO, Ludolph AC et al. Laryngospasm: an underdiagnosed symptom of X-linked spinobulbar muscular atrophy. Neurology 2005; 64(4):753-754.
8. Dejager S, Bry-Gauillard H, Bruckert E et al. A comprehensive endocrine description of Kennedy's disease revealing androgen insensitivity linked to CAG repeat length. J Clin Endocrinol Metab 2002; 87(8):3893-3901.
9. Zajac JD, MacLean HE. Kennedy's disease: clinical aspects. In: Wells RD, Warren ST, Sarmiento M, eds. Genetic Instabilities and Hereditary Neurological Diseases. San Diego: Academic Press; 1998:87-100.
10. Kennedy WR, Alter M, Sung JH. Progressive proximal spinal and bulbar muscular atrophy of late onset. A sex-linked recessive trait. Neurology 1968; 18(7):671-680.
11. Lee JH, Shin JH, Park KP et al. Phenotypic variability in Kennedy's disease: implication of the early diagnostic features. Acta Neurol Scand 2005; 112(1):57-63.
12. Kassubek J, Juengling FD, Sperfeld AD. Widespread white matter changes in Kennedy disease: a voxel based morphometry study. J Neurol Neurosurg Psychiatry 2007; 78(11):1209-1212.
13. Soragna D, Messa C, Mochi M et al. Dopaminergic pathways involvement in Kennedy's disease: neurophysiological and. Journal of Neurology 2001; 248(8):710-712.
14. Shaw PJ, Thagesen H, Tomkins J et al. Kennedy's disease: unusual molecular pathologic and clinical features. Neurology 1998; 51(1):252-255.
15. Ishihara H, Kanda F, Nishio H et al. Clinical features and skewed X-chromosome inactivation in female carriers of X-linked recessive spinal and bulbar muscular atrophy. J Neurol 2001; 248(10):856-860.
16. Schmidt BJ, Greenberg CR, Allingham-Hawkins DJ et al. Expression of X-linked bulbospinal muscular atrophy (Kennedy disease) in two homozygous women. Neurology 2002; 59(5):770-772.
17. Johansen JA, Yu Z, Mo K et al. Recovery of function in a myogenic mouse model of spinal bulbar muscular atrophy. Neurobiol Dis 2009; 34(1):113-120.

18. Ferrante MA, Wilbourn AJ. The characteristic electrodiagnostic features of Kennedy's disease. Muscle Nerve 1997; 20(3):323-329.
19. Suzuki K, Katsuno M, Banno H et al. CAG repeat size correlates to electrophysiological motor and sensory phenotypes in SBMA. Brain 2008; 131(Pt 1):229-239.
20. Sobue G, Hashizume Y, Mukai E et al. X-linked recessive bulbospinal neuronopathy. A clinicopathological study. Brain 1989; 112(Pt 1):209-232.
21. Harding AE, Thomas PK, Baraitser M et al. X-linked recessive bulbospinal neuronopathy: a report of ten cases. J Neurol Neurosurg Psychiatry 1982; 45(11):1012-1019.
22. Li M, Sobue G, Doyu M et al. Primary sensory neurons in X-linked recessive bulbospinal neuropathy: histopathology and androgen receptor gene expression. Muscle Nerve 1995; 18(3):301-308.
23. Soraru G, D'Ascenzo C, Polo A et al. Spinal and bulbar muscular atrophy: skeletal muscle pathology in male patients and heterozygous females. Journal of the Neurological Sciences 2008; 264(1-2):100-105.
24. Soukup GR, Sperfeld AD, Uttner I et al. Frontotemporal cognitive function in X-linked spinal and bulbar muscular atrophy (SBMA): a controlled neuropsychological study of 20 patients. Journal of Neurology 2009; 256(11):1869-1875.
25. Logroscino G, Traynor BJ, Hardiman O et al. Descriptive epidemiology of amyotrophic lateral sclerosis: new evidence and unsolved issues. J Neurol Neurosurg Psychiatry 2008; 79(1):6-11.
26. McDermott CJ, Shaw PJ. Diagnosis and management of motor neurone disease. BMJ 2008; 336(7645): 658-662.
27. Corcia P, Meininger V. Management of amyotrophic lateral sclerosis. Drugs 2008; 68(8):1037-1048.
28. van der Graaff MM, de Jong JM, Baas F et al. Upper motor neuron and extra-motor neuron involvement in amyotrophic lateral sclerosis: a clinical and brain imaging review. Neuromuscul Disord 2009; 19(1):53-58.
29. Pugdahl K, Fuglsang-Frederiksen A, de Carvalho M et al. Generalised sensory system abnormalities in amyotrophic lateral sclerosis: a European multicentre study. J Neurol Neurosurg Psychiatry 2007; 78(7):746-749.
30. Amoiridis G, Tsimoulis D, Ameridou I. Clinical, electrophysiologic and pathologic evidence for sensory abnormalities in ALS. Neurology 2008; 71(10):779.
31. Suzuki K, Katsuno M, Banno H et al. The profile of motor unit number estimation (MUNE) in spinal and bulbar muscular atrophy. J Neurol Neurosurg Psychiatry 2010; 81(5):567-571.
32. La Spada AR, Wilson EM, Lubahn DB et al. Androgen receptor gene mutations in X-linked spinal and bulbar muscular atrophy. Nature 1991; 352(6330):77-79.
33. Greenland KJ, Zajac JD. Kennedy's disease: pathogenesis and clinical approaches. Intern Med J 2004; 34(5):279-286.
34. Adachi H, Katsuno M, Minamiyama M et al. Widespread nuclear and cytoplasmic accumulation of mutant androgen receptor in SBMA patients. Brain 2005; 128(Pt 3):659-670.
35. Li M, Miwa S, Kobayashi Y et al. Nuclear inclusions of the androgen receptor protein in spinal and bulbar muscular atrophy. Annals of Neurology 1998; 44(2):249-254.
36. Grossmann M, Thomas MC, Panagiotopoulos S et al. Low testosterone levels are common and associated with insulin resistance in men with diabetes. J Clin Endocrinol Metab 2008; 93(5):1834-1840.
37. MacLean HE, Ball EM, Rekaris G et al. Novel androgen receptor gene mutations in Australian patients with complete androgen insensitivity syndrome. Hum Mutat 2004; 23(3):287.
38. Yeh S, Tsai MY, Xu Q et al. Generation and characterization of androgen receptor knockout (ARKO) mice: an in vivo model for the study of androgen functions in selective tissues. Proc Natl Acad Sci USA 2002; 99(21):13498-13503.
39. Notini AJ, Davey RA, McManus JF et al. Genomic actions of the androgen receptor are required for normal male sexual differentiation in a mouse model. J Mol Endocrinol 2005; 35(3):547-555.
40. Mariotti C, Castellotti B, Pareyson D et al. Phenotypic manifestations associated with CAG-repeat expansion in the androgen receptor gene in male patients and heterozygous females: a clinical and molecular study of 30 families. Neuromuscul Disord 2000; 10(6):391-397.
41. Thomas PS Jr., Fraley GS, Damian V et al. Loss of endogenous androgen receptor protein accelerates motor neuron degeneration and accentuates androgen insensitivity in a mouse model of X-linked spinal and bulbar muscular atrophy. Hum Mol Genet 2006; 15(14):2225-2238.
42. Cary GA, La Spada AR. Androgen receptor function in motor neuron survival and degeneration. Phys Med Rehabil Clin N Am 2008; 19(3):479-494, viii.
43. Poletti A, Negri-Cesi P, Martini L. Reflections on the diseases linked to mutations of the androgen receptor. Endocrine 2005; 28(3):243-262.
44. Walsh R, Storey E, Stefani D et al. The roles of proteolysis and nuclear localisation in the toxicity of the polyglutamine diseases. A review. Neurotox Res 2005; 7(1-2):43-57.
45. McCampbell A, Taylor JP, Taye AA et al. CREB-binding protein sequestration by expanded polyglutamine. Hum Mol Genet 2000; 9(14):2197-2202.

46. Minamiyama M, Katsuno M, Adachi H et al. Sodium butyrate ameliorates phenotypic expression in a transgenic mouse model of spinal and bulbar muscular atrophy. Hum Mol Genet 2004; 13(11):1183-1192.

47. Walcott JL, Merry DE. Ligand promotes intranuclear inclusions in a novel cell model of spinal and bulbar muscular atrophy. J Biol Chem 2002; 277(52):50855-50859.

48. Chan HY, Warrick JM, Andriola I et al. Genetic modulation of polyglutamine toxicity by protein conjugation pathways in Drosophila. Human Molecular Genetics 2002; 11(23):2895-2904.

49. Morfini G, Pigino G, Szebenyi G et al. JNK mediates pathogenic effects of polyglutamine-expanded androgen receptor on fast axonal transport. Nat Neurosci 2006; 9(7):907-916.

50. Katsuno M, Adachi H, Minamiyama M et al. Disrupted transforming growth factor-beta signaling in spinal and bulbar muscular atrophy. Journal of Neuroscience 2010; 30(16):5702-5712.

51. Ellerby LM, Hackam AS, Propp SS et al. Kennedy's disease: caspase cleavage of the androgen receptor is a crucial event in cytotoxicity. J Neurochem 1999; 72(1):185-195.

52. Adachi H, Katsuno M, Minamiyama M et al. Heat shock protein 70 chaperone overexpression ameliorates phenotypes of the spinal and bulbar muscular atrophy transgenic mouse model by reducing nuclear-localized mutant androgen receptor protein. J Neurosci 2003; 23(6):2203-2211.

53. Paulson HL, Perez MK, Trottier Y et al. Intranuclear inclusions of expanded polyglutamine protein in spinocerebellar ataxia type 3. Neuron 1997; 19(2):333-344.

54. Li M, Nakagomi Y, Kobayashi Y et al. Nonneural nuclear inclusions of androgen receptor protein in spinal and bulbar muscular atrophy. Am J Pathol 1998; 153(3):695-701.

55. Beitel LK, Scanlon T, Gottlieb B et al. Progress in spinobulbar muscular atrophy research: insights into neuronal dysfunction caused by the polyglutamine-expanded androgen receptor. Neurotox Res 2005; 7(3):219-230.

56. Chevalier-Larsen ES, O'Brien CJ, Wang H et al. Castration restores function and neurofilament alterations of aged symptomatic males in a transgenic mouse model of spinal and bulbar muscular atrophy. J Neurosci 2004; 24(20):4778-4786.

57. Katsuno M, Banno H, Suzuki K et al. Efficacy and safety of leuprorelin in patients with spinal and bulbar muscular atrophy (JASMITT study): a multicentre, randomised, double-blind, placebo-controlled trial. Lancet Neurol 2010; 9(9):875-884.

58. Matsumoto A, Micevych PE, Arnold AP. Androgen regulates synaptic input to motoneurons of the adult rat spinal cord. J Neurosci 1988; 8(11):4168-4176.

59. Goldstein LA, Sengelaub DR. Timing and duration of dihydrotestosterone treatment affect the development of motoneuron number and morphology in a sexually dimorphic rat spinal nucleus. J Comp Neurol 1992; 326(1):147-157.

60. Watson NV, Freeman LM, Breedlove SM. Neuronal size in the spinal nucleus of the bulbocavernosus: direct modulation by androgen in rats with mosaic androgen insensitivity. J Neurosci 2001; 21(3):1062-1066.

61. Schroder HD, Reske-Nielsen E. Preservation of the nucleus X-pelvic floor motosystem in amyotrophic lateral sclerosis. Clin Neuropathol 1984; 3(5):210-216.

62. Katsuno M, Adachi H, Minamiyama M et al. Reversible disruption of dynactin 1-mediated retrograde axonal transport in polyglutamine-induced motor neuron degeneration. Journal of Neuroscience 2006; 26(47):12106-12117.

63. Takeyama K, Ito S, Yamamoto A et al. Androgen-dependent neurodegeneration by polyglutamine-expanded human androgen receptor in Drosophila. Neuron 2002; 35(5):855-864.

64. McManamny P, Chy HS, Finkelstein DI et al. A mouse model of spinal and bulbar muscular atrophy. Hum Mol Genet 2002; 11(18):2103-2111.

65. Abel A, Walcott J, Woods J et al. Expression of expanded repeat androgen receptor produces neurologic disease in transgenic mice. Hum Mol Genet 2001; 10(2):107-116.

66. Kinirons P, Rouleau GA. Administration of testosterone results in reversible deterioration in Kennedy's disease. J Neurol Neurosurg Psychiatry 2008; 79(1):106-107.

67. Goldenberg JN, Bradley WG. Testosterone therapy and the pathogenesis of Kennedy's disease (X-linked bulbospinal muscular atrophy). Journal of the Neurological Sciences 1996; 135(2):158-161.

68. Katsuno M, Adachi H, Kume A et al. Testosterone reduction prevents phenotypic expression in a transgenic mouse model of spinal and bulbar muscular atrophy. Neuron 2002; 35(5):843-854.

69. Katsuno M, Adachi H, Doyu M et al. Leuprorelin rescues polyglutamine-dependent phenotypes in a transgenic mouse model of spinal and bulbar muscular atrophy. Nat Med 2003; 9(6):768-773.

70. Yang Z, Chang YJ, Yu IC et al. ASC-J9 ameliorates spinal and bulbar muscular atrophy phenotype via degradation of androgen receptor. Nat Med 2007; 13(3):348-353.

71. Tokui K, Adachi H, Waza M et al. 17-DMAG ameliorates polyglutamine-mediated motor neuron degeneration through well-preserved proteasome function in an SBMA model mouse. Human Molecular Genetics 2009; 18(5):898-910.

72. Adachi H, Waza M, Tokui K et al. CHIP overexpression reduces mutant androgen receptor protein and ameliorates phenotypes of the spinal and bulbar muscular atrophy transgenic mouse model. Journal of Neuroscience 2007; 27(19):5115-5126.

73. Palazzolo I, Stack C, Kong L et al. Overexpression of IGF-1 in muscle attenuates disease in a mouse model of spinal and bulbar muscular atrophy. Neuron 2009; 63(3):316-328.

74. Banno H, Katsuno M, Suzuki K et al. Phase 2 trial of leuprorelin in patients with spinal and bulbar muscular atrophy. Annals of Neurology 2009; 65(2):140-150.

75. Shahani S, Braga-Basaria M, Basaria S. Androgen deprivation therapy in prostate cancer and metabolic risk for atherosclerosis. J Clin Endocrinol Metab 2008; 93(6):2042-2049.

76. Greenspan SL. Approach to the prostate cancer patient with bone disease. J Clin Endocrinol Metab 2008; 93(1):2-7.

77. Grossmann M. Bone and metabolic health in patients with nonmetastatic prostate cancer receiving androgen deprivation therapy—management guidelines. Med J Aust. In Press.

78. Waza M, Adachi H, Katsuno M et al. 17-AAG, an Hsp90 inhibitor, ameliorates polyglutamine-mediated motor neuron degeneration. Nat Med 2005; 11(10):1088-1095.

79. Banerji U, O'Donnell A, Scurr M et al. Phase I pharmacokinetic and pharmacodynamic study of 17-allylamino, 17-demethoxygeldanamycin in patients with advanced malignancies. J Clin Oncol 2005; 23(18):4152-4161.

80. Piccioni F, Roman BR, Fischbeck KH et al. A screen for drugs that protect against the cytotoxicity of polyglutamine-expanded androgen receptor. Hum Mol Genet 2004; 13(4):437-446.

81. Caplen NJ, Taylor JP, Statham VS et al. Rescue of polyglutamine-mediated cytotoxicity by double-stranded RNA-mediated RNA interference. Hum Mol Genet 2002; 11(2):175-184.

82. Takeuchi Y, Katsuno M, Banno H et al. Walking capacity evaluated by the 6-minute walk test in spinal and bulbar muscular atrophy. Muscle Nerve 2008; 38(2):964-971.

83. Banno H, Adachi H, Katsuno M et al. Mutant androgen receptor accumulation in spinal and bulbar muscular atrophy scrotal skin: a pathogenic marker. Annals of Neurology 2006; 59(3):520-526.

CHARACTERISING THE NEUROPATHOLOGY AND NEUROBEHAVIOURAL PHENOTYPE IN FRIEDREICH ATAXIA

A Systematic Review

Louise A. Corben,*,[1] Nellie Georgiou-Karistianis,[2]
John L. Bradshaw,[2] Marguerite V. Evans-Galea,[1]
Andrew J. Churchyard,[3] and Martin B. Delatycki[1,4,5]

[1]*Bruce Lefroy Centre for Genetic Health Research, Murdoch Childrens Research Institute, The Royal Children's Hospital, Parkville, Victoria, Australia;* [2]*Experimental Neuropsychology Research Unit, School of Psychology and Psychiatry, Monash University, Clayton, Victoria, Australia;* [3]*Monash Neurology, Monash Medical Centre, Clayton, Victoria, Australia;* [4]*Department of Clinical Genetics, Austin Health, Heidelberg, Victoria, Australia;* [5]*Department of Medicine, University of Melbourne at Austin Health, Heidelberg, Victoria, Australia*
Corresponding Author: Louise Corben—Email: louise.corben@ghsv.org.au

Abstract: Friedreich ataxia (FRDA), the most common of the hereditary ataxias, is an autosomal recessive, multisystem disorder characterised by progressive ataxia, sensory symptoms, weakness, scoliosis and cardiomyopathy. FRDA is caused by a GAA expansion in intron one of the *FXN* gene, leading to reduced levels of the encoded protein frataxin, which is thought to regulate cellular iron homeostasis. The cerebellar and spinocerebellar dysfunction seen in FRDA has known effects on motor function; however until recently slowed information processing has been the main feature consistently reported by the limited studies addressing cognitive function in FRDA. This chapter will systematically review the current literature regarding the neuropathological and neurobehavioural phenotype associated with FRDA. It will evaluate more recent evidence adopting systematic experimental methodologies that postulate that the neurobehavioural phenotype associated with FRDA is likely to involve impairment in cerebello-cortico connectivity.

Tandem Repeat Polymorphisms: Genetic Plasticity, Neural Diversity and Disease,
edited by Anthony J. Hannan ©2012 Landes Bioscience and Springer Science+Business Media.

INTRODUCTION

Friedreich ataxia (FRDA), a multisystem autosomal-recessive disease, is the most common inherited ataxia, affecting approximately 1 in 29,000 individuals.[1] FRDA was first described by Nicholaus Friedreich, Professor of Medicine in Heidelberg, Germany, in 1863 as a distinct clinical syndrome characterized by ataxia that affected several progeny of unaffected parents.[2] The hallmark clinical features of FRDA include progressive ataxia, spasticity, absent lower limb reflexes, impaired vibration sense and proprioception, scoliosis, foot deformity and cardiomyopathy.[1,3] At present there is no proven treatment that can slow the progression or eventual outcome of this life-shortening condition.

CLINICAL FEATURES

Since the discovery by Campuzano and colleagues in 1996 of the *FXN* gene and the mutations therein associated with FRDA, molecular diagnosis of affected individuals and heterozygotes has been possible. Until this time FRDA was generally diagnosed according the criteria of Harding which included progressive ataxia of all four limbs, absent tendon reflexes in the lower limbs, extensor plantar responses and an onset of disease prior to 25 years of age.[3] Molecular diagnosis has now enabled thorough documentation of the phenotype associated with this condition.

Onset of symptoms in FRDA is typically around the age of 10 (±6.4) years; however this can vary considerably, even between siblings.[4,5] The hallmark symptoms tend to be gait instability or generalised clumsiness;[4] however in rare cases cardiac manifestation may precede neurological signs.[5] Mixed cerebellar and sensory gait ataxia results in a broad based ataxic gait associated with frequent falls. As the gait ataxia progresses the individual requires increased support, usually by furniture, walls, another person and eventually a gait aid. They eventually lose the ability to mobilise and thus become fully dependent on a wheelchair for mobility on average 10 to 15 years after disease onset.[3,4] Progressive limb weakness initially affecting proximal muscles further compounds the capacity to walk and, in some cases, also affects hand function.[6] Spasticity, though little studied in FRDA, also has a significant effect on mobility and hand function. Many individuals complain of painful spasms in the legs, particularly at night, as the condition progresses.[7] The disease affects many motor pathways as evident in hand function by dysmetria, dysdiadiokinesia and intention tremor; speech and swallowing by dysarthria and in the latter stages of the disease, dysphagia; and vision in terms of instability of fixation with square wave jerks.[5,7-8] As previously mentioned, most individuals with FRDA present with absent lower limb reflexes, extensor plantar responses (Babinski sign) and a reduction or loss of vibration sense and proprioception.[1] Perception of light touch, pain and temperature is essentially normal in the early stages of the disease, but may become disturbed as the disease progresses[7] with some stimuli perceived as quite noxious. Autonomic disturbance is evident in cold and cyanosed extremities, particularly feet and lower legs, and becomes more apparent and painful as the condition unfolds.[4] A considerable number of individuals with FRDA report neurogenic bladder disturbance as evidenced by urinary frequency and occasionally incontinence.[4,9] Bowel disturbance is rarely reported.[5]

Skeletal changes including pes cavus and scoliosis are also considered cardinal clinical signs of FRDA.[1,10] Pes cavus occurs frequently in individuals with FRDA[1,10] though the role this deformity has in the decline in mobility associated with FRDA is unknown. Scoliosis occurs in most individuals with FRDA[11] with a high prevalence of double thoracic or lumbar curves. Scoliosis may be progressive, that is usually occurring before the age of 10 years and often exhibiting curve greater than 60 degrees; or less severe and non progressive with a curve less than 40 degrees.[12]

Speech disturbance or dysarthria is a universal symptom of FRDA, usually becoming apparent within 2 years of disease onset.[5] Dysarthria associated with FRDA manifests as impaired articulation though is not always ataxic in nature. It may include a blend of spastic, ataxic and/or flaccid components depending on the stage of the disease.[13] Dysarthria invariably progresses and in the latter stages of the disease is associated with dysphagia.

Cardiac changes are apparent in the majority of individuals with FRDA, although few have reported symptoms.[14] Clinically there is considerable variation in the type and extent of impairment; however 90% of people with FRDA have an abnormal ECG with the commonest abnormality being T wave inversion.[15] Cardiac hypertrophy and changes in septal and lateral long-axis LV function,[16] ventricular mass index, posterior wall and septal wall thickness are common.[14] In severe cases of cardiomyopathy, dilation can lead to progressive deterioration of left ventricular ejection fraction and severe cardiac failure.[7] Various arrhythmias including atrial fibrillation and supraventricular arrhythmia may be present.[5,15]

Visual changes in individuals with FRDA may occur and include prolonged saccadic latency and fixation abnormalities such as square wave jerks and ocular flutter. Brainstem dysfunction can result in diminished vestibulo-ocular reflex.[8] Approximately 30% of people with FRDA develop optic atrophy, though this is not often symptomatic.[5] Some individuals with FRDA also show evidence of auditory neuropathy and a disturbed auditory brainstem response reflecting auditory pathway abnormality.[17,18] This impairment is apparent in reduced auditory processing in the setting of normal sound detection and may become more evident as the disease progresses. Diabetes mellitus is more common in people with FRDA, affecting 10% of diagnosed individuals.[5,6]

FRDA is a heterogeneous condition and as such the extent of morbidity and mortality associated with the condition varies. Historically the average age of death from FRDA was 37 years;[3] however more recent studies have found the average age of death to be in the fourth and fifth decade[7] with heart failure and arrhythmias being the most common reported cause.[19,20]

MOLECULAR GENETICS

In 98% of cases, FRDA is due to homozygosity for an expansion of a GAA trinucleotide repeat in intron one of the *FXN* gene.[21] The other 2% are compound heterozygous for a GAA expansion and a unique mutation in *FXN*.[22] These unique mutations include missense, nonsense or splice-site point mutations, or deletion/insertion mutations which can result in a typical or atypical phenotype as evidence by altered presence or absence of reflexes and increased lower limb spasticity, slower disease progression, absence of dysarthria and mild cerebellar ataxia.[23-25] Homozygosity for a point mutation has been found in one family (Schmucker et al, 2011 unpublished data).[26] Unaffected chromosomes contain GAA repeats <30.[27] Individuals with FRDA have expansions ranging from 66 to >1000, most often within the range of 600-900.[6,27] Carriers of one expanded allele are clinically healthy.[21] The FRDA related expansion demonstrates instability when transmitted from parent to child.[21,28] Paternal transmission usually results in contractions; however maternal transmission can result in both expansion and contraction of the repeat size in offspring.[28,29]

FXN expression is highest in the heart and spinal cord, lower in the cerebellum, liver, pancreas and skeletal muscle and lowest in the cerebral cortex.[30] Whilst these areas are consistent with those most affected by the disease process, diminished frataxin levels are noted beyond those sites affected by FRDA.[31] It has been proposed that the areas specifically affected by FRDA are highly dependent on mitochondrial metabolism, and moreover, are areas where there is little or no cellular regeneration and thus exhibit pathology associated with FRDA.[31]

The protein encoded by *FXN* is frataxin. Frataxin was believed to only reside in the mitochondria; however it is now also seen to be present in the cytosol.[27] The GAA expansion results in partial silencing of *FXN* thus reducing the production of frataxin. The exact mechanism(s) of *FXN* silencing in FRDA is not fully understood, but recent studies indicate a role for epigenetic regulation. Epigenetics is a rapidly expanding field of research that examines mechanisms of phenotypic change that do not alter the DNA sequence itself, such as chromatin remodelling and DNA methylation. There is evidence that the GAA expansions in the disease-causing range induce heterochromatin formation which can impede elongation during transcription of *FXN*.[32,33] Histone deacetylase (HDAC) inhibitors are able to increase *FXN* expression in both cells from individuals with FRDA and an FRDA mouse model.[34,35] DNA methylation upstream of the GAA expansion also correlates with *FXN* expression in both affected and non-affected individuals, indicating that *FXN* may normally be regulated in this manner, but is altered in FRDA.[36]

Individuals homozygous for GAA expansions are able to synthesize between 4 and 29% of normal levels of structurally and functionally normal frataxin;[30] however as demonstrated in mouse studies, complete lack of frataxin production leads to embryonic lethality.[37] The exact role of frataxin is still not fully elucidated, but it is largely seen as a mitochondrial membrane protein involved in iron sulphur protein production and storage, iron chaperone activity and reducing the production of reactive oxygen species.[30,31,38] Frataxin is essential for overall cellular iron homeostasis.[31] Therefore impaired cellular iron homeostasis in FRDA results in mitochondrial dysfunction, defects in energy metabolism, oxidative stress response and associated oxidative damage.[6,38] Whilst the exact sequence of molecular and cellular events caused by the silencing of *FXN* is still uncertain, impaired intramitochondrial metabolism with raised free iron levels and a defective mitochondrial respiratory chain may result in increased free radical generation causing oxidative stress. This ultimately causes cell death as evident in neuronal die back.[39] Whilst FRDA is seen as a neurodegenerative disease implying normal development until the onset of symptoms,[40] studies of several different organisms have demonstrated the metabolic functions regulated by frataxin are fundamental to normal development[31] and as such FRDA could also be viewed partly as a neurodevelopmental disorder.

PATHOLOGY

The major sites of pathology in FRDA include the dorsal root ganglia (DRG) and posterior columns of the spinal cord, spinocerebellar tracts, corticospinal tracts, dentate nucleus of the cerebellum and the heart.[6,40,41] Post mortem studies reveal a reduction in size and discolouration of the dorsal columns.[40,42] Early in the course of the disease there is an overall reduction in the size of DRG due to atrophy, with active ganglion cell destruction in addition to loss of larger myelinated fibres in the dorsal roots.[42-44] Smaller ventral roots (VR) remain largely unaffected in FRDA, though Koeppen et al[43] report some thinning of efferent motor fibres, the clinical significance of which is still not apparent. The loss of DRG neurons has an ancillary effect resulting in the significant reduction of dorsal column fibres and afferent connections to the dorsal nuclei of Clarke as originally noted in postmortem studies by Friedreich.[43] The loss of the primary sensory neurons in the DRG and associated atrophy of the posterior spinal columns result in loss of vibration and proprioceptive sense; hence the use of the term "sensory ataxia" to describe the gait impairment associated with FRDA.[5]

Imaging studies in FRDA have provided evidence of mild cerebellar atrophy, but little cortical atrophy has been demonstrated.[6,45-48] De Michele et al[49] utilized single photon emission tomography (SPET) to compare rates of regional cerebral perfusion (rCP) with brain atrophy (as measured by MRI) in people with spinocerebellar degenenerative conditions (including six individuals with FRDA). All individuals with FRDA had cerebellar atrophy and a direct correlation between the degree of hypoperfusion and the severity of cerebellar atrophy was found.[49] Gilman et al[50] examined the local cerebral metabolic rate for glucose (LCMRG) in 22 individuals with FRDA and 23 control subjects and observed that the metabolic rate for glucose increases throughout the CNS (beyond that of control subjects) early in the disease process, perhaps as a consequence of initial compensatory mechanisms that subsequently prove ineffective. However as the disease progressed, the rate decreased in all areas except the basal ganglia. Gilman et al[50] suggested that decline in metabolic rate later in the disease process might be a function of a reduction in neurons and synaptic terminals, evidenced by atrophy of related regions. A further study by Junck and colleagues[41] found significantly greater atrophy in all regions in people with FRDA (including the cerebral hemispheres) when compared with controls. Moreover, generalized brain atrophy correlated with clinical severity, as did lower LCMRG in the thalamus, cerebellum and brain stem. Junck and colleagues[41] also attributed the decline in LCMRG to loss of neurons and synaptic terminals and decreased activity in the remaining synaptic terminals, suggesting elevation of LCMRG early in the disease process may result in increased glycolysis.

More recent studies utilizing transcranial magnetic stimulation,[51] magnetic resonance imaging (MRI) including tractography,[48,52,53] transcranial sonography[54] and whole brain voxel-based morphometry (VBM)[55] have elucidated the specific cortical and cerebellar regions affected in FRDA. Della Nave and colleagues[52] used VBM and identified gray matter (GM) loss in the dorsal medulla and rostral cerebellar vermis, inferomedial cerebellar hemispheres and white matter (WM) loss in the dentate region of the cerebellum. These results were confirmed in a subsequent study that used tract based spatial statistics (TBSS) to examine degeneration of the brainstem and cerebellar WM tracts.[48] Whilst WM degeneration in spinal cord, brainstem and cerebellum is well documented, there is a growing interest in the involvement of WM tracts in the brain of people with FRDA.[52] Pagani and colleagues[56] used anisotropy maps derived from diffusion-tensor MRI and a nonlinear registration, to map the extent of WM changes in the brain of individuals with FRDA. WM changes were noted in the brainstem, including the medial lemnisci and medial longitudinal fasciculi, bilateral superior cerebellar peduncles, cerebellar peridentate region and the optic chiasma. All areas of WM changes correlated with the clinical symptoms exhibited in FRDA.[56] Furthermore, there was an unexpected finding of atrophy in the left frontopontine tract and reticular formation. França and colleagues[55] also employed VBM analysis in conjunction with proton magnetic resonance spectroscopy to assess structural changes and assess neuronal changes in people with FRDA. Areas of atrophy were generally consistent with previous studies, although this study also indentified axonal damage in the deep cerebral WM of people with FRDA. França and colleagues[55] postulated that such cerebral changes may be due to changes in the developing brain rather than true degenerative atrophy.

Morphologically the cerebellar cortex is essentially unaffected and Purkinje cells are essentially preserved in FRDA.[44] In sharp contrast, the dentate nucleus suffers severe loss of large neurons with associated pathological modification of synaptic terminals termed "grumose degeneration".[40,57,58] Loss of large neurons contributes to atrophy of the dentate nucleus which is believed to be the major cause of the dysmetria, dysarthria and dysphagia that characterizes FRDA.[58] Dentate nucleus projections to motor, premotor, oculomotor and prefrontal (DLPFC) regions form feedforward loops that pass through the superior cerebellar peduncle (SCP) and the thalamus.[59] A recent structural imaging study of people with FRDA found the SCP is approximately 60% smaller in people with FRDA compared to matched controls.[53] Anatomical changes to this structure imply cerebellar connections to cortical structures such as the DLPFC are likely impaired in individuals with FRDA and as such functions dependent on an intact feedforward dentate-cortical loop may be compromised in affected individuals.

GENOTYPE-PHENOTYPE CORRELATION

Given that smaller GAA repeat expansions allow higher expression of *FXN* and thus increase the amount of available frataxin,[60] it is not surprising that the severity of the phenotype is correlated with repeat size.[6] An inverse correlation has been found between the number of GAA repeats on the smaller allele (GAA1) and age at onset;[1,14,61] more rapid disease progression;[4,62] time to wheelchair use;[14] cardiomyopathy and other clinical features including diabetes, optic atrophy and hearing loss.[63] Delatycki and colleagues[4] also found a significant correlation between GAA1 and presence of scoliosis, wheelchair dependence, impaired vibration sense and presence of foot deformity. Koeppen and colleagues[44] found that the longer allele (GAA2), particularly in the case of late onset FRDA correlates with disease duration and phenotype. Nevertheless there is still considerable variability in the FRDA phenotype that is not explained by GAA repeat length. For example the size of the GAA1 allele only accounts for approximately 47% of age of onset.[6,31,64] Hence it is not possible to accurately predict disease severity or rate of progression based on repeat size only.[31] Further studies have examined the predictive power of variables other than GAA repeat size on disease progression and conclude that age at disease onset is more predictive of progression to wheelchair use, inability to complete activities of daily living[62,65] and cardiomyopathy.[66] The molecular mechanism that underlies such clinical variability is still unclear, but it has been proposed that mitotic instability resulting in somatic mosaicism of the expansion size may be a contributing factor.[67]

A series of recent studies have demonstrated that epigenetic factors also contribute to clinical outcome in FRDA. Treatment with HDAC inhibitors can alleviate disease phenotype in an FRDA mouse model.[68] A characteristic DNA methylation pattern has also been identified flanking the site of expansion in FRDA primary human samples compared to controls—increased methylation upstream and decreased methylation downstream.[32] Significant correlations between the size of the expansion, the degree of methylation and age of onset have also been observed.[69]

VARIANTS

In about 25% of presentations individuals with FRDA fall outside the diagnostic criteria as described by Harding in 1981.[25,70,71] Late Onset Friedreich Ataxia (LOFA) is defined as onset between 25 and 29 years and Very Late Onset Friedreich Ataxia (VLOFA) is defined as onset 30 years and beyond. Such individuals have smaller GAA1 alleles than those with typical (<25 years) onset.[72,73] A recent study by Koeppen and colleagues[44] indicated individuals with LOFA have significantly shorter repeat expansions on both alleles when compared to those with usual onset. Individuals with LOFA or VLOFA have slower disease progression and fewer associated symptoms such as pes cavus and scoliosis.[63,72,74]

A further clinical variant is Friedreich Ataxia with Retained Reflexes (FARR) associated with retained deep tendon reflexes, slower and milder disease progression.[75] Finally, Acadian ataxia, seen in families from Louisiana and Eastern Canada, also presents as a slower, milder form of the disease with a decreased incidence of hypertrophic cardiomyopathy and abnormal glucose metabolism.[76,77]

COGNITIVE AND BEHAVIOURAL IMPLICATIONS OF FRDA

Until recent years documentation of cognitive function in FRDA was scarce, possibly because FRDA was widely held to predominantly affect the spinal cord, peripheral sensory nerves and cerebellum[78] and therefore not affect cognition. Traditionally, the cerebellum has been thought to coordinate voluntary movement and control of motor tone, posture and gait.[79] The last few decades has seen a significant shift from a corticocentric view of cognitive function to a view that embraces the significant contribution of the cerebellum in cognitive, language and affective regulation.[80-87] This notion is underpinned by neuroanatomical[88] and neuroimaging studies (for a review see ref. 89) that demonstrate cerebellar connectivity with cerebral areas involved in nonmotor functions.[86] In particular, Strick et al[86] noted 75% of cerebellar projections to the motor cortex originate from the dentate nucleus; however in terms of projections from the dentate nucleus only 30% target the motor cortex, suggesting that a significant proportion of dentate nucleus projections are to regions supporting nonmotor functions.[86] In tandem with this shift in awareness of the role of the cerebellum in cognitive function has come an associated interest in examining the effect of FRDA on cognitive development and function. This has been important for a number of reasons not the least of which is the importance of characterizing the cognitive behavioural profile associated with FRDA in order to understand the functional consequences of such a profile and develop interventions that ensure individuals are able to maximize their potential in terms of cognitive function. Furthermore, understanding the effect of FRDA on the developing brain is clinically important in the context of the effect of age of disease onset on subsequent cognitive and motor capacities. Finally, examination of FRDA has contributed to the increasing body of knowledge regarding the nonmotor function of the cerebellum.

The focus of early studies examining neurobehaviour in FRDA was based on psychiatric impairment,[90,91] speed of information processing[78,92-95] and general cognitive function.[96,97] Slowed information processing was a consistent finding of these studies, whereas changes to other cognitive functions were not consistently reported. Reviews of the early empirical literature highlighted the methodological limitations of these studies including poor matching with control participants, the motor confound associated with

FRDA, small numbers and importantly the lack of consistency in the neuropsychological assessments used.[98,99] Furthermore there was little examination of the correlation between possible important determinants of disease severity, such as the size of the GAA expansion, age of onset and the potential effect of FRDA on the developing brain. As previously mentioned, the neuropathology in FRDA is associated with a deficiency in frataxin.[30,100,40] Consistent with the distribution of pathology within the central nervous system (CNS), frataxin expression is highest in the spinal cord, lower in the cerebellum and lowest in the cerebral cortex, a somewhat surprising finding in the context of apparent cognitive impairment associated with FRDA.[64] Neurons with high energy demands may specifically display dysfunction when frataxin is reduced and accordingly may exhibit such dysfunction when placed under higher metabolic loads.[101] Moreover, reduced frataxin may have deleterious effects on the development of cerebro-ponto-cerebello-thalamo-cerebral pathways necessary for cognitive function. If the development of neural circuitry is particularly prone to the effects of reduced frataxin then parameters such as the age of disease onset would correlate with neurobehavioural performance. Given cognitive deficit is a potential source of morbidity in the young individual with FRDA this is of significant developmental importance.[98] It is critical that studies addressing the cognitive and behaviour profile of FRDA also correlate cognitive and behavioural outcomes with clinical indicators of disease severity. Earlier studies acknowledged the difficulty of dissociating a deficit in motor function from a cognitive deficit,[102] however were in general agreement that changes to cognitive function in people with FRDA were secondary to cerebellar pathology.

Recent years have seen an increment in studies aimed at understanding the neurobehavioural profile associated with FRDA, often utilizing traditional neuropsychological assessments incorporating tests of intelligence.[101,103,104] Mantovan et al[101] identified individuals with FRDA who were, compared to control participants, impaired in tasks related to verbal fluency, visuoconstructive and visuoperceptual capacity, motor and mental reaction times. Whilst these results did not represent a major impairment in intelligence, it did appear the intelligence of people with FRDA was characterized by concrete thinking resulting in reduced concept formation and visuospatial reasoning.[101] De Nóbrega et al[103] also examined performance in individuals with FRDA and matched controls on phonemic, semantic and action verbal fluency tasks and in so doing also attempted to address the difficult issue of dissecting cognitive performance from the motor impairment associated with FRDA. Individuals with FRDA and control participants performed at a similar level on the semantic fluency task; however individuals with FRDA performed significantly poorly on the phonemic and action fluency tests when compared to control participants. Lack of correlation between the results of the verbal fluency tasks, dysarthria tests and reaction time task indicated the source of the group difference lay not with the motor speech impairment associated with FRDA, but rather with possible impairment in the neural circuitry associated with greater executive control. It should be noted that the use of standardized assessments of intelligence in people with FRDA, whilst providing some information regarding intelligence per se, provides little understanding of the capacity of people with FRDA to manage a functional situation demanding flexible and comprehensive cognitive abilities. People with FRDA often become quite physically impaired at a critical time of their neural development and education. It has also been recently identified that people with FRDA have a concurrent hearing impairment that if undetected in the school environment may have an impact on the capacity to learn.[17] Poor performance on specific tests of intelligence may well reflect compromised educational opportunities rather than direct impairment.[104,105] Alternatively, as pointed out by Mantovan and colleagues[101] the neuropsychological and affective changes in people with FRDA may reflect primary cerebellar pathology or alternatively the consequences of frataxin deficiency disrupting function in other parts of the central nervous system. Functional changes to cortical activation may be the result of disruption in cerebello-cortical loops or may be indicative of primary cortical activation disturbance.[101]

The area of personality changes as a consequence of FRDA is little understood. The most significant albeit small study has been that by Mantovan and colleagues,[101] who administered the Minnesota Multiphasic Personality Inventory (MMPI) to eight participants with FRDA. Of these eight participants, half (n = 4) returned personality profiles that were outside the normal range. In particular, these profiles included increased irritability, poor impulse control and blunting of affect. None of the control participants demonstrated similar profiles. Curiously such impairments may also be indicative of inhibitory dysfunction. Despite comprising a small cohort, the results of this investigation provide a fascinating insight into the little investigated area of personality changes related to FRDA. Mantovan and colleagues[101] postulated these changes may be the consequence of the onset of symptomatology of FRDA in adolescence, a critical time for personality formation or perhaps the daily living and participatory restrictions imposed by the condition that provide a modifying influence on otherwise adaptive behaviour. Either proposal seems plausible and provides an interesting basis for future examination.[101]

FRDA, whilst being largely heterogeneous, does inherently present as a disorder of movement. The most effective avenue to explore cognitive function is usually through the use of paradigms that assess reaction time (RT). However, attempting to characterize cognitive function in FRDA utilizing RT is likely to be impacted by the inherent motor impairment and careful consideration of experimental characteristics is warranted. Corben and colleagues[106-109] employed a number of studies to examine psychomotor function in people with FRDA whilst controlling for issues related to RT. These studies examined the capacity to respond to unexpected change in movement, to utilize advance information to plan movement, to generate a motoric response to incongruent stimuli and to accommodate changes in moving to targets differing in size and distance apart. Importantly each study correlated movement outcomes with clinical parameters. Individuals with FRDA had difficulty accommodating unexpected movement, were disadvantaged by conditions requiring initiation of movement without a direct visual cue and were differentially affected in reaction time to incongruent, compared with congruent stimuli. Moreover, examination of the kinematic profile indicated people with FRDA were unable to modulate the time spent planning and preparing movement according to task parameters. In the majority of studies there was a significant negative correlation between age at disease onset and the movement outcomes, suggesting the impact of FRDA on the developmental unfolding of motor cognition. These studies used a range of differing analytic methodologies that provided a consistent pattern of results ensuring confidence that the confounding issue of using RT to examine psychomotor function in FRDA was accounted for.[110]

A number of other studies have characterized impairment of cognitive function in FRDA. Fielding and colleagues[111] examined antisaccade and memory-guided saccade characteristics in 13 individuals with FRDA and matched control participants. Individuals with FRDA demonstrated disproportionally greater and significantly different antisaccade and memory-guided saccades compared to control participants. Importantly the difference lay in the cognitive control of ocular movements, rather than being a generalized slowing of eye movement. Hocking and colleagues[112] extended the study of Fielding and colleagues[111] by employing a gap overlap task to examine attentional engagement and disengagement of eye movements in FRDA. People with FRDA were able to move attention promptly in the absence of a distracter; however as evidenced by prolonged saccadic latencies they had difficulty disengaging attention from fixation to move to a target in the presence of a distracter. As with the previous study,[111] scores on a measure of clinical severity, the Friedreich Ataxia Rating Scale (FARS) significantly correlated with saccadic latencies.[112] The capacity to successfully complete attentional, antisaccade and memory-guided saccades requires inhibition of a prepotent response and subsequent selection of a suitable response. Cortical areas responsible for conflict resolution and response selection include the posterior and superior parietal areas, DLPFC and the dorsal ACC,[113] hence successful completion of these ocular motor tasks requires cerebellar access to cortical areas. Saccadic abnormalities in FRDA may reflect a disruption to the cerebro-ponto-cerebello-thalamo-cerebral circuitry accessing these areas. Also, consistent with the series of studies by Corben and colleagues,[106-109] significant correlations were found between a number of measures of ocular motor control and clinical parameters including scores on the FARS and disease duration. These studies support the premise that disruption to the cerebro-ponto-cerebello-thalamo-cerebral circuitry results in disturbed cognitive function, in this case the cognitive control of eye movements.

Klopper and colleagues[114] examined sustained volitional attention and working memory in a cohort of 16 individuals with FRDA and matched control participants utilizing subtests from the Test of Everyday Attention (TEA). Individuals with FRDA demonstrated significant impairment on four out of five subtests (a ceiling effect consistent with that of the controls tested was noted in the subtest that failed to reach significance). These results, the first to document attentional and working memory capacity in people with FRDA, provided further evidence for a disruption to the network of brain regions that modulate and underlie working memory and attentional capacity. Again, significant correlation with some clinical parameters in particular, the repeat length of GAA1, implicated the effect of reduced frataxin on developing brain function.

Results from these empirical studies confirm the presence of cognitive impairment in people with FRDA. One or more of three explanations could account for such impairment in people with FRDA: (1) cortical pathology, (2) dysfunction in the cerebellum interrupting cerebro-ponto-cerebellar-thalamo-cerebral loops, or (3) a combination of the two. Consistent with related studies into the role of the cerebellum in cognition, such dysfunction in people with FRDA may be indicative of disruption to the corticocerebellar neural network reflecting a failure to access prefrontal, frontal and parietal regions necessary for effective cognitive control.[84,85,99,115]

What is the unique role of the cerebellum in modulating cognitive functions? In motor function, through inhibitory mechanisms, the role of the cerebellum is thought to determine rate, rhythm and force of motor behaviour.[116] Conceivably the role of the cerebellum is similar for cognitive function. The inhibitory function of the cerebellum can be viewed as a modulator or regulator of behaviour through its interactive feedforward and inverse loops. As proposed by Ito,[117] the role of the cerebellum may be to predict the outcome of a cognitive activity as it does with a motor task.[87,118] In this way the cerebellum takes the available information, anticipates and adapts it to the requirement of the current task and feeds forward the blueprint for the task to be achieved.[116] Such a predictive role can be expressed as accurately perceiving and producing movement in an appropriate temporal order.[86] Interference in this capacity may result in the loss of the selection and sequencing of motor and cognitive commands as apparent by reduced motor control or problems with task-shifting and other executive functions.[84] Leggio and colleagues[119] also highlight the role of the cerebellum in sequencing, an integral component of predicting optimum states in movement or cognitive control via a feedforward mechanism. All of the above roles in terms of predicting optimum sensory and motor states, timing and sequencing, combine to underscore the significant role of the cerebellum in internal models of control. Hence, anticipating the outcome of the task and feeding forward the appropriate adjustments required to achieve the outcome, is a fundamental component of cerebellar control in nonmotor activity.[116] As in motor tasks, error detection and modulation enables constant updating of cognitive information. Interference in this capacity may result in the loss of the selection and sequencing of appropriate cognitive commands as apparent by difficulty with task-shifting and other executive functions.[86] The cerebellum completes this essential role through neural networks involving feedforward and feedback links via cerebello-thalamic-cortical projections. As previously stated, people with FRDA sustain significant neuropathological changes to the dentate nucleus, the source and target of extensive projections to motor, premotor, oculomotor and prefrontal regions.[59] It is inevitable that in people with FRDA, such changes will interfere with feedforward loops fundamental to many nonmotor tasks.[105,99] Inexorable subsequent impairment is the logical outcome of such neuropathology.

CONCLUSION

To summarize, there is now established evidence of psychomotor dysfunction in individuals with FRDA. If the source of this dysfunction was purely related to cerebellar dysfunction then a more uniform pattern of dysfunction would be observed across all cerebellar ataxias.[105] Since this is not the case, the source of this dysfunction in FRDA is likely disruption to cerebro-ponto-cerebello-thalamo-cerebral circuitry interfering with access to more anterior structures essential to the feedforward and inverse model of cognitive and motor control.[99,105,120] The neural dysfunction in FRDA results from deficiency of frataxin. Neuronal death is a consequence of reduced frataxin and precedes the first symptoms; hence pathological changes have already occurred prior to diagnosis.[31] Since the identification of the *FXN* gene in 1996 much has been learnt regarding the mechanism of the gene dysfunction and the role of frataxin in the neuropathology associated with FRDA. Reported correlations between age of disease onset and FARS with neurobehavioural measures support the hypothesis that pathological changes to essential neuroanatomical structures may underscore the severity of cognitive dysfunction associated with FRDA. However, there is still a considerable way to go to fully understand the effect of the neurobehavioural phenotype and the underlying neuropathology associated with FRDA. This is critical if we are to effectively evaluate the impact of new treatments in forestalling the condition. The adoption of more direct multimodal neuroimaging techniques will provide a more comprehensive understanding of the neurobehavioural profile and neuropathology associated with FRDA. This mode of investigation has proved extremely informative in other neurological diseases such as Huntington disease[121] and may be expanded to include use of other technologies such as transcranial-magnetic stimulation (TMS) to further elucidate areas of cortical activation in individuals with FRDA.[122,123]

Of particular interest for future investigation is possible compensatory reorganisation of the cerebellar-cortical network in individuals with FRDA. Multi-modal neuroimaging assessment in people with FRDA will provide the essential next-steps in identifying sensitive biomarkers of both neuropathology and progression for consideration in future therapeutic trials.

REFERENCES

1. Delatycki MB, Williamson R, Forrest SM. Friedreich ataxia: an overview. J Med Gen 2000; 37(1):1-8.
2. Friedreich N. Uber degenerative Atrophie der Spinalen Hinterstrange. Virchows Arch Path Anat 1863; 26:433-459.
3. Harding AE. Friedreich's ataxia: a clinical and genetic study of 90 families with an analysis of early diagnostic criteria and intrafamilial clustering of clinical features. Brain 1981; 104(3):589-620.
4. Delatycki MB, Paris DB, Gardner RJ et al. Clinical and genetic study of Friedreich ataxia in an Australian population. Am J Med Gen 1999; 87(2):168-174.
5. Pandolfo M. Friedreich Ataxia: the clinical picture. J Neurol 2009; 256:3-8.
6. Pandolfo M. Friedreich ataxia. Arch Neurol 2008; 65(10):1296-1303.
7. Shultz JB, Boesch S, Bürk K et al. Diagnosis and treatment of Friedreich ataxia: a European perspective. Nat Rev Neurol 2009; 5:222-234.
8. Fahey MC, Cremer PD, Swee TA et al. Vestibular, saccadic and fixation abnormalities in genetically confirmed Friedreich ataxia. Brain 2008; 131:1035-1045.
9. Nardulli R, Monitillo V, Losavio E et al. Urodynamic evaluation of 12 ataxic subjects: neurophysiopathologic considerations. Func Neurol 1992; 7(3):223-235.
10. Campanella G, Filla A, DeFalco F et al. Friedreich's ataxia in the south of Italy: a clinical and biochemical survey of 23 patients. Can J Neurol Sci 1980; 7(4):351-357.
11. Shapiro F, Specht L. The diagnosis and orthopaedic treatment of childhood spinal muscular atrophy, peripheral neuropathy, Friedreich ataxia and arthrogryposis. J Bone Joint Surg—Series A 1993; 75(11):1699-1714.
12. Labelle H, Tohme S, Duhaime M et al. Natural history of scoliosis in Friedreich's Ataxia. J Bone Joint Surg 1986; 68:564-572.
13. Folker J, Murdoch B, Cahill L et al. Dysarthria in Friedreich's Ataxia: a perceptual analysis. Folia Phon Log 2010; 62:97-103.
14. Ribaï P, Pousset F, Tanguy M et al. Neurological, cardiological and oculomotor progression in 104 patients with Friedreich ataxia during long-term follow-up. Arch Neurol 2007; 64:558-564.
15. Dutka DP, Donnelly JE, Palka P et al. Echocardiographic characterization of cardiomyopathy in Friedreich's ataxia with tissue Doppler echocardiographically derived myocardial velocity gradients. Circulation 2000; 102(11):1276-1282.
16. Mottram PM, Delatycki MB, Donelan L et al. Early changes in left ventricular long axis function in Friedreich ataxia—relation with the FXN gene mutation and cardiac structural change. J Am Soc Echo 2011; 24(7):782-789.
17. Rance G, Fava R, Baldock H et al. Speech perception ability in individuals with Friedreich ataxia. Brain 2008; 131:2002-2012.
18. Rance G, Corben L, Barker E et al. Auditory perception in individuals with Friedreich's ataxia. Audiol Neurotol 2010; 15:229-240.
19. Meyer C, Schmid G, Görlitz S et al. Cardiomyopathy in Friedreich ataxia: Assessment by cardiac MRI. Mov Dis 2007; 22(11):1615-1622.
20. Tsou AY, Paulsen EK, Lagedrost SJ et al. Mortality in Friedreich ataxia. J Neurol Sci 2011; 307:46-49.
21. Campuzano V, Montermini L, Molto MD et al. Friedreich's ataxia: autosomal recessive disease caused by an intronic GAA triplet repeat expansion. Science 1996; 271(5254):1423-1427.
22. Voncken M, Ioannou P, Delatycki MB. Friedreich ataxia-update on pathogenesis and possible therapies. Neurogenetics 2004; 5(1):1-8.
23. Evans-Galea MV, Corben LA, Hasell J et al. A novel deletion-insertion mutation identified in exon 3 of FXN in two siblings with a severe Friedreich ataxia phenotype. Neurogenetics 2011.
24. Forrest SM, Knight M, Delatycki MB et al. The correlation of clinical phenotype in Friedreich ataxia with the site of point mutations in the FRDA gene. Neurogenetics 1998; 1(4):253-257.
25. Cossée M, Dürr A, Schmitt M et al. Friedreich's ataxia: point mutations and clinical presentation of compound heterozygotes. Ann Neurol 1999; 45(2):200-206.

26. Schmucker S, Reutenauer L, Devos et al. Identification of an atypical Friedreich ataxia patient with no GAA expansion but with a homozygous point mutation in the mitochondrial targeting sequence of frataxin. Presented at the Friedreich ataxia scientific meeting; Strasbourg, France. 2011.
27. Boehm T, Scheiber-Mojdehkar B, Kluge B et al. Variations of frataxin protein levels in normal individuals. Neurol Sci 2010:327-330.
28. Delatycki MB, Paris D, Gardner RJ et al. Sperm DNA analysis in a Friedreich ataxia premutation carrier suggests both meiotic and mitotic expansion in the FRDA gene. J Med Gen 1998; 35(9):713-716.
29. Pianese L, Cavalcanti F, De Michele G et al. The effect of parental gender on the GAA dynamic mutation in the FRDA gene. Am J Hum Gen 1997; 60(2):460-463.
30. Campuzano V, Montermini L, Lutz Y et al. Frataxin is reduced in Friedreich ataxia patients and is associated with mitochondrial membranes. Hum Mol Gen 1997; 6(11):1771-1780.
31. Santos R, Lefevre S, Sliwa S et al. Friedreich Ataxia: Molecular mechanisms, redox considerations and therapeutic opportunities. Ant Red Sig 2010; 13(5):651-690.
32. Al-Mahdawi S, Pinto RM, Ismail O et al. The Friedreich ataxia GAA repeat expansion mutation induces comparable epigenetic changes in human and transgenic mouse brain and heart tissues. Hum Mol Gen 2008; 17(5):735-746.
33. Punga T, Bühler M. Long intronic GAA repeats causing Friedreich ataxia impede transcription elongation. EMBO Mol Med 2010; 2:120-129.
34. Herman D, Jenssen K, Burnett R et al. Histone deacetylase inhibitors reverse gene silencing in Friedreich's ataxia. Nat Chem Biol 2007; 2(10):551-558.
35. Rai M, Soragni E, Jenssen K et al. HDAC inhibitors correct frataxin deficiency in a Friedreich ataxia mouse model. PloS One 2008; 3:e1958.
36. Evans-Galea MV, Carrodus N, Rowley SN et al. FXN methylation predicts expression and clinical outcome in Friedreich ataxia. Ann Neurol 2011 In press.
37. Cossee M, Puccio H, Gansmuller A et al. Inactivation of the Friedreich ataxia mouse gene leads to early embryonic lethality without iron accumulation. Hum Mol Gen 2000; 9(8):1219-1226.
38. Pandolfo M, Pastore A. The pathogenesis of Friedreich ataxia and the structure and function of frataxin. J Neurol 2009; 256(Suppl 1):9-17.
39. Calabrese V, Lodi R, Tonon C et al. Oxidative stress, mitochondrial dysfunction and cellular stress response in Friedreich's Ataxia. J Neurol Sci 2005; 233(1-2):145-162.
40. Koeppen AH. Friedreich's ataxia: pathology, pathogenesis and molecular genetics. J Neurol Sci 2011; 303(1-2):1-12.
41. Junck L, Gilman S, Gebarski SS et al. Structural and functional brain imaging in Friedreich's ataxia. Arch Neurol 1994; 51(4):349-355.
42. Koeppen AH. Neuropathology of the inherited ataxias. In: Manto U-M, Pandolfo M, eds. The Cerebellum and its Disorders. Cambridge: Cambridge University Press, 2002: 387-409.
43. Koeppen AH, Morral JA, Davis AN et al. The dorsal root ganglion in Friedreich's ataxia. Acta Neuropath 2009; 118(6):763-776.
44. Koeppen AH, Morral JA, McComb RD et al. The neuropathology of late-onset Friedreich's ataxia. Cerebellum 2011; 10(1):96-103.
45. Botez MI, Leveille J, Lambert R et al. Single photon emission computed tomography (SPECT) in cerebellar disease: cerebello-cerebral diaschisis. Eur Neurol 1991; 31(6):405-412.
46. Ormerod IE, Harding AE, Miller DH et al. Magnetic resonance imaging in degenerative ataxic disorders. J Neurol Neurosurg Psych 1994; 57(1):51-57.
47. Klockgether T, Zuhlke C, Schulz JB et al. Friedreich's ataxia with retained tendon reflexes: molecular genetics, clinical neurophysiology and magnetic resonance imaging. Neurology 1996; 46(1):118-121.
48. Della Nave R, Ginestroni A, Giannelli M et al. Brain structural damage in Friedreich's ataxia. J Neurol, Neurosurg, Psych 2008; 79:82-85.
49. De Michele G, Mainenti PP, Soricelli A et al. Cerebral blood flow in spinocerebellar degenerations: a single photon emission tomography study in 28 patients. J Neurol 1998; 245(9):603-608.
50. Gilman S, Junck L, Markel DS et al. Cerebral glucose hypermetabolism in Friedreich's ataxia detected with positron emission tomography. Ann Neurol 1990; 28(6):750-757.
51. Brighina F, Scalia S, Gennuso M et al. Hypo-excitability of cortical areas in patients affected by Friedreich ataxia: A TMS study. J Neurol Sci 2005; 235:19-22.
52. Della Nave R, Ginestroni A, Tessa C et al. Brain white matter tracts degeneration in Friedreich ataxia. An in vivo MRI study using tract-based spatial statistics and voxel-based morphometry NeuroImage 2008; 40(1):19-35.
53. Akhlaghi H, Corben LA, Georgiou-Karistianis N et al. Superior cerebellar peduncle atrophy in Friedreich's Ataxia correlates with disease symptoms. Cerebellum 2011; 10:81-87.
54. Synofzik M, Godau J, Lindig T et al. Transcranial sonography reveals cerebellar, nigral and forebrain abnormalities in Friedreich's ataxia. Neurodeg Dis 2011; 8(6):470-475.

55. França AE, D'Abreu A, Yasuda CL et al. A combined voxel-based morphometry and 1H-MRS study in patients with Friedreich's ataxia. J Neurol 2009; 256:1114-1120.
56. Pagani E, Ginestroni A, Della Nave R et al. Assessment of brain white matter fiber bundle atrophy in patients with Friedreich ataxia. Radiology 2010; 255(3):882-889.
57. Koeppen AH, Michael SC, Knutson MD et al. The dentate nucleus in Friedreich's ataxia: the role of nonresponsive proteins. Acta Neuropath 2007; 114:163-173.
58. Koeppen AH, Davis AN, Morral JA. The cerebellar component of Friedreich's ataxia. Acta Neuropath 2011; 122(3):323-330.
59. Middleton FA, Strick PL. Basal ganglia and cerebellar loops: motor and cognitive circuits. Brain Res Rev 2000; 31:236-250.
60. Bidichandani SI, Ashizawa T, Patel PI. The GAA triplet-repeat expansion in Friedreich ataxia interferes with transcription and may be associated with an unusual DNA structure. Am J Hum Gen 1998; 62(1):111-121.
61. De Michele G, Filla A, Criscuolo C et al. Determinants of onset age in Friedreich's ataxia. J Neurol 1998; 245(3):166-168.
62. Mateo I, Llorca J, Volpini V et al. Expanded GAA repeats and clinical variation in Friedreich's ataxia. Acta Neurol Scand 2004; 109(1):75-78.
63. Montermini L, Richter A, Morgan K et al. Phenotypic variability in Friedreich ataxia: role of the associated GAA triplet repeat expansion. Ann Neurol 1997; 41(5):675-682.
64. Montermini L, Andermann E, Labuda M et al. The Friedreich ataxia GAA triplet repeat: premutation and normal alleles. Hum Mol Gen 1997; 6(8):1261-1266.
65. La Pean A, Jeffries N, Grow C et al. Predictors of progression in patients with Friedreich ataxia. Mov Dis 2008; 23(14):2026-2032.
66. Maione S, Giunta A, Filla A et al. May age onset be relevant in the occurrence of left ventricular hypertrophy in Friedreich's ataxia? Clin Cardiol 1997; 20(2):141-145.
67. Montermini L, Kish SJ, Jiralerspong S et al. Somatic mosaicism for Friedreich's ataxia GAA triplet repeat expansions in the central nervous system. Neurology 1997; 49(2):606-610.
68. Sandi C, Pinto RM, Al-Mahdawi S et al. Prolonged treatment with pimelic o-aminobenzamide HDAC inhibitors ameliorates the disease phenotype of a Friedreich ataxia mouse model. Neurobiol Dis 2011; 42:496-505.
69. Castaldo I, Pinelli M, Monticelli A et al. DNA methylation in intron 1 of the frataxin gene is related to GAA repeat length and age of onset in Friedreich ataxia patients. J Med Gen 2008; 45(12):808-812.
70. Dürr A, Cossee M, Agid Y et al. Clinical and genetic abnormalities in patients with Friedreich's ataxia. New Eng J Med 1996; 335(16):1169-1175.
71. Filla A, De Michele G, Coppola G et al. Accuracy of clinical diagnostic criteria for Friedreich's ataxia. Mov Dis 2000; 15(6):1255-1258.
72. Bidichandani SI, Garcia CA, Patel PI et al. Very late-onset Friedreich ataxia despite large GAA triplet repeat expansions. Arch Neurol 2000; 57(2):246-251.
73. Bhidayasiri SP, P Stefan, D Geschwind. Late onset Friedreich Ataxia. Phenotypic analysis, magnetic resonance imaging findings and review of the literature. Arch Neurol 2005; 62:1865-1869.
74. De Michele G, Filla A, Cavalcanti F et al. Late onset Friedreich's disease: clinical features and mapping of mutation to the FRDA locus. J Neurol Neurosurg Psych 1994; 57(8):977-979.
75. Palau F, De Michele G, Vilchez JJ et al. Early-onset ataxia with cardiomyopathy and retained tendon reflexes maps to the Friedreich's ataxia locus on chromosome 9q. Ann Neurol 1995; 37(3):359-362.
76. Barbeau A, Roy M, Sadibelouiz M et al. Recessive ataxia in Acadians and "Cajuns". Can J Neurol Sci 1984; 11(4 Suppl):526-533.
77. Richter A, Poirier J, Mercier J et al. Friedreich ataxia in Acadian families from eastern Canada: clinical diversity with conserved haplotypes. Am J Med Gen 1996; 64(4):594-601.
78. Wollmann T, Barroso J, Monton F et al. Neuropsychological test performance of patients with Friedreich's ataxia. J Clin Exp Neuropsych 2002; 24(5):677-686.
79. Holmes G. The cerebellum of man (The Hughlings Jackson memorial lecture). Brain 1939; 62:1-30.
80. Schmahmann JD. An emerging concept. The cerebellar contribution to higher function. Arch Neurol 1991; 48(11):1178-1187.
81. Grafman J, Litvan I, Massaquoi S et al. Cognitive planning deficit in patients with cerebellar atrophy. Neurology 1992; 42(8):1493-1496.
82. Schmahmann JD, Sherman JC. The cerebellar cognitive affective syndrome. Brain 1998; 121(Pt 4):561-579.
83. Botez-Marquard T, Bard C, Leveille J et al. A severe frontal-parietal lobe syndrome following cerebellar damage. Eur J Neurol 2001; 8(4):347-353.
84. Thach WT. On the mechanism of cerebellar contributions to cognition. Cerebellum 2007; 6:163-167.
85. Baillieux H, De Smet HJ, Paquier PF et al. Cerebellar neurocognition: Insights into the bottom of the brain. Clin Neurol Neurosurg 2008; 110:763-773.
86. Strick PL, Dum RP, Fiez JA. Cerebellum and nonmotor function. Ann Rev Neurosci 2009; 32:413-434.

87. Timmann D, Drepper J, Frings M et al. The human cerebellum contributes to motor, emotional and cognitive associative learning. A review. Cortex 2010; 46(7):845-857.
88. Dum RP, Strick PL. An unfolded map of the cerebellar dentate nucleus and its projections to the cerebral cortex. J Neurophysiol 2003; 89(1):634-639.
89. Stoodley CJ, Schmahmann JD. Evidence for topographic organization in the cerebellum of motor control versus cognitive and affective processing. Cortex 2010; 46(7):831-844.
90. Flood MK, Perlman SL. The mental status of patients with Friedreich's ataxia. J Neurosci Nurs 1987; 19(5):251-255.
91. Giordani B, Boivan M, Berent S et al. Cognitive and emotional function in Friedreich's Ataxia. J Clin Expl Neuropsychol 1989; 11:53-54.
92. Botez-Marquard T, Botez MI. Cognitive behavior in heredodegenerative ataxias. Eur Neurol 1993; 33(5):351-357.
93. Botez-Marquard T, Botez MI. Olivopontocerebellar atrophy and Friedreich's ataxia: neuropsychological consequences of bilateral versus unilateral cerebellar lesions. Int Rev Neurobiol 1997; 41:387-410.
94. Hart RP, Henry GK, Kwentus JA et al. Information processing speed of children with Friedreich's ataxia. Dev Med Child Neurol 1986; 28(3):310-313.
95. Hart RP, Kwentus JA, Leshner RT et al. Information processing speed in Friedreich's ataxia. Ann Neurol 1985; 17(6):612-614.
96. White M, Lalonde R, Botez-Marquard T. Neuropsychologic and neuropsychiatric characteristics of patients with Friedreich's ataxia. Acta Neurol Scand 2000; 102(4):222-226.
97. Wollmann T, Nieto-Barco A, Monton-Alvarez F et al. Ataxia de Friedreich: analisis de parametros de resonancia magnetica y correlatos con el enlentecimiento cognitivo y motor. Rev Neurol 2004; 38(3):217-222.
98. Corben LA, Georgiou-Karistianis N, Fahey MC et al. Towards an understanding of cognitive function in Friedreich Ataxia. Brain Res Bull 2006; 70:197-202.
99. Manto M, Lorivel T. Cognitive repercussions of hereditary cerebellar disorders. Cortex 2011; 47(1):81-100.
100. Lynch DR, Farmer JM, Balcer LJ et al. Friedreich ataxia: effects of genetic understanding on clinical evaluation and therapy. Arch Neurol 2002; 59(5):743-747.
101. Mantovan MC, Martinuzzi A, Squarzanti F et al. Exploring mental status in Friedreich's ataxia: a combined neuropsychological, behavioural and neuroimaging study. Eur J Neurol 2006; 13:827-835.
102. Lalonde R, Botez T, Botez MI. Methodologic considerations in neuropsychologic testing of ataxic patients. Arch Neurol 1992; 49(3):218-219.
103. de Nóbrega E, Nieto A, Barrosso J et al. Differential impairment in semantic, phonemic and action fluency performance in Friedreich's ataxia: Possible evidence of prefrontal dysfunction. J Int Neuropsych Soc 2007; 13:944-952.
104. Ciancarelli I, Cofini V, Carolei A. Evaluation of neuropsychological functions in patients with Friedreich ataxia before and after cognitive therapy. Func Neurol 2010; 25(2):81-85.
105. Bürk K. Cognition in hereditary ataxia. Cerebellum 2007; 6:280-286.
106. Corben LA, Delatycki MB, Bradshaw JL et al. Impairment in motor reprogramming in Friedreich ataxia reflecting possible cerebellar dysfunction. J Neurol 2010; 257(5):782-791.
107. Corben LA, Akhlaghi H, Georgiou-Karistianis N et al. Impaired inhibition of prepotent motor tendencies in Friedreich ataxia demonstrated by the Simon interference task. Brain Cog 2011; 76 (1):140-145.
108. Corben LA, Georgiou-Karistianis N, Bradshaw JL et al. The Fitts task reveals impairments in planning and online control of movement in Friedreich ataxia: reduced cerebellar-cortico connectivity? Neuroscience 2011; 192:382-390.
109. Corben LA, Delatycki MB, Bradshaw JL et al. Utilisation of advance motor information is impaired in Friedreich ataxia. Cerebellum 2011: June 2. DOI 10.1007/s12311-011-0289-7.
110. Salthouse TA, Heddon T. Interpreting reaction time measures in between-group comparisons. J Clin Exp Neuropsychol 2002; 24(7):858-872.
111. Fielding J, Corben L, Cremer P et al. Disruption to higher order processes in Friedreich ataxia. Neuropsychologia 2010; 48(1):235-42.
112. Hocking DR, Fielding J, Corben L A et al. Ocular Motor Fixation Deficits in Friedreich Ataxia. Cerebellum 2010; 9:411-418.
113. Liu X, Banich M, Jacobson B et al. Common and distinct neural substrates of attentional control in an integrated Simon and spatial Stroop task as assessed by event-related fMRI. Neuroimage 2004; 22(3):1097-1106.
114. Klopper F, Delatycki M.B, Corben LA et al. The test of everyday attention reveals significant sustained volitional attention and working memory deficits in Friedreich Ataxia. J Int Neuropsy Soc 2011; 17:196-200.
115. Timmann D, Drepper J, Maschke M et al. Motor deficits cannot explain impaired cognitive associative learning in cerebellar patients. Neuropsychologia 2002; 40(7):788-800.

116. Koziel LF, Budding DE. The cerebellum: Quality control, creativity, intuition and unconscious working memory. Subcortical structures and cognition: Implication for neuropsychological assessment. New York: Springer; 2009; pp. 124-65.
117. Ito M. Control of mental activities by internal models in the cerebellum. Nat Rev Neurosci 2008; 9(4):304-313.
118. Courchesne E, Allen G. Prediction and preparation, fundamental functions of the cerebellum. Learn Mem 1997; 4(1):1-35.
119. Leggio MG, Chiricozzi FR, Clausi S et al. The neuropsychological profile of cerebellar damage: the sequencing hypothesis. Cortex 2011; 47(1):137-144.
120. Ebner TJ, Pasalar S. Cerebellum predicts the future motor state. Cerebellum 2008; 7:583-588.
121. Thiruvady D, Georgiou-Karistianis N, Egan G et al. Functional connectivity of the prefrontal cortex in Huntington's disease. J Neurol, Neurosurg, Psych 2007; 78(2):127-133.
122. Medina FJ, Tunez I. Huntington's disease: the value of transcranial meganetic stimulation. Curr Med Chem 2010; 17(23):2482-2491.
123. Stagg CJ, O'Shea J, Johansen-Berg H. Imaging the effects of rTMS-induced cortical plasticity. Res Neurol Neurosci 2010; 28(4):425-436.

CHAPTER 12

POLYALANINE TRACT DISORDERS AND NEUROCOGNITIVE PHENOTYPES

Cheryl Shoubridge* and Jozef Gecz

Department of Genetics and Molecular Pathology, SA Pathology at the Women's and Children's Hospital, North Adelaide, South Australia, Australia; and Department of Pediatrics, University of Adelaide, Adelaide, South Australia, Australia
Corresponding Author: Cheryl Shoubridge—Email: cheryl.shoubridge@adelaide.edu.au

Abstract: Expansion of polyalanine tracts cause at least 9 inherited human diseases. Eight of these nine diseases are due to expansions in transcription factors and give rise to congenital disorders, many with neurocognitive phenotypes. Disease-causing expansions vary in length depending upon the gene in question, with the severity of the associated clinical phenotype generally increasing with length of the polyalanine tract. The past decade has seen considerable progress in the understanding on how these mutations may arise and the functional effect of expanded polyalanine tracts on the resulting protein. Despite this progress, the pathogenic mechanism of expanded polyalanine tracts contributing to the associated disease states remains poorly understood. Gaining insights into the mechanisms that underlie the pathogenesis of different expanded polyalanine tract mutations will be a necessary step on the path to the design of potential treatment strategies for the associated diseases.

INTRODUCTION

Expansions of tri-nucleotides sequence in the genome above a certain length have been linked to a growing number of human diseases. Unstable tri-nucleotide repeats can be located in either noncoding sequences, transcribed but not translated sequences or within translated sequences for homomeric stretch of either glutamine or alanine amino acids. There are nine hereditable disorders caused by the expansion of polyalanine tracts.[1] In the human proteome approximately 500 proteins contain polyalanine tracts of greater than 4 residues, with a quarter of these containing a tract of seven or more

Tandem Repeat Polymorphisms: Genetic Plasticity, Neural Diversity and Disease,
edited by Anthony J. Hannan ©2012 Landes Bioscience and Springer Science+Business Media.

uninterrupted alanines but not exceeding 20 residues.[1,2] These proteins are particularly enriched for transcription regulators and in particular homeobox containing proteins, which account for the largest single functional group of proteins containing polyalanine tracts. Eight out of nine polyalanine tract expansion disorders are associated with transcription factors, four of which are homeobox genes (Table 1). The disease causing expanded polyalanine tracts are stably transmitted across multiple generations[1] and generally lead to congenital disorders arising from a loss of function of transcription factors during development.

Expansions of polyglutamine tracts also form the basis of several human diseases.[3] Unlike the congenital disorders of the expanded polyalanine tracts, these diseases are characterised by expansion of unstable tri-nucleotide repeats[4] and result in neuronal dysfunction from mid-life progressing to severe neurodegeneration.[5] The large expansions of polyglutamine tracts are thought to contribute to the generation of mis-folded protein intermediates that eventually lead to aggregates in susceptible neuronal sub-types, a hallmark of the associated disease states.[6,7] In contrast to polyglutamine disorders the human data on in vivo aggregation of expanded polyalanine tract mutant proteins are not yet available. The only exception is PABPN1, a protein involved in mRNA polyadenylation, in which expanded polyalanine tract mutations cause later onset oculo-pharyngeal myotonic dystrophy (OPMD). OPMD is clinically associated with the presence of pathological filamentous inclusions in muscle fibre nuclei.[8,9] The challenge remains to identify the pathogenic mechanism(s) underlying expanded polyalanine tracts that contribute to associated disease states, knowledge required for the design of potential therapeutic strategies.

CONGENITAL DISORDERS CAUSED BY MUTATIONS IN GENES WITH EXPANDED POLYALANINE TRACTS

Of the nine hereditary diseases caused by expansions of polyalanine tracts, eight are congenital disorders (Table 1).[1] The diseases range from disturbances to the body plan, incorrect development of reproductive structures including the ovaries, limbs, bones and the autonomous ventilation system as well as the incorrect development and function of the brain including the hypothalamic—pituitary axis (Table 1). Two of the six transcription factors are located on the X-chromosome and as such the hemizygous males are affected and heterozygote female carriers are normally asymptomatic. The remaining six of the eight transcription factors and the remaining member, a poly(A) binding protein PABPN1, mutated in these diseases are located on autosomes. The mode of transmission for these is generally autosomal dominant (with varying degrees of penetrance) meaning heterozygotes carrying the mutant allele present clinically with the disease.

Complete penetrance of autosomal dominant diseases occurs due to mutations in *HOXA13, ZIC2* and *PABPN1*. Expansion mutations have been identified in each of the three large polyalanine tracts (14, 12 and 18 residues) of *HOXA13*, with all mutations leading to hand-foot-genital syndrome.[10-14] The same clinical outcome is also observed due to mutations deleting or truncating *HOXA13*. Similarly, heterozygous loss of function mutations due to frameshift and truncation mutations indicate haploinsufficiency of *ZIC2* contributes to the aetiology of Holoprosencephaly (HPE).[15] The same clinical phenotype also manifests in patients with a loss or partial loss-of-function of the transcription factor due to expanded polyalanine tract mutations interfering with DNA binding and transcriptional activation.[16] The only adult onset disease caused by expanded polyalanine tract disorders is OPMD due to mutations in *PABPN1* (Table 1). Moreover, unlike all the other genes with disease causing expanded polyalanine tracts, PABPN1 is not a transcription factor but rather a protein involved in polyadenylation of mRNA precursors.[17] This protein contains a single polyalanine tract of 10 residues that when expanded cause OPMD[18] (Table 1). Inheritance of OPMD is generally in an autosomal dominant fashion with heterozygous mutation carriers displaying two to seven additional alanines in this tract. A polymorphism of one additional alanine residue is noted in two percent of the population. In the homozygous state this polymorphism results in autosomal recessive OPMD, while as a compound heterozygote with a pathogenic mutation of two additional alanines, this polymorphism results in a more severe phenotype.[18] The only other type of disease causing mutation in this gene also results in an expanded polyalanine tract, although by a 35G-C transversion.[19] This mutation alters the glycine residue immediately 3-prime of the alanine tract to an alanine. As this residue is followed by a further two alanine residues in the normal protein, this mutation essentially results in an uninterrupted tract of 13 alanines.

Individuals with expanded polyalanine tract mutations often display milder phenotypic consequences compared to individuals with complete loss of function of a given gene. However, the polyalanine tract mutation phenotype is often coupled with broader clinical features. In some cases these mild features may be difficult to recognise on the more severe background of a complete loss of function. However, in some instances these findings imply that in addition to the loss (or partial loss) of function of the mutant allele, these expanded polyalanine tracts may in some cases confer a gain of abnormal function on the protein. One such example is a deletion of the *RUNX2* locus which cause the severe skeletal abnormality Cleidocranial Dysplasia (CCD).[20,21] RUNX2 has a tract of 23 uninterrupted glutamine residues followed by 17 uninterrupted alanine residues on the N-terminal side of the functional Runt domain.[20] When an alanine tract is expanded by recurrent 30 bp duplication resulting in 10 additional alanine residues, the affected individuals have a phenotype of minor craniofacial features of CCD, with the additional features of brachydactyly of hands and feet.[20] In the case of *FOXL2*, 30% of all mutations in this gene are expansions of the single 14-residue polyalanine tract to 19 or 24 residues.[22] Although phenotypic variability is noted in patients with these mutations, the predominate disease outcome is Blepharophimosis ptosis epicantus inversus syndrome (BPES-II). This is a less severe form than BPES-I more commonly caused by mutations leading to PTC and or loss of function of the forkhead domain of FOXL2.[23] Essentially, BPES-I is the likely outcome of a null allele of *FOXL2* where BPES-II is likely due to a hypomorphic allele.[24] Further to this, despite BPES being an autosomal dominant disorder, a recent report observed autosomal recessive inheritance in homozygotes with the Ala19 allele.[25]

The penetrance of the disease and severity of the clinical phenotype increases with the length of the expanded polyalanine tract in several genes including *HOXD13* and *PHOX2B*.[26,27] HOXD13 has three polyalanine tracts, two less than seven residues and one with 15 residues. Expansions to the longest tract were first identified in families with Synpolydactyly (SPD).[28,29] Subsequent analysis identifying a range of mutations expanding the longest polyalanine tract with variable clinical phenotypes and truncation and amino acid substitution mutations in patients with SPD1 (Table 1).[26,30-32] In patients with the smallest expansion mutations, phenotypic variation is not only seen within

Table 1. Diseases caused by expanded polyalanine tract mutations

Gene (MIM)	Gene Name*	Gene Locus	Disease	Disease Inheritance
ARX (300382)	Aristaless related homeobox	Xp22.13	NS-XLID; PRT; MR/RTS/dys; ISSX/WS; IEDE; OS	X-linked Multiple clinical syndromes arise due to expanded PA tract mutations. Phenotypic severity increases with length of expanded PA tract.
FOXL2 (110100)	Forkhead box L2	3q23	BPES I and II −/+ POF	Autosomal dominant Autosomal recessive—homozygous for Ala19 expansion allele.
HOXA13 (142959)	Homeobox A13	7p15-p14.2	HFGS	Autosomal dominant—fully penetrant. Little variation in severity of phenotype.
HOXD13 (142989)	Homeobox D13	2q31-q32	Syndactyly type 5 SPD1	Autosomal dominant—reduced penetrance (semidominant). Increased penetrance with increasing expansion of Ala tract
PABPN1 (602279)	Poly(A) binding protein, nuclear 1	14q11.2-q13	OPMD	Autosomal dominant—complete penetrance. Phenotype more severe in compound heterozygotes with pathogenic (GCG)9 mutation and (GCG)7 polymorphism. Autosomal recessive—homozygous for polymorphic (GCG)7 allele.

continued on next page

Table 1. Continued

Gene (MIM)	Gene Name*	Gene Locus	Disease	Disease Inheritance
PHOX2B (603851)	Paired-like homeobox 2b	4p12	CCHS	Autosomal dominant with reduced penetrance
RUNX2 (600211)	Runt-related transcription factor 2	6p21	Bracydactyly and minor clinical features CCD	Autosomal dominant Expansion of PA tract cause less severe phenotypes than mutations causing homozygous loss of RUNX2 or stop codons in runt or C-terminal transactivating domain all leading to CCD.
SOX3 (313430)	SRY (sex determining region Y)-box 3	Xq26.3	XH	X-linked Phenotypic severity increases with length of expanded PA tract.
ZIC2 (603073)	Zinc family member 2 (odd-paired homolog, drosophila)	13q32	HPE5	Autosomal dominant Aetiology of HPE very heterogenous. Haploinsufficiency.

*Gene names are official full names provided by HGNC.

BPES, Blepharophimosis ptosis epicanthus inversus syndrome (MIM 110100); CCD, Cleidocranial dysplasia (MIM 119600); CCHS, Congenital central hypoventilation syndrome (MIM 209880); HFGS, Hand-Foot-Genital syndrome (MIM 140000); HPE5, Holoprosencephaly 5 (MIM 60937); IEDE, Infantile epileptic-dyskinetic encephalopathy (MIM 308350); ISSX (WS), Infantile spasms X-linked (West syndrome) (MIM 308350); NS-XLID, nonsyndromic X-linked intellectual disability; MR/TS/Dys, mental retardation with tonic seizures with dystonia; OPMD, Oculopharyngeal muscular dystrophy (MIM 164300); OS, Ohtahara syndrome—Early infantile epileptic encephalopathy (MIM 308350); POF, Premature Ovarian Failure; PRTS, Partington syndrome—Intellectual disability with dystonic movements, Ataxia and seizures (MIM 309510); SPD1, Synpolydactyly 1 (MIM 186000); Syndactyly type 5 (113200), XH, X-linked hypopituitarism (MIM 300123).

families (intrafamilial) but also within individuals, such that one hand or foot can be affected while the other is completely normal.[26] In the case of *PHOX2B*, frameshift mutations and a variety of mutations expanding the longest of two C-terminal polyalanine tracts (Table 2) are associated with Congenital Central Hypoventilation syndrome (CCHS).[27,33-35] The severity of the respiratory phenotype, related symptoms and age of onset, all correlate with increasing length of expanded polyalanine tract in PHOX2B.[27]

The remaining two transcription factors, *ARX* and *SOX3* are located on the X-chromosome. With only one X chromosome active in any given cell, X-linked inheritance is more complex than simply grouping as either dominant or recessive.[36] In males, all cells have the same copy of the maternal X-chromosome, while in females dosage compensation occurs as a result of random inactivation of one of the two X-chromosomes in every cell. Disease phenotypes due to polyalanine tract expansion mutations in both these X chromosome genes are predominantly seen in males while the heterozygous female carriers are generally not affected or have very mild clinical presentations.[37-39]

When looking at the X-linked genes *ARX* and *SOX3* we can see an emerging genotype-phenotype correlation between the length of the polyalanine tract expansion and the severity of the disorder. Of the four polyalanine tracts in ARX, the two N-terminal tracts are expanded contributing to ~60% of all mutations reported in this gene.[40] Expansions of the first polyalanine tract from 16 to 17, 19, 23 and 27 alanines result in increasingly severe phenotypes from nonsyndromic intellectual disability, to X-linked infantile spasms and Ohtahara syndrome.[40] The second polyalanine tract is also expanded by mutations increasing the 12-residue tract to 20 and 27 residues.[40-42] Expansion of the second polyalanine tract by a 24 bp duplication is the most frequent mutation reported for *ARX* and leads to an unusual breadth of variation in the clinical presentations, both within and between families.[37] Interestingly, a duplication of 33 bp in the same tract gives rise to a less severe phenotype than would otherwise be expected.[43] While the overall number of alanine residues increases from 12 to 22, the tract is interrupted by a glycine after the tenth alanine residue. This glycine interruption likely ameliorates the severity of this polyalanine tract expansion by a yet to be defined mechanism. It is difficult to determine whether the expanded polyalanine tract mutations in *ARX* cause disease due to a dominant gain of function or due to a loss of function. A complete loss of function is unlikely given the comparison with loss of *ARX* function mutations that result in severe brain malformation phenotypes, including lissencephaly, hydranencephaly and agensis of the corpus callosum.[44,45] Although partial loss of function is implied by the formation of aggregates in the nucleus and cytoplasm,[46,47] a loss of transcriptional activity due to the two most frequent expanded polyalanine tracts in ARX is not supported by cell based assays.[48]

In the case of *SOX3*, over-dosage due to a duplication of the gene, as well as expansion of the first of four polyalanine tracts lead to X-linked hypopituitarism (XH), with variable pituitary deficiency and incompletely penetrant mental retardation.[49,50] When the 15-residue tract is expanded by seven alanines, the patients have XH with normal cognitive function. Expansion by 11 residues leads to XH with isolated growth hormone deficiency and X-linked intellectual disability. Experimental evidence indicates that both expansions of the polyalanine tract in *SOX3* are associated with decreased activity of the transcription factor.[50,51] Hence, disturbance to the activity levels of SOX3, either by partial loss of function of the protein due to expanded polyalanine tract mutations or duplication of the whole gene both contribute to similar disease phenotypes.

DNA MUTATION MECHANISM(S) UNDERLYING POLYALANINE TRACT EXPANSION

The size of disease-causing polyalanine tract expansions in the nine genes reported to date varies across these genes (Table 2). The fine balance between the normal and disease-associated variation in function might be demonstrated by expansion of just one extra alanine of the 16-residue polyalanine tract of ARX, which causes nonsyndromic X-linked intellectual disability.[40,52] The largest expansions of 14 residues occur in both HOXA13 and HOXD13 and lead to tracts of 32 and 29 uninterrupted alanines, respectively.[14,26] The longest polyalanine tract reported to date is a 33-residue tract in PHOX2B in patients with congenital central hypoventilation syndrome.[53]

Both replication slippage and nonhomologous recombination (or unequal cross over of mis-paired alleles) have been proposed to explain the increase in tract length. Understanding the mechanism responsible for causing the expansion of these tracts has often been confounded by the sequence of these tracts themselves. In all nine genes, the polyalanine tracts themselves are coded, to at least some degree, by imperfect tri-nucleotide repeats. The majority of the mutations leading to disease causing expansions have been identified as partial direct repeats of these imperfect tri-nucleotide repeats, often with several sites of insertion (Table 2). Mis-pairing of normal alleles followed by unequal cross-over events[54] provide a rational explanation for these types of expanded tracts. Further to this, some cases of a recurrent alanine tract expansion in *ZIC2* are suggested to arise due to errors in somatic recombination (i.e., mitotic rather than meiotic) in fathers of affected individuals.[55]

On the other hand, a portion of polyalanine tracts are encoded by repeats of a single codon, often (GCG). When disease-causing mutations lead to in-frame duplication, the underlying mechanism is more difficult to attribute. For example, the first polyalanine tract of ARX is made up of a $(GCG)(GCA)(GCG)_{10}(GCA)(GCG)(GCG)_2$ coding for a 16 polyalanine tract.[41] This tract is expanded due to an in frame insertion of seven (GCG) codons resulting in the tract of 23 alanines.[56] Replication slippage, although possible, is an unlikely mechanism in this type of case. The mutation is stable across multiple generations in affected families, implying this expanded tract is stable during both meiosis and mitosis. Furthermore, the sequence this insertion arises from is much shorter than the 34-38 uninterrupted repeats that would be expected to cause replication slippage.[57] A recent mutation in this polyalanine tract of *ARX* identified the inclusion of two imperfect tri-nucleotide repeats as well as nine of the 10 GCG repeats ((GCG) $(GCA)(GCG)_9$) leading to a 27 polyalanine tract.[58] This type of mutation is consistent with unequal cross over between mis-paired normal alleles and suggests the insertion of seven (GCG) codons could just as likely arise due to this type of mechanism.

A similar scenario is observed for the expansion of the only polyalanine tract in PABPN1. The 10 alanines within this tract are encoded by $(GCG)_6(GCA)_4$ with the first mutations identified as expansion of the (GCG) codon by as little as two and up to eight leading to (GCG)8-13 range of expansion mutations.[18] Again, replication slippage has been touted as a possible mechanism underlying the mutations in this gene. Diagnostically, mutations in this gene have often been analyzed as an increased size of a specific PCR product containing the expanded region. However, closer scrutiny of these expansion mutations in a cohort of OPMD patients by direct sequencing has revealed that a third of these mutations were interspersed with a GCA codon[19] and not just expansions of the GCG repeat previously reported.[18] In agreement with the case for some mutations in *ARX,* the

Table 2. Proposed mechanisms of polyalanine tract mutations

Gene	Alanine Tracts	Expanded PA Tracts	Type of Repeat/Insertion	Putative Mechanism of Mutation Formation
ARX	PA1 = 16 (GCN)2+ (GCG)10+ (GCN)4 PA2 = 12 PA3 = 7 PA4 = 9	(GCG)11 = 17 (GCG)13 = 19 (GCG)17 = 23 $\underline{16 + 11 = 27}$ 12 + 8 = 20 10 + G + 12 = 23 12 + 9 = 21	PA1 17 to 23 Ala—PA tract 1 Ala 3 to 13 coded by GCG repeats, which are expanded 1 to 7 times. PA1 27 Ala and all PA2 mutations are partial direct repeats of imperfect tri-nucleotide repeat with several sites of insertion.	Replication slippage or unequal crossover for PA1 mutations 17 to 23 Ala. Unequal crossover between mis-paired normal alleles for PA1 27 Ala and all PA2 mutations.
FOXL2	PA1 = 14	+5 =19 +10 =24	Partial direct repeat(s) of imperfect tri-nucleotide repeat. Several sites of insertion.	Unequal crossover between mis-paired normal alleles
HOXA13	PA1 = 14 PA2 = 12 PA3 = 18	$\underline{14 + 8 = 22}$ $\underline{12 + 6 = 18}$ 18 + 6 = 24 +8 = 26 +9 = 27 +10 = 28 +11 = 29 +12 = 30 +14 = 32	In frame duplication of cryptic tri-nucleotide repeat.	Recurrent replication slippage with point mutations in alanine codons = increase in tract length across evolution. Unequal crossover between mis-paired normal alleles.
HOXD13	PA1 = 6 PA2 = 15 PA3 = 5	15 + 7 =22 +8 = 23 +9 = 24 +10 = 25 +14 = 29	Partial direct repeat of imperfect tri-nucleotide repeat with several sites of insertion.	Unequal crossover between mis-paired normal alleles. +9 by FoTeS or unequal cross over coupled with polymorphism or mutation within the expansion.

continued on next page

Table 2. Continued

Gene	Alanine Tracts	Expanded PA Tracts	Type of Repeat/Insertion	Putative Mechanism of Mutation Formation
PABPN1	PA1 = 10 (GCG)6+ (GCA)4	(GCG)8 = 12 (GCG)9 = 13 (GCG)10 = 14 (GCG)11 = 15 (GCG)12 = 16 (GCG)13 = 17	First 6 of 10 alanines are coded by GCG repeats which are expanded 2 to 7 times.	Replication slippage or Unequal crossover between mis-paired normal alleles.
PHOX2B	PA1 = 9 PA2 = 20	+5 = 25 +6 = 26 +7 = 27 +8 = 28 +9 = 29 +10 = 30 +13 = 33	Partial direct repeat of imperfect tri-nucleotide repeat. Several sites of insertion within the tri-nucleotide tract.	Unequal crossover between mis-paired normal alleles.
RUNX2	PA1 = 17	+10 = 27	Partial direct repeat of imperfect tri-nucleotide repeat.	Unequal crossover between mis-paired normal alleles.
SOX3	PA1 = 15 PA2 = 7 PA3 = 8 PA4 = 12	+7 = 22 +11 = 26	Partial direct repeat of imperfect tri-nucleotide repeat.	Unequal crossover between mis-paired normal alleles.
ZIC2	PA1 = 9 PA2 = 9 PA3 = 5 PA4 = 15	+10 = 25	Partial direct repeat of imperfect tri-nucleotide repeat inserted Head-to-Tail.	Unequal crossover between mis-paired normal alleles. Somatic recombination in mosaic fathers.

FoTeS, Fork stalling and template switching.

expansion of *PABPN1* is likely due to rare, but stable events of unequal recombination of mis-aligned normal alleles for both GCG alone and GCA containing expansions.

Several polyalanine tract expansion mutations, however, can not be adequately explained by the process of unequal cross-over.[26,59-61] For example, in 86 OPMD patients with expansion mutations in *PABPN1* there were 13 different types of expansions, seven of which were uninterrupted GCG repeats and six had GCA interspersed in the repeat. Using a theoretical model of unequal cross over of mis-paired normal alleles based on the seven mutations with an interspersed GCA, 12 of the 13 different types of expansion mutations could be accounted for.[60] The $(GCG)_{13}$ allele comprises an addition of seven (GCG) codons to the $(GCG)_6(GCA)_4$ sequence and as such is not readily explained by unequal cross over. This mutation has been suggested presumably to arise due to slippage or a combination of unequal cross over and slippage.[60] Similarly, there are several mutations for which recombination events cannot be deduced from the observed sequence in patients with CCHS due to expanded polyalanine tracts in *PHOX2B*.[53] These particular mutations and the presence of somatic mosaicism in some cases support a mutational mechanism that may involve misalignment (either template or nascent strand) within a slippage model, resulting in either contraction or expansion due to replication if the unpaired sequence is not repaired.[61]

Fork stalling and template switching (FosTeS)[62] has been suggested as an alternative mechanism to account for expansion mutations that do not fit with recombination events. Essentially, FosTeS utilizes microhomologies of just a few base pairs that act as bridges for the DNA replication fork to skip (forward or backward) along the chromosome when encountering complex genomic architecture or DNA lesions.[63] This FosTeS mechanism has been suggested from the sequence of a polyalanine expansion identified in a mouse with a spontaneous mutation in the *Hoxd13* gene.[64] *Spdh* mice modeling the most common mutation in human SPD have a 21 bp in frame duplication in the polyalanine stretch of exon-1.[65-67] This expansion is a straightforward reduplication of a short segment of the imperfect tri-nucleotide repeat. In contrast, the expansion mutation in *Dyc* mice appears to have resulted from two smaller, more complex duplications.[64] When applied to the expansion in *Hoxd13* in *Dyc* mice, FosTeS could potentially explain the sequence of the expansion. Moreover, polyalanine expansions not explained by unequal cross-over in other genes could, with only one exception, be accounted for by this complex mechanism[64] (Table 2). The one exception not explained by any one of the mechanisms discussed so far is the expansion of the 15 polyalanine tract in HOXD13 by an additional 9 residues. As suggested when this mutation was first identified this particular expansion could only arise if a point mutation in one of the alanine codons has occurred in addition to either unequal crossing over or FosTeS to give rise to the duplication.[26,64]

PATHOGENIC MECHANISMS OF POLYALANINE TRACT EXPANSIONS

The predicted mechanism of protein dysfunction for these autosomal dominant and X-linked disorders arising from expanded polyalanine tract mutations ranges from complete or partial loss of function, dominant negative effect or a gain of function.[68] In the case of autosomal genes, dominant transmission of disease traits due to expanded polyalanine tracts means that heterozygotes with one wild-type allele and one mutant allele will be affected. Hence, cells specifically expressing the genes in question will essentially have both wild-type and mutant protein present in the same cell. The potential exists for expanded polyalanine tracts in the mutant protein to sequester the wild-type protein and/or other factors. This interference with either the localisation or function of these normal proteins is likely to confound the pathogenic effect of the mutation and contribute to the disease phenotype due to a dominant negative effect. This scenario is observed with longest expansion of *PHOX2B,* which exerts a partial dominant negative effect over the wild-type protein in addition to functional haploinsufficiency.[69] A dominant negative mechanism of these expanded polyalanine tract mutations may extend to include sequestration of other proteins required for cell function. Expanded polyalanine tract mutations in *HOXD13* are suggested to have a dominant negative effect not only over wild-type HOXD13, but over other wild-type HOX proteins as well.[66] In support of this suggestion, mice with inactivation of the *Hoxd13* alleles have a milder phenotype than encountered in human patients heterozygous for expanded polyalanine tract mutations. However, mice with inactivation of multiple *Hoxd* genes together (*Hoxd11, Hoxd12* and *Hoxd13*) have SPD-like malformations in keeping with the associated clinical outcomes in patients with expanded polyalanine tract mutations in *HOXD13.*[70]

An emerging theme over the last few years has held that expansion of polyalanine tracts above a certain threshold results in degradation of the mutant protein. Depending upon the efficiency of this process, the length of the expansion in the mutant protein and overall expression levels, aggregation of the protein may occur. The threshold at which these events occur differ between proteins, but are a common finding in over-expression studies in routine and explant cell culture for *FOXL2,*[71] *HOXD13, SOX3, RUNX2, HOXA13,*[72] *PHOX2B*[69] and *ARX.*[46,47] Although these types of studies indicate aggregation is a key functional consequence of polyalanine expansion mutations, the contribution of these aggregates to the pathogenesis of disease remains to be demonstrated.

As more and more disease-causing expansions to polyalanine tracts have been identified many have begun to investigate the impact of tract length on localisation, aggregation and transcriptional activity of the mutant proteins. Using a range of different length polyalanine tracts generated as fusion constructs with GFP, Moumne and colleagues[73] examined a range of constructs with normal and expanded tracts of FOXL2 identified in patients, as well as tracts expanded well above the number seen even in other disease states. The wild-type tract of 14 residues in FOXL2 is expanded to 19 and 24 in human disease (Table 2). The data showed the longer the tract, the higher the propensity of the mutant protein to aggregate and mis-localise to the cytoplasm. When the longest tract of 37 alanines in total was tested, all cells transfected had abnormal cytoplasmic sub-cellular localisation of the mutant protein.[73] Functional analysis of a range of expanded polyalanine tracts in FOXL2 revealed a clear correlation between transcriptional activity of FOXL2 and the resulting clinical phenotype.[74] Variants that cause the more severe Type 1 BPES had significantly reduced transcriptional activity in Luciferase reporter assays using two specific reporter systems. In contrast, mutations leading to the less severe Type II BPES had activity levels not different from wild-type levels. The mutation that is known to give rise to both types of BPES syndromes interestingly displayed intermediate effects on the transcriptional activity in this analysis.[74] In addition to this functional haploinsufficiency, co-aggregation of wild-type protein is also noted for expanded polyalanine tract mutations in FOXL2,[71] indicating a partial dominant negative effect may contribute to the dominant inheritance of BPES.

A similar disruption to transcriptional activity has been reported in several other proteins containing expanded polyalanine tracts including *SOX3,*[51] *PHOX2B*[69] and *ZIC2*[16] (Table 3). In the case of mutations in PHOX2B, the transcriptional activity on the regulatory regions of *DHB* and *PHOX2A* target genes was increasingly impaired as

Table 3. Effects on protein activity due to expanded polyalanine tract mutations

Protein/Function	Functional Domains	Impact of Expanded PA Tract on Protein Function	Aggregation/Mis-Localisation of Mutant Protein—In Vitro/In Vivo
ARX Crucial role in development, in particular of the brain	Paired-type Homeodomain (DNA binding) Octapeptide, aristaless	Predicted partial loss-of-function compared to severe phenotype of complete loss-of function mutations[44,45] and aggregation and mis-localisation of mutant protein.[46,47] Conversely, no loss of transcriptional activity due to the two most frequent expanded polyalanine tracts.[48]	Aggregation and mis-localisation to cytoplasm increases with expansion length.[47] Reduced Arx positive cells in striatum and mis-localised mutant protein in interneurons of mice modelling the PA1 23 alanine tract.[80] Conversely, normal localisation of PA1 23 alanine mutant protein in brain regions of knock-in mice.[81]
FOXL2 Crucial role in ovarian development	Forkhead (DNA binding)	24 Ala retains wild-type protein in aggregates = possible DNE.[73]	19 and up Ala-Aggregation and mis-localisation[71,73]
HOXA13 Critical to the development of the autopod, reproductive and extraembryonic structures	Homeodomain (DNA binding)	28 Ala transgenic mice indistinguishable from null mice, i.e., loss-of-function. Reduced levels of steady state protein indicate likely degradationof mutant protein resulting in reduced in vivo protein.[13]	28 and 32 Ala form cytoplasmic aggregates, with distribution between nucleus and cytoplasm dependant on size of tract. Can sequester Wt in aggregates.[14]
HOXD13 Crucial regulatory role in development of the Anterior-posterio polarity of tetrapod limb	Homeodomain (DNA binding)	Expansion of alanine tract leads to a specific gain-of-function, likely acting in a DNE over other Hoxgenes (except HoxA13).[66]	Aggregation and mis-localisation to cytoplasm increases with expansion length.[67] Reduced levels of 22 Ala mutant proteins in Spdh mice. Protein mis-localised to cytoplasm.[67]
PABPN1 Binds with high affinity to nascent poly(A) tails of 3′ eukaryotic mRNA	RNP-type (RNA binding) domain	Predicted loss-of-function of PABPN1 in OPMD due to decreased availability of mutant protein.[78,79]	PABP2 is a component of the hallmark filamentous inclusions (Intranuclear inclusions) in muscle fibre nuclei of patients with OPMD.[8]

continued on next page

Table 3. Continued

Protein/Function	Functional Domains	Impact of Expanded PA Tract on Protein Function	Aggregation/Mis-Localisation of Mutant Protein—In Vitro/In Vivo
PHOX2B Essential role in normal patterning of Autonomous ventilation system and ANS	Paired-type Homeodomain (DNA binding)	Expansion of alanine tract leads to a functional haploinsufficiency and a progressive partial a DNE against wild-type protein with increasing length of the tract. Correlation between increased expansion length and reduced transcriptional activity.[69]	Aggregation and mis-localisation from nucleus to cytoplasm correlates with increases in expansion length. Only partial sequestration of Wt in nuclear aggregates.[69]
RUNX2 Master regulator of oesteoblast differentiation	Runt domain (DNA binding) C-terminal transactivating	Expansion of alanine tract likely causes a gain-of-function given the distinct phenotype compared to loss of function mutations causing CCD.[20]	Aggregation and mis-localisation to cytoplasm increases with expansion length.[8]
SOX3 Hypothalamic-pituitary axis; neuronal differentiation	HMG box (DNA binding)	Predicted partial loss-of-function; aggregation and mis-localisation and transcriptional activity of mutant protein is reduced.[50,51]	Aggresome formation and mis-localisation from nucleus to cytoplasm correlates with increases in expansion length.[50,51]
ZIC2 Transcriptional activation of *APOE*	5 x Zinc Fingers (DNA binding)	Predicted loss-of-function/functional haploinsufficiency. Near complete loss of transactivation activity and reduces DNA binding.[16]	No aggregation or mis-localisation compared to wild-type.[16]

the length of the polyalanine tract expanded, causing both cytoplasmic retention and aggregate formation.[69] Subsequent studies indicate upregulation of heat shock response prevents the formation and induces clearance of existing cytoplasmic aggregates of mutant protein.[75] Components of the heat shock response pathway and members of the proteasomal machinery have been colocalised to aggregates of PABPN1, ARX and PHOX2B with expanded polyalanine tract mutations.[9,46,47,53,76] Increased expression of HSP70 and recruitment to the nucleus has been implicated in reducing aggregation of ARX and PABPN1 expanded polyalanine mutants and alleviating the incidence of aggregate related cell death.[46,76] Similarly, up-regulation of the heat-shock response by the application of geldanamycin prevented formation of aggregates, was able to clear preformed aggregates and partially restored transactivation activity mutant PHOX2b.[75]

A potential sequence of events has been described by the work of Albrecht et al[65] outlining that although mutant protein may initially be translocated to the nucleus, prolonged exposure or high expression levels may cause the formation of small aggregates in perinuclear region. In turn these can grow to large inclusions around the nucleus as they trap mutant and in the case of some autosomal genes, wild-type proteins alike. Eventually, these events will hinder adequate levels of these transcriptional factors to reach the nuclear environment and interact with specific targets, leading to partial or complete loss-of-activity. This model easily accommodates the observation that mutations leading to larger expanded tract length often result in more severe phenotypes as well as the apparent dominant negative aspect of some autosomal diseases. Although appealing, this common pathogenic mechanism needs to be considered in the context that human data on in vivo aggregation of proteins with expanded polyalanine tracts is not available.

The only exception is the presence of pathological filamentous inclusions in muscle fibre nuclei in patients with OPMD due to expanded polyalanine tract mutations in PABPN1.[8,9,18,77] Mis-folding and aggregation of mutant PABPN1 protein underlies the formation of insoluble inclusions which in turn sequester poly(A) RNA.[8] A range of cellular factors including molecular chaperones, components of Ubiquitin-proteasome pathway and transcription cofactors are also caught within these intranuclear inclusions. Despite the possible contribution of this sequestration to disease outcomes, a loss of function in OPMD due to decreased availability of PABPN1 has been suggested.[78] Expansions to the polyalanine tract in PABPN1 alter the protein conformation and change protein binding properties of interacting proteins.[79] Hence, the likely pathogenic mechanism causing OPMD is a loss of function of PABPN1. This loss of function may be due to the expanded polyalanine tract leading to confirmation changes of the mutant protein and contributing to a combination of aggregation and altered protein-protein interactions.[79]

A growing number of the polyalanine tract expansion disorders are being modelled in mice, providing an avenue to address the role of aggregates or mis-localised mutant proteins and investigate mechanisms contributing to the disease phenotype. Heterozygous and homozygous mice generated to express a patient relevant expansion to the polyalanine tract of Hoxa13 showed no phenotypic differences to mice with deletion of this gene,[13] supporting the observations in affected patients.[10] Limb buds in mice with the expanded allele had normal mRNA expression and splicing but had reduced protein levels, suggesting the loss of function of the mutant protein is due to reduced levels of expanded protein, likely due to increased degradation of the mutant protein.[13] However, the sub-cellular localisation or aggregation of the mutant protein in these animals was not reported. In the case of *ARX* there are currently two, independently generated mouse models of the expansion in the first polyalanine tract from 16 to 23Ala.[80,81] Both of these genetic mouse models recapitulate many of the phenotypic features of affected human individuals and support a partial loss-of-function of the mutant Arx protein. Closer examination identified a selective reduction of Arx-positive GABAergic interneurons in the striatum of mutant mice. At the cellular level mutant Arx protein was mis-localised to the cytoplasm in 55% of interneurons throughout the cortex when examined in mature/adult mice,[80] supporting previous in vitro findings.[47] In contrast, no specific formation of intranuclear inclusions (or cytoplasmic accumulation) occurred in migratory cells positive for *Arx* expression in the ganglionic eminence at E12 and in the cortical cells at P0.[81] Moreover, no specific aggregation of Arx protein was noted when the more frequent c.429_452dup was modeled.[81] If aggregate formation is not an obvious pathogenic hallmark explaining the loss of function of expanded polyalanine tract mutations of *ARX* then alternative mechanisms such as partial loss of function, aberrant protein-protein interactions, or protein-degradation remain to be investigated.

There are examples supporting alternative pathogenic mechanisms to aggregation of expanded polyalanine tract containing proteins contributing to disease. Increases in the polyalanine tract of ZIC2 (expanded from 15 to 25 alanines) result in no difference in localisation of the mutant protein compared to the wild-type ZIC2 protein.[16] Instead of aggregation leading to reduced levels of protein reaching the specific DNA target the reduced transcriptional activity of mutant ZIC2 is instead linked to altered binding of the protein to specific DNA targets.[16] An elegant study recently identified, in mice homozygous for expanded polyalanine tract in *hoxd13*, the phenotypic consequences that arise from mutant Hoxd13 altering the regulation of a rate-limiting enzyme involved in the production of retinoic acid (RA). The subsequent, reduced levels of RA in the limb buds do not adequately repress chondrogenesis, leading to accelerated and uncontrolled differentiation of interdigital cells into chondrocytes and hence, fused digits.[82]

CONCLUSION

Despite the considerable progress made in recent years of understanding how expanded polyalanine tract mutations may arise, the contribution of these mutations to disease pathogenesis still remains poorly understood. Although there is a range of mechanisms potentially contributing to diseases associated with expanded polyalanine tract mutations, the loss of function appears to be a common cause. The mechanisms underlying this loss of function, however, remain to be demonstrated in most cases. A dominant negative effect in addition to functional haploinsufficiency is also likely to underpin the pathogenesis of several expanded polyalanine tract diseases. Expanded polyalanine tracts may cause increased degradation of the mutant protein, interference in the efficient binding to protein partners and/or binding to specific DNA targets required for normal transcription factor activity, thereby contributing to the disease pathogenesis.

Increasingly sophisticated techniques modelling expanded polyalanine tract mutations in the correct cellular context are likely to be required to reveal at least part of the pathogenic mechanisms contributing to the associated diseases. Unfortunately, this gene-specific strategy means a 'magic bullet' solution to these debilitating diseases caused by expanded polyalanine tract mutations is unlikely. Despite this rather bleak outlook, some hope can be derived from the early success of approaches in other neurocognitive disorders, often focusing on common sets of targets or pathways impacted by a range of disease states. Although not the focus of this chapter, several recent examples help to illustrate this concept. In the case of Fragile X syndrome (FXS), the loss of FMRP activity is predicted to result in unchecked mGluR-dependent protein synthesis, which in turn contributes to the pathogenesis of FXS.[83] The understanding of the basic disease pathogenesis of FXS lead to a potential therapeutic avenue of chronically down-regulating Gp1 mGluR signalling to correct, at least in part, phenotypic outcomes of FXS.[84] Similarly, the superfamily of histone deacetylases have also been put forward as potential targets of therapeutic intervention in diseases such as cancer and more recently disorders of the central nervous system, in particular Fragile X syndrome and polyglutamine diseases such as Huntington's disease and spinocerebellar ataxia.[85] One lesson to be taken from these studies is that a physiologically integrated picture surrounding the expanded polyalanine tract mutations in each gene will be necessary for potential therapeutic approaches to be devised. A potential caveat would be that understanding the pathogenesis of each disease in turn might highlight areas of commonality that may be amenable to intervention, similar to the interrogation of the current data on molecular mechanisms to identify therapeutic strategies for the polyglutamine diseases.[7]

REFERENCES

1. Albrecht A, Mundlos S. The other trinucleotide repeat: polyalanine expansion disorders. Curr Opin Genet Dev 2005; 15(3):285-293.
2. Lavoie H, Debeane F, Trinh QD et al. Polymorphism, shared functions and convergent evolution of genes with sequences coding for polyalanine domains. Hum Mol Genet 2003; 12(22):2967-2979.
3. La Spada AR, Taylor JP. Repeat expansion disease: progress and puzzles in disease pathogenesis. Nat Rev Genet 11(4):247-258.
4. McMurray CT. Mechanisms of trinucleotide repeat instability during human development. Nat Rev Genet 11(11):786-799.
5. Bauer PO, Nukina N. The pathogenic mechanisms of polyglutamine diseases and current therapeutic strategies. J Neurochem 2009; 110(6):1737-1765.
6. Li LB, Bonini NM. Roles of trinucleotide-repeat RNA in neurological disease and degeneration. Trends Neurosci 33(6):292-298.
7. Takahashi T, Katada S, Onodera O. Polyglutamine diseases: where does toxicity come from? What is toxicity? Where are we going? J Mol Cell Biol 2(4):180-191.
8. Calado A, Tome FM, Brais B et al. Nuclear inclusions in oculopharyngeal muscular dystrophy consist of poly(A) binding protein 2 aggregates which sequester poly(A) RNA. Hum Mol Genet 2000; 9(15): 2321-2328.
9. Abu-Baker A, Messaed C, Laganiere J et al. Involvement of the ubiquitin-proteasome pathway and molecular chaperones in oculopharyngeal muscular dystrophy. Hum Mol Genet 2003; 12(20):2609-2623.
10. Goodman FR, Bacchelli C, Brady AF et al. Novel HOXA13 mutations and the phenotypic spectrum of hand-foot-genital syndrome. Am J Hum Genet 2000; 67(1):197-202.
11. Utsch B, Becker K, Brock D et al. A novel stable polyalanine [poly(A)] expansion in the HOXA13 gene associated with hand-foot-genital syndrome: proper function of poly(A)-harbouring transcription factors depends on a critical repeat length? Hum Genet 2002; 110(5):488-494.
12. Debeer P, Bacchelli C, Scambler PJ et al. Severe digital abnormalities in a patient heterozygous for both a novel missense mutation in HOXD13 and a polyalanine tract expansion in HOXA13. J Med Genet 2002; 39(11):852-856.
13. Innis JW, Mortlock D, Chen Z et al. Polyalanine expansion in HOXA13: three new affected families and the molecular consequences in a mouse model. Hum Mol Genet 2004; 13(22):2841-2851.
14. Utsch B, McCabe CD, Galbraith K et al. Molecular characterization of HOXA13 polyalanine expansion proteins in hand-foot-genital syndrome. Am J Med Genet A 2007; 143A(24):3161-3168.
15. Roessler E, Lacbawan F, Dubourg C et al. The full spectrum of holoprosencephaly-associated mutations within the ZIC2 gene in humans predicts loss-of-function as the predominant disease mechanism. Hum Mutat 2009; 30(4):E541-E554.
16. Brown L, Paraso M, Arkell R et al. In vitro analysis of partial loss-of-function ZIC2 mutations in holoprosencephaly: alanine tract expansion modulates DNA binding and transactivation. Hum Mol Genet 2005; 14(3):411-420.

17. Danckwardt S, Hentze MW, Kulozik AE. 3' end mRNA processing: molecular mechanisms and implications for health and disease. EMBO J 2008; 27(3):482-498.

18. Brais B, Bouchard JP, Xie YG et al. Short GCG expansions in the PABP2 gene cause oculopharyngeal muscular dystrophy. Nat Genet 1998; 18(2):164-167.

19. Robinson DO, Wills AJ, Hammans SR et al. Oculopharyngeal muscular dystrophy: a point mutation which mimics the effect of the PABPN1 gene triplet repeat expansion mutation. J Med Genet 2006; 43(5):e23.

20. Mundlos S, Otto F, Mundlos C et al. Mutations involving the transcription factor CBFA1 cause cleidocranial dysplasia. Cell 1997; 89(5):773-779.

21. Cunningham ML, Seto ML, Hing AV et al. Cleidocranial dysplasia with severe parietal bone dysplasia: C-terminal RUNX2 mutations. Birth Defects Res A Clin Mol Teratol 2006; 76(2):78-85.

22. Beysen D, Moumne L, Veitia R et al. Missense mutations in the forkhead domain of FOXL2 lead to subcellular mislocalization, protein aggregation and impaired transactivation. Hum Mol Genet 2008; 17(13):2030-2038.

23. Crisponi L, Deiana M, Loi A et al. The putative forkhead transcription factor FOXL2 is mutated in blepharophimosis/ptosis/epicanthus inversus syndrome. Nat Genet 2001; 27(2):159-166.

24. De Baere E, Dixon MJ, Small KW et al. Spectrum of FOXL2 gene mutations in blepharophimosis-ptosis-epicanthus inversus (BPES) families demonstrates a genotype—phenotype correlation. Hum Mol Genet 2001; 10(15):1591-1600.

25. Nallathambi J, Moumne L, De Baere E et al. A novel polyalanine expansion in FOXL2: the first evidence for a recessive form of the blepharophimosis syndrome (BPES) associated with ovarian dysfunction. Hum Genet 2007; 121(1):107-112.

26. Goodman FR, Mundlos S, Muragaki Y et al. Synpolydactyly phenotypes correlate with size of expansions in HOXD13 polyalanine tract. Proc Natl Acad Sci USA 1997; 94(14):7458-7463.

27. Matera I, Bachetti T, Puppo F et al. PHOX2B mutations and polyalanine expansions correlate with the severity of the respiratory phenotype and associated symptoms in both congenital and late onset Central Hypoventilation syndrome. J Med Genet 2004; 41(5):373-380.

28. Akarsu AN, Stoilov I, Yilmaz E et al. Genomic structure of HOXD13 gene: a nine polyalanine duplication causes synpolydactyly in two unrelated families. Hum Mol Genet 1996; 5(7):945-952.

29. Muragaki Y, Mundlos S, Upton J et al. Altered growth and branching patterns in synpolydactyly caused by mutations in HOXD13. Science 1996; 272(5261):548-551.

30. Kjaer KW, Hedeboe J, Bugge M et al. HOXD13 polyalanine tract expansion in classical synpolydactyly type Vordingborg. Am J Med Genet 2002; 110(2):116-121.

31. Johnson D, Kan SH, Oldridge M et al. Missense mutations in the homeodomain of HOXD13 are associated with brachydactyly types D and E. Am J Hum Genet 2003; 72(4):984-997.

32. Kjaer KW, Hansen L, Eiberg H et al. A 72-year-old Danish puzzle resolved—comparative analysis of phenotypes in families with different-sized HOXD13 polyalanine expansions. Am J Med Genet A 2005; 138(4):328-339.

33. Amiel J, Laudier B, Attie-Bitach T et al. Polyalanine expansion and frameshift mutations of the paired-like homeobox gene PHOX2B in congenital central hypoventilation syndrome. Nat Genet 2003; 33(4):459-461.

34. Sasaki A, Kanai M, Kijima K et al. Molecular analysis of congenital central hypoventilation syndrome. Hum Genet 2003; 114(1):22-26.

35. Weese-Mayer DE, Berry-Kravis EM, Zhou L et al. Idiopathic congenital central hypoventilation syndrome: analysis of genes pertinent to early autonomic nervous system embryologic development and identification of mutations in PHOX2b. Am J Med Genet A 2003; 123A(3):267-278.

36. Dobyns WB. The pattern of inheritance of X-linked traits is not dominant or recessive, just X-linked. Acta Paediatr Suppl 2006; 95(451):11-15.

37. Turner G, Partington M, Kerr B et al. Variable expression of mental retardation, autism, seizures and dystonic hand movements in two families with an identical ARX gene mutation. Am J Med Genet 2002; 112(4):405-411.

38. Partington MW, Turner G, Boyle J et al. Three new families with X-linked mental retardation caused by the 428-451dup(24bp) mutation in ARX. Clin Genet 2004; 66(1):39-45.

39. Gecz J, Cloosterman D, Partington M. ARX: a gene for all seasons. Curr Opin Genet Dev 2006; 16(3):308-316.

40. Shoubridge C, Fullston T, Gecz J. ARX spectrum disorders: making inroads into the molecular pathology. Hum Mutat 31(8):889-900.

41. Stromme P, Mangelsdorf ME, Shaw MA et al. Mutations in the human ortholog of Aristaless cause X-linked mental retardation and epilepsy. Nat Genet 2002; 30(4):441-445.

42. Reish O, Fullston T, Regev M et al. A novel de novo 27 bp duplication of the ARX gene, resulting from postzygotic mosaicism and leading to three severely affected males in two generations. Am J Med Genet A 2009; 149A(8):1655-1660.

43. Demos MK, Fullston T, Partington MW et al. Clinical study of two brothers with a novel 33 bp duplication in the ARX gene. Am J Med Genet A 2009; 149A(7):1482-1486.
44. Kitamura K, Yanazawa M, Sugiyama N et al. Mutation of ARX causes abnormal development of forebrain and testes in mice and X-linked lissencephaly with abnormal genitalia in humans. Nat Genet 2002; 32(3):359-369.
45. Kato M, Das S, Petras K et al. Mutations of ARX are associated with striking pleiotropy and consistent genotype-phenotype correlation. Hum Mutat 2004; 23(2):147-159.
46. Nasrallah IM, Minarcik JC, Golden JA. A polyalanine tract expansion in Arx forms intranuclear inclusions and results in increased cell death. J Cell Biol 2004; 167(3):411-416.
47. Shoubridge C, Cloosterman D, Parkinson-Lawerence E et al. Molecular pathology of expanded polyalanine tract mutations in the Aristaless-related homeobox gene. Genomics 2007; 90(1):59-71.
48. McKenzie O, Ponte I, Mangelsdorf M et al. Aristaless-related homeobox gene, the gene responsible for West syndrome and related disorders, is a Groucho/transducin-like enhancer of split dependent transcriptional repressor. Neuroscience 2007; 146(1):236-247.
49. Laumonnier F, Ronce N, Hamel BC et al. Transcription factor SOX3 is involved in X-linked mental retardation with growth hormone deficiency. Am J Hum Genet 2002; 71(6):1450-1455.
50. Woods KS, Cundall M, Turton J et al. Over- and underdosage of SOX3 is associated with infundibular hypoplasia and hypopituitarism. Am J Hum Genet 2005; 76(5):833-849.
51. Wong J, Farlie P, Holbert S et al. Polyalanine expansion mutations in the X-linked hypopituitarism gene SOX3 result in aggresome formation and impaired transactivation. Front Biosci 2007; 12:2085-2095.
52. Bienvenu T, Poirier K, Friocourt G et al. ARX, a novel Prd-class-homeobox gene highly expressed in the telencephalon, is mutated in X-linked mental retardation. Hum Mol Genet 2002; 11(8):981-991.
53. Trochet D, Hong SJ, Lim JK et al. Molecular consequences of PHOX2B missense, frameshift and alanine expansion mutations leading to autonomic dysfunction. Hum Mol Genet 2005; 14(23):3697-3708.
54. Warren ST. Polyalanine expansion in synpolydactyly might result from unequal crossing-over of HOXD13. Science 1997; 275(5298):408-409.
55. Brown LY, Odent S, David V et al. Holoprosencephaly due to mutations in ZIC2: alanine tract expansion mutations may be caused by parental somatic recombination. Hum Mol Genet 2001; 10(8):791-796.
56. Scheffer IE, Wallace RH, Phillips FL et al. X-linked myoclonic epilepsy with spasticity and intellectual disability: mutation in the homeobox gene ARX. Neurology 2002; 59(3):348-356.
57. Eichler EE, Holden JJ, Popovich BW et al. Length of uninterrupted CGG repeats determines instability in the FMR1 gene. Nat Genet 1994; 8(1):88-94.
58. Kato M, Saitoh S, Kamei A et al. A longer polyalanine expansion mutation in the ARX gene causes early infantile epileptic encephalopathy with suppression-burst pattern (Ohtahara syndrome). Am J Hum Genet 2007; 81(2):361-366.
59. De Baere E, Beysen D, Oley C et al. FOXL2 and BPES: mutational hotspots, phenotypic variability and revision of the genotype-phenotype correlation. Am J Hum Genet 2003; 72(2):478-487.
60. Robinson DO, Hammans SR, Read SP et al. Oculopharyngeal muscular dystrophy (OPMD): analysis of the PABPN1 gene expansion sequence in 86 patients reveals 13 different expansion types and further evidence for unequal recombination as the mutational mechanism. Hum Genet 2005; 116(4):267-271.
61. Trochet D, de Pontual L, Keren B et al. Polyalanine expansions might not result from unequal crossing-over. Hum Mutat 2007; 28(10):1043-1044.
62. Lee JA, Carvalho CM, Lupski JR. A DNA replication mechanism for generating nonrecurrent rearrangements associated with genomic disorders. Cell 2007; 131(7):1235-1247.
63. Gu W, Zhang F, Lupski JR. Mechanisms for human genomic rearrangements. Pathogenetics 2008; 1(1):4.
64. Cocquempot O, Brault V, Babinet C et al. Fork stalling and template switching as a mechanism for polyalanine tract expansion affecting the DYC mutant of HOXD13, a new murine model of synpolydactyly. Genetics 2009; 183(1):23-30.
65. Johnson KR, Sweet HO, Donahue LR et al. A new spontaneous mouse mutation of Hoxd13 with a polyalanine expansion and phenotype similar to human synpolydactyly. Hum Mol Genet 1998; 7(6):1033-1038.
66. Bruneau S, Johnson KR, Yamamoto M et al. The mouse Hoxd13(spdh) mutation, a polyalanine expansion similar to human type II synpolydactyly (SPD), disrupts the function but not the expression of other Hoxd genes. Dev Biol 2001; 237(2):345-353.
67. Albrecht AN, Schwabe GC, Stricker S et al. The synpolydactyly homolog (spdh) mutation in the mouse—a defect in patterning and growth of limb cartilage elements. Mech Dev 2002; 112(1-2):53-67.
68. Messaed C, Rouleau GA. Molecular mechanisms underlying polyalanine diseases. Neurobiol Dis 2009; 34(3):397-405.
69. Bachetti T, Matera I, Borghini S et al. Distinct pathogenetic mechanisms for PHOX2B associated polyalanine expansions and frameshift mutations in congenital central hypoventilation syndrome. Hum Mol Genet 2005; 14(13):1815-1824.
70. Zakany J, Duboule D. Hox genes in digit development and evolution. Cell Tissue Res 1999; 296(1):19-25.

71. Caburet S, Demarez A, Moumne L et al. A recurrent polyalanine expansion in the transcription factor FOXL2 induces extensive nuclear and cytoplasmic protein aggregation. J Med Genet 2004; 41(12):932-936.

72. Albrecht AN, Kornak U, Boddrich A et al. A molecular pathogenesis for transcription factor associated poly-alanine tract expansions. Hum Mol Genet 2004; 13(20):2351-2359.

73. Moumne L, Dipietromaria A, Batista F et al. Differential aggregation and functional impairment induced by polyalanine expansions in FOXL2, a transcription factor involved in cranio-facial and ovarian development. Hum Mol Genet 2008; 17(7):1010-1019.

74. Dipietromaria A, Benayoun BA, Todeschini AL et al. Towards a functional classification of pathogenic FOXL2 mutations using transactivation reporter systems. Hum Mol Genet 2009; 18(17):3324-3333.

75. Bachetti T, Bocca P, Borghini S et al. Geldanamycin promotes nuclear localisation and clearance of PHOX2B misfolded proteins containing polyalanine expansions. Int J Biochem Cell Biol 2007; 39(2):327-339.

76. Wang Q, Mosser DD, Bag J. Induction of HSP70 expression and recruitment of HSC70 and HSP70 in the nucleus reduce aggregation of a polyalanine expansion mutant of PABPN1 in HeLa cells. Hum Mol Genet 2005; 14(23):3673-3684.

77. Tome FM, Fardeau M. Nuclear inclusions in oculopharyngeal dystrophy. Acta Neuropathol 1980; 49(1):85-87.

78. Klein AF, Ebihara M, Alexander C et al. PABPN1 polyalanine tract deletion and long expansions modify its aggregation pattern and expression. Exp Cell Res 2008; 314(8):1652-1666.

79. Tavanez JP, Bengoechea R, Berciano MT et al. Hsp70 chaperones and type I PRMTs are sequestered at intranuclear inclusions caused by polyalanine expansions in PABPN1. PLoS ONE 2009; 4(7):e6418.

80. Price MG, Yoo JW, Burgess DL et al. A triplet repeat expansion genetic mouse model of infantile spasms syndrome, Arx(GCG)10+7, with interneuronopathy, spasms in infancy, persistent seizures and adult cognitive and behavioral impairment. J Neurosci 2009; 29(27):8752-8763.

81. Kitamura K, Itou Y, Yanazawa M et al. Three human ARX mutations cause the lissencephaly-like and mental retardation with epilepsy-like pleiotropic phenotypes in mice. Hum Mol Genet 2009; 18(19):3708-3724.

82. Kuss P, Villavicencio-Lorini P, Witte F et al. Mutant Hoxd13 induces extra digits in a mouse model of synpolydactyly directly and by decreasing retinoic acid synthesis. J Clin Invest 2009; 119(1):146-156.

83. Bear MF, Huber KM, Warren ST. The mGluR theory of fragile X mental retardation. Trends Neurosci 2004; 27(7):370-377.

84. Dolen G, Osterweil E, Rao BS et al. Correction of fragile X syndrome in mice. Neuron 2007; 56(6):955-962.

85. Kazantsev AG, Thompson LM. Therapeutic application of histone deacetylase inhibitors for central nervous system disorders. Nat Rev Drug Discov 2008; 7(10):854-868.

INDEX

A

Aggregation 31-34, 60-63, 115-122, 125-134, 161, 164, 186, 195-199
in cell 128
Aggresome 129, 197
Androgen receptor (AR) 59, 60, 86, 93, 97, 120, 121, 125, 153-156, 159-164
Ataxia 1-3, 32, 35, 55, 57, 59, 61, 62, 64, 65, 67-69, 78, 79, 87, 88, 90-94, 125, 169, 170, 171, 173, 175, 178, 179, 189, 200
Ataxin-3 32, 59, 61, 63, 116, 120, 121, 133
Autism 79, 80, 84, 100, 102, 103
Autophagy 131, 132
Autosomal dominant inheritance 159

B

Basal ganglia 142, 149, 150, 173
Behavior 1, 3, 4, 20, 21, 42, 43, 89, 102, 103, 118, 120, 129, 142, 176-178
Behavioral phenotype 80
Brain atrophy 90, 173
Brain function 3, 4, 178

C

CAG repeat 2-4, 56, 59-64, 67, 69, 70, 88, 133, 141, 142, 144, 146-148, 153, 155-157, 159, 160, 164
Cardiomyopathy 169-171, 174, 175
Carrier 45, 78, 79, 81-103, 171, 186, 187, 190
Cerebellum 33, 35, 60-62, 80, 87, 88, 93, 94, 160, 172, 173, 175, 176, 178, 179
CGG repeat 64, 67, 70, 78-82, 86, 88, 91-101
Chaperone 115, 116, 121, 131, 132, 163, 164, 172, 198
Chromatin remodeling 45, 47, 70
Cingulate cortex 144, 146, 148
Cognitive decline 67, 79, 85, 92, 94, 100, 101, 103
Cognitive function 4, 60, 89, 169, 175-178, 190
Compartment 115, 129, 141, 143-145, 149
Congenital disorder 185, 186
Corticospinal tract 172